OPTICAL INTERCONNECTIONS AND PARALLEL PROCESSING: TRENDS AT THE INTERFACE

OPTICAL INTERCONNECTIONS AND PARALLEL PROCESSING: TRENDS AT THE INTERFACE

EDITED BY

Pascal BERTHOMÉ
LRI, Université Paris XI
Orsay, France

Afonso FERREIRA
CNRS, LIP, ENS-Lyon
Lyon, France

with a preface by Pierre Chavel

KLUWER ACADEMIC PUBLISHERS
DORDRECHT / BOSTON / LONDON

A C.I.P. Catalogue record for this book is available from the Library of Congress.

ISBN 978-1-4419-4782-6

Published by Kluwer Academic Publishers,
P.O. Box 17, 3300 AA Dordrecht, The Netherlands.

Sold and distributed in the U.S.A. and Canada
by Kluwer Academic Publishers,
101 Philip Drive, Norwell, MA 02061, U.S.A.

In all other countries, sold and distributed
by Kluwer Academic Publishers Group,
P.O. Box 322, 3300 AH Dordrecht, The Netherlands.

Printed on acid-free paper

CONTENTS

CONTRIBUTORS

Contributors

Pascal Berthomé
LRI CNRS
Bât 490
Université Paris-Sud
91405 Orsay Cedex, France
berthome@lri.fr

J.-P. Bouzinac
CERT-ONERA / DERO
2,avenue E. Belin
Complexe Aérospatial
BP 4025
31055 Toulouse Cedex, France

R. Buhleier
LAAS CNRS
7, avenue du Colonel Roche
31077 Toulouse Cedex, France

Pascal Churoux
CERT-ONERA / DERO
2, avenue E. Belin
Complexe Aérospatial
BP 4025
31055 Toulouse Cedex, France
churoux@onecert.fr

Jacques Collet,
LAAS CNRS
7, avenue du Colonel Roche
31077 Toulouse Cedex, France
collet@laas.fr

Thierry Collette
LETI (CEA - Technologies Avancées)
DEIN - CEA Saclay
F-91191 Gif sur Yvette Cedex, France
collette@gaap.saclay.cea.fr

D. Comte
CERT-ONERA
Computing Department
2,avenue E. Belin
Complexe Aérospatial
BP 4025
31055 Toulouse Cedex, France

Marc Desmulliez
Heriot-Watt University
Department of Computing & Electrical Engineering
Riccarton
Edinburgh EH14 4AS
Scotland, UK
marc@hw.ac.uk

Sadik C. Esener
University of California San Diego
ECE Department
9500 Gilman Drive
La Jolla CA 92093-0407, USA
sadik@ucsd.edu

Afonso Ferreira
LIP CNRS ENS-Lyon
46, allée d'Italie
69364 Lyon Cedex 07, France
ferreira@ens-lyon.fr

Laurent Fesquest
LAAS CNRS,
7, avenue du Colonel Roche
31077 Toulouse Cedex, France
fesquest@laas.fr

M. Fracès
CERT-ONERA / DERO
2, avenue E. Belin
Complexe Aérospatial
BP 4025
31055 Toulouse Cedex, France

N. Hifdi
CERT-ONERA
Computing Department
2, avenue E. Belin
Complexe Aérospatial
BP 4025
31055 Toulouse Cedex, France

Yoshiki Ichioka
Osaka University
Faculty of Engineering
Department of Material and Life Science
2-1 Yamadaoka, Suita 565, Japan

S. Kocon
CERT-ONERA / DERO
2, avenue E. Belin
Complexe Aérospatial
BP 4025
31055 Toulouse Cedex, France

Philippe J. Marchand
University of California San Diego
ECE Department
9500 Gilman Drive
La Jolla CA 92093-0407, USA
pmarchand@ucsd.edu

Rami Melhem
Department of Computer Science
and Electrical Engineering
University of Pittsburgh
Pittsburgh Pensylvannia 15260, USA
melhem@cs.pitt.edu

S. Thomas Obenaus
Department of Electrical Engineering
McGill University
Montreal
PQ Canada H3A 2A7, Canada

Haldun M. Özaktas
Bilkent University
Department of Electrical Engineering
TR-06533 Bilkent, Ankara, Turkey
haldun@ee.bilkent.edu.tr

Charles Salisbury
Department of Computer Science
and Electrical Engineering
University of Pittsburgh
Pittsburgh Pensylvannia 15260, USA
salisbur@cs.pitt.edu

Patrick Scheer
LETI (CEA - Technologies Avancées)
DEIN - CEA Saclay
F-91191 Gif sur Yvette Cedex, France
scheer@gaap.saclay.cea.fr

Assaf Schuster
Computer Department
Technion
Israel Institute of Technology
Technion City
Haifa 32000, Israel
assaf@cs.technion.ac.il

Michel Syska
I3S CNRS
Université Nice Sopia-Antipolis
BP 145, 930 Route des Colles
06903 Sophia-Antipolis FRANCE
and
INRIA
2004 Route des Lucioles
BP 93
06902 Sophia-Antipolis, France
Michel.Syska@sophia.inria.fr

Ted Szymanski
Department of Electrical Engineering
McGill University
Montreal
PQ Canada H3A 2A7, Canada
teds@macs.ee.mcgill.ca

Jun Tanida
Osaka University
Faculty of Engineering
Department of Material and Life Science
2-1 Yamadaoka, Suita 565, Japan
tanida@ap.eng.osaka-u.ac.jp

Brian S. Wherrett
Heriot-Watt University
Department of Physics
Riccarton
Edinburgh EH14 4AS
Scotland, UK
phybsw@hw.ac.uk

Gökçe I. Yayla
University of California San Diego
ECE Department
9500 Gilman Drive
La Jolla CA 92093-0407, USA

PREFACE

As stated by the two editors in their introduction, "the purpose of this book is to bring the optical interconnects research into the mainstream research in parallel processing". It is particularly delightful for an optical scientist who has been following the field since the "first International Optical Computing Conference" to read this sentence written by two computer scientists and thereby measure the evolution in 23 years of research. While early mentions of optical logic can be traced to the beginning of the lasers, the mainstream of interest was initially devoted to the analog parallel processing of images still a research subject on its own. With the advent of optical bistability, which was first observed in gas cells and in bulk semiconductors before specific devices were developed, it became fashionable to think in terms of all-optical processors. In a few cases, mostly in sketchy reports for the general public, the optical computer was presented as an alternative and even a competitor to its microelectronic counterpart. A small number of electro-optic and acousto-optic processing units were demonstrated as demanding exercises on the way to a clear analysis of the real role of light in computing. In spite of them, it has been one common mistake to underestimate the number of layers in the design of a computer and the complexity of the construct that goes from the physical effect behind the switch or the logic gate to the complete machine. In this context, some confusion prevailed for a while before it was widely recognized that the potential for optics rests on the dense and fast transmission of information. Dense transmission at moderate rate has been common practice for decades in imaging and video applications; high bandwidth, non parallel transmission has determined the current explosion of telecommunications. Parallel and fast transmission with potential impact on computing is still an open challenge.

For a long time, the optical computing community was composed almost exclusively of optical scientists, perhaps more generally physicists and electrical engineers. Its interaction with computer scientists, although much desired, was in fact restricted to inviting renowned computer architects at basically every conference and hear them invariably say that they did not see a role for optics, that there was no present need for new technologies and that market conditions were incompatible with exotic ideas. In the last few years however, the situation has started changing. From what I gather about the field of high performance computing, dedicated parallel processor development has been limited by hardware and economic issues. The availability of very large communication bandwidths in telecommunication networks has opened new perspectives in "distributed computing" with machines far away from each other joining their strengths to perform a given task. Downscaling the distances, the approach of building parallel machines around sophisticated interconnection networks is perceived as a promising avenue by at least some computer scientists. Attention is being devoted to the optical nature of information transmission in these networks. Among others, the fact that telecommunication protocols have added wavelength as one new dimension in multiplexing gives rise to new models and this may be a place to note parenthetically that the same mistake of underestimating the number of layers between the physics and the computer should be avoided. In this context, Pascal Berthomé and Afonso Ferreira took the initiative of a Workshop on Optics and Computer Science (WOCS 1), which in spite of icy road conditions and of the worst traffic strike France has experienced in many years led to very stimulating informal discussions during two days at Georgia Tech Lorraine, Metz in December 1995. Then they both proposed to edit a book that I was unfortunately unable to contribute to because of overcommitments. Now that I have agreed to participate anyway with this preface, I can see that the project evolved from a "WOCS book" to a collection of chapters that, although mostly written by participants of WOCS 1, ambitiously and effectively covers a broad range of subjects of common interest to all researchers active in optical interconnects.

Discussions at the workshop had emphasized that interactions between optical and computer scientists needed to be extended to computer engineers: I still feel this as a prerequisite for progress. Anyway, the fact is that the two communities have started interacting. I do not know who first described the evolution of effective human relationships in a four stage scheme: stage one, talk about the weather and the current sports events; stage two, exchange information about each other; stage three, start talking to the partner; stage four, start talking with the partner and then you can interact with him and get something done jointly. Although the division of this book does not follow the division between the two communities, it is still quite clear if one given chapter has been written by members of one or of the other. I would therefore guess that optical and computer scientists are somewhere between the end of stage two and the beginning of stage three. The conclusion is obvious: much more work is ahead of us, success will follow.

Pierre Chavel, Centre National de la Recherche Scientifique
Institut d'Optique, Orsay (France)

INTRODUCTION

P. Berthomé and A. Ferreira, Editors

Advances in semiconductor technologies coupled with progress in high performance computing and communication systems and multi-computing are placing stringent requirements on inter-system and intra-system communications. Demands for high density, high bandwidth, and low power interconnections are already present in a wide variety of computing and switching applications, including, for example, multiprocessing and parallel computing (simulations of real problems, monitoring of parallel programming, etc), and enhanced digital telecommunications services (broadcast TV, video on demand, video conferencing, wireless communication, etc.). Furthermore, with advances in silicon and Ga-As technologies, processor speed will soon reach the gigahertz (GHz) range. Thus, the communication technology is becoming and will remain a potential bottleneck in many systems. This dictates that significant progress needs to be made in the traditional metal-based interconnects, and/or that new interconnect technologies, such as optics, be introduced in these systems.

Optical media are now widely used in telecommunication networks and the evolution of optical and optoelectronic technologies tends to show that their wide range of techniques could be successfully introduced in shorter distance interconnection systems. However, introducing optics in interconnect systems also means that specific problems have yet to be solved while some unique features of the technology must be taken into account in order to design optimal systems. Such problems and features include device characteristics, network topologies, packaging issues, compatibility with silicon processors, system level modeling, and others.

The purpose of this book is to bridge the existing gap between the communities working with optical interconnects and all those concerned with modeling, and computing and communication issues in parallel processing. Our aim is to bring research in optical interconnects into the mainstream research in high performance computing and communication systems, of which parallel processing is just an example, while at the same time providing the latter community with a more comprehensive understanding of the advantages and limitations of optics as applied to high-speed communications. Therefore, we assembled a group of major researchers, from the optical, architecture and communication research communities alike, who deal with the field of optical interconnects for assessing its current status, and identifying future directions to pursue. This book is thus divided in three parts, reflecting the fact that it is intended to provide the reader with insights on the interface of Optical Interconnections, Computer Architectures, and Computing. It is our hope that it will motivate the readers to pursue their research in this very promising domain.

Part I: Optical Interconnections

The first part of the book is composed of four chapters studying technology questions related to optical interconnections. Consider, for instance, the design of a computing machine. It takes place at several levels of abstraction ranging from materials and device engineering to system architecture to high-level software. With this system of nested abstraction levels, the design problem can be broken down into manageable subproblems, much as in a procedural programming language. On the other hand, it makes difficult the introduction of novel concepts and technologies such as optoelectronic device planes, which do not readily fit in the existing scheme of things. In Chapter 1, Haldun Ozaktas develops an understanding of this system of levels of abstraction, why and how it resists the introduction of optical technology, and how one can modify it so as to successfully house optical technology.

Smart-pixel technology or hybrid-VLSI electronics may be seen as a serious candidate for the fabrication of massively parallel, large throughput bandwidth optoelectronics systems. The chapter by Desmulliez and Wherrett reviews the current state of the art and the future trends in this field. Basic considerations show that the technology depends on few parameters in the optical and electronic domains. A system-specific calculation on the optimum throughput is carried out in the case of the bitonic sorter demonstrator from Heriot-Watt University. The methodology developed emphasizes the need for more compact, high sensitivity, high gain and low power consumption photoreceiver amplifiers as well as the development of short pulse duration, high modulation rate, and high energy laser sources.

As it often happens with the advent of a new technology, here also it is essential and to identify the regions of superiority between electrical and optical interconnects, and identify some of the important trade-offs. Scaling of VLSI technology has been dramatically increasing microelectronic device densities and speeds. However, the interconnection technology between devices does not advance proportionally. Limited available interconnect materials compatible with VLSI and packaging technologies, increased wire resistance as a result of scaling, residual wire capacitance due to fringing fields and fields between interconnect wires are among the factors that prohibit drastic improvement of the electrical interconnect performance. As a result, the performance of VLSI systems becomes increasingly dominated by the performance of long interconnects. To overcome this limitation, free-space optical interconnects have been suggested, where long electrical interconnects are replaced by an optical transmitter, a photodetector, and interconnection optics between them. This scheme, although devoid of electrical interconnection drawbacks, has its own difficulties.

This would help system designers choose the proper interconnect technology for the application in question. Therefore, Yayla, Marchand and Esener compare, in Chapter 3, the speed performance and energy cost of a class of electrical and optical interconnections for their use in large scale computing systems.

The contribution of Fesquet, Collet and Buhleier presents an investigation on how to reduce the internode communication latency in distributed multiprocessor systems. The strategy consists of replacing the "store and forward" mechanism that is usually carried out on each electronic node by a free diffusion of optical packets along an optical bus. No routing, no extraction of optical packets and no access arbitration into the bus are needed. Intermediate nodes between two processors engaged in a communication are optically transparent. These specifications considerably simplify the optoelectronic interface of the node and reduce the latency of communication. The authors discuss the utilization of all-optical logical gates and the specifications for the requested photonic devices.

Part II: Architectures for Optically Interconnected Systems

In the second part of the book, Chapters 5, 6 and 7 present computer architectures implementing optical interconnections. Churoux, Bouzinac, Kocon, Fracès, Comte, Hifdi, Collette and Scheer, present projects being carried out at the ONERA-CERT by way of the MILORD and OEDIPE architectures. The SYNOPTIQUE project is described in Chapter 6, by Scheer, Collette and Chouroux. In order to fully take advantage of powerful components within parallel architectures, they started this research project, in which they study the advantages of optical interconnects. In Chapter 7, Tanida and Ichioka show the architecture of an optoelectronic computing system, which is called *Optical Array Logic Network Computing* (OAL-NC). Based on the idea of *optical compunection*, which means merge of optical interconnection and computation, the concept and implementation methods of the OAL-NC are described, along with one possible system construction of the OAL-NC.

Part III: Computing and Communication Issues

The third and final part of this book, composed of five chapters, addresses computing issues related to optically interconnected multiprocessor systems, e.g. modeling, control and routing. It starts with the study of models for one of the main techniques used to implement distributed memory lightwave networks, the Optical Passive Star (OPS) coupler. It is a very efficient medium for transmitting information, offering multiple access channels that allow a substantial reduction in the latencies for one-to-many communications, since every processor can access all its neighbors in a single step through an OPS. Ferreira, in Chapter 8, gives an overview of effective models for OPS-based lightwave networks, which can capture most aspects related to the required resources (single or multiple OPS's), technology used (single or multiple wavelengths) and diameter (single or multiple hops in pairwise communications).

Then, Szymanski and Obenaus describe interesting embedding properties of some emerging reconfigurable and partitionable optical networks, and motivate and formalize several combinatorial optimization problems associated with embeddings in these networks. In particular, their chapter dicusses the embedding properties of a reconfigurable multichannel free-space optical backplane called the "HyperPlane". By partitioning the backplane optical channels appropriately, the optical backplane can be dynamically reconfigured to embed arbitrary networks in real time. The optical backplane can thus provide terabits of low latency bandwidth for message-passing multiprocessors based upon graphs, and shared memory multiprocessors based upon broadcast busses. It is also shown that partitionable optical networks exhibit a significant improvement in performance over non-partitionable optical networks.

With respect to routing, circuit switching techniques are preferred in optical multiprocessor interconnection networks because they do not require any optical to electronic signal conversion to route messages. Thus, they can provide all-optical paths between message sources and destinations and the large communication bandwidth of optical signals is available to the nodes attached to the network. However, circuit switching techniques increase the complexity of managing the network, and thus may increase the communication delay. In a massively parallel processor, this added delay may affect the performance of a parallel program. Therefore, Chapter 10, by Salisbury and Melhem, models the performance impact of multiplexed circuit switched network management techniques. They show how program performance is affected by the choice of circuit switching techniques, by the multiplexing degree, and by the characteristics of the network.

In addition, Schuster considers a type of packet routing known as hot-potato routing, in Chapter 11. In hot-potato routing there is no intermediate storage for the packets (messages) that are on their way to their destinations, which is an important feature for communication networks that are based on optical hardware and for which the messages are composed of beams of light. In particular a "practical" mode of routing, known as greedy routing, is considered. In greedy routing, unless some local congestion forbids it, an intermediate network node always attempts to send packets towards their destinations. He presents several algorithms and analysis methods that were recently suggested for greedy routing, along with some negative results by means of a general lower bound.

The chapter by Berthomé and Syska closes the book with an overview of the optical models used in the literature. For each model, new specific problems arise, making this field very attractive for researchers in many areas of Computer Science. The interested reader will find in that chapter the basic framework of models for optical interconnections and pointers to more specialized results.

Acknowledgments

We wish to thank all the contributors of this book for their efforts in preparing these state-of-the-art chapters. The edition of this book was highly motivated by very fruitful discussions held with the participants of the French Working Group ROI - Rencontres Optique-Informatique, and of the First Workshop in Optics and Computer Science, Metz, December 1995. We are particularly grateful to Pierre Chavel, Philippe Marchand and Karina Marcus, for their endless support.

1

LEVELS OF ABSTRACTION IN COMPUTING SYSTEMS AND OPTICAL INTERCONNECTION TECHNOLOGY

Haldun M. Ozaktas

Bilkent University
Department of Electrical Engineering
Turkey

ABSTRACT

The design of a computing machine takes place at several levels of abstraction ranging from materials and device engineering to system architecture to high-level software. This system of levels of abstraction enables the design problem to be broken down into manageable subproblems, much as in a procedural programming language. On the other hand, it makes difficult the introduction of novel concepts and technologies such as optoelectronic device planes ("smart pixels"), which do not readily fit in the existing scheme of things. We try to develop an understanding of this system of levels of abstraction, why and how it resists the introduction of optical technology, and how one can modify it so as to successfully house optical technology. We argue that in the near future, optoelectronic technology can be successfully introduced if: (i) changing technology or applications create a significant bottleneck in the existing system of levels of abstraction that can be removed by the introduction of optical technology (e.g. interconnections, memory access); (ii) special purpose applications involving very few levels of abstraction can be identified (e.g. sensing, image processing); (iii) it is possible to modify a few levels of abstraction above the level that optical technology is introduced, so that the optical technology is smoothly "grafted" to the existing system of levels of abstraction (e.g. modifying communications schemes or standards so as to match the capabilities of optical switching systems, employing parallel architectures to match the parallel flow of information generated by optical subsystems).

P. Berthomé and A. Ferreira (eds.), Optical Interconnections and Parallel Processing: Trends at the Interface, 1-18.

1 INTRODUCTION

The integration of larger numbers of primitive computing elements (switches, transistors, gates, processors, etc.) to produce computers of greater processing power requires the use of interconnections with greater length/width ratios.[1]

As the length of an interconnection is increased, the time it takes for a signal to propagate to the other end also increases, at least as much as dictated by the speed of light. While this limitation holds for all types of interconnections, normally conducting electrical interconnections have much more severe limitations. The signal delay is a quadratic function of the length/width ratio beyond a certain length/width ratio, since the line becomes too lossy to allow pulse propagation. The energy per transmitted bit also increases with line length, even when repeaters are used [8, 11].

For these and other reasons (e.g., the possibility of non-planar interconnections, voltage isolation, very little or no frequency dependent crosstalk and distortion, no impedance matching problems even with multiple taps, etc.) that have been extensively discussed elsewhere, it has been suggested to use optical interconnections for implementing the longer connections in computing systems, especially when an electrical line to be used instead would have a high length/width ratio. (See [1–6, 8, 9, 11–14] and the references therein.)

From a fundamental perspective, the advantages of optical interconnections in comparison to conducting interconnections is almost obvious to anyone with a basic understanding of the physics involved. Many analytical and quantitative studies, as well as some technology demonstrations also confirm the advantages to be gained by employing optical interconnections. Despite this, the penetration of optical technology into mainstream computing systems has been disappointingly slow. In this essay, we would like to take the opportunity to explore the reasons for this in a qualitative manner. Given the nature of the problem we try to address and the subjective and descriptive style of this essay, we would like to acknowledge its limitations, incompleteness, and need for refinement at the outset.

[1] This can be avoided by resorting to architectures with local connections only, but for problems which intrinsically require global flow of information this merely amounts to breaking down the necessary long distance communication paths into a large number of short hops, which is not necessarily optimal [13].

2 LEVELS OF ABSTRACTION IN PROBLEM SOLVING

A lot of attention has been paid to determining at what level of the interconnection hierarchy optics should be employed (board to board? chip to chip? etc.). On the other hand, it seems that little attention has been paid to the issue of determining at what level of abstraction of the computational process optics should be introduced. (It will be evident that the two issues are not unrelated, the levels of abstraction to some degree corresponding to the levels in the packaging hierarchy.)

To understand what this means, let us reflect on how one usually accomplishes a task by using a computer. Say that we wish to modify a high resolution image in some way so that it is more pleasing to the eye. It might be possible to come up with certain operations involving convolutions, matrix operations etc. that enable us to do this. These mathematical operations will have to be broken down into repetitive or recursive sequences of more elementary operations. In doing this, we will have developed an algorithm for solving the problem. The algorithm will be written down in the form of a high-level programming language, which will be translated into assembly language, which will be run by a microprocessor, which is essentially a high-level logic system, which is made up of lower-level logic functions (such as shift, add, etc.), which are made of gates, which are made of transistors.

Given a certain number of transistors or logic gates, there is no reason to think that this way of doing image enhancement is optimal, but at least it is possible. Realization of the techniques associated with each level of abstraction can be posed as self-contained problems which can be solved by specialists, with some care being necessary to ensure successful interface to the levels immediately above and below. Some degree of optimization within each level is usually performed, but it is often not possible to optimize over several levels. No central committee has ever decided on what these levels of abstraction are either; they are the outcome of historical developments.

Device physicists try to minimize the switching time and energy, and computer scientists try to minimize the consecutive number of steps required for the completion of a task. This suggests that the combined effort of both camps will result in an optimal machine, but closer examination reveals that things are not so simple. For instance, perhaps it is the case that for larger systems, performance saturates with increasing device speed, and devices beyond a certain speed offer no further increase in performance.

The controversy over the relative virtues of global and local computation cannot be resolved unless optimization over several levels of abstraction are performed (see [13] and references therein). Globally connected systems allow fewer steps of computation but result in longer duration per step, whereas locally connected systems require a larger number of steps with shorter duration per step; these considerations being closely related to the choice of algorithm, architecture, and interconnection media. To find the optimum degree of globality or locality, one must optimize jointly over possible algorithms and physical realizations of the machine.

The difficulty of introducing optical technology despite its clear fundamental physical advantages can to some extent be explained in the light of the above discussion. If we had a theory of computing which allowed joint optimization over all levels of abstraction, we could throw in the possibility of optical interconnections and switching into the parameter space. Then, given a computing task, we would perform the optimization, which would not only clearly indicate whether and when we should use optics, but also the architectures and paradigms that must be used. Since we cannot do this, we instead try to show that, say, a globally connected interconnection network is faster if implemented optically. But what if a locally connected network, which can be implemented electrically, allows the same task to be done in overall less time by running a different algorithm? (We have argued that this is not the case in [13], but not definitively.)

3 ALTERNATIVE SYSTEMS OF LEVELS OF ABSTRACTION FOR GENERAL PURPOSE COMPUTING?

It is clear that a very fast, large, and low switching energy array of optical switches or "smart pixels" has tremendous computing potential. However, it is too difficult a task to start from this array and arrive at a general purpose system in a single leap. If we are interested in designing a general purpose computer, we must guide our efforts by some system of levels of abstraction. It is first necessary to show how certain elementary functional units (in the abstract sense) can be formed, and then how these can form higher-level units and so on, until we arrive at some kind of high-level "programming language" enabling the problem description to be formulated. (In most cases, the burden of providing a higher-level platform, which must rest on intermediate-level platforms, will belong to whoever provides the computer. People will certainly be reluctant if they are presented with an array of optoelectronic devices, no matter how fast or large, if we cannot show them where to plug the keyboard and monitor, and where to buy a C Language compiler. Without any registers, accumulators, microprocessors, assembly or C Languages, no user will want to program or configure their systems at such a low level.)

We could try to come forward with a new system of levels of abstraction complete with the techniques necessary for realization of each level, and then build machines including optical components based on this system. Taking the array of optical switches as our starting point, and without being biased by the mainstream system of abstraction, we may try to work our way up to the level of problem description.

Some alternative systems of abstraction do already exist, such as cellular automata, connectionist systems, and most significantly parallel computing. There is some reason to think that these might house optical technology better, but unfortunately these "paradigms" (which differ from the mainstream in varying degrees) are not that well developed. For instance, the techniques for only the lowest levels of abstraction are developed for cellular automata; nobody has a high-level programming language which they can compile into some kind of "assembly language" which will run on some kind of cellular automata hardware (which consists of several levels of abstraction down to the level of a single cell). The state of development of techniques for doing things with cellular automata is comparable to that of low-level logic in the mainstream system, such as shift registers etc.[2]

In conclusion, it seems difficult to come forward with a general purpose optically interconnected computer based on such novel paradigms; the development of the mainstream system of abstraction having spanned at least a century.

---

[2] It has been shown how to simulate conventional logic operations in cellular automata, so that one can in principle do anything with a cellular automaton that one can do with conventional logic. However, this is a meaningless approach if the cellular automata is implemented using logic gates in the first place, or simulated on a workstation. But things may change if cellular automata are implemented by virtue of some atomic scale physical phenomenon.

4 GENERAL PURPOSE SYSTEMS VERSUS SPECIAL PURPOSE SYSTEMS

Unlike general purpose systems which can be programmed to do any task with reasonable efficiency,[3] special purpose systems are "hardwired" and can do only certain prescribed tasks. Broadly speaking, the more general purpose a computer is—that is, the greater the range of tasks it can do with reasonable efficiency—the greater the number of levels of abstraction it has. The more special purpose a computer is, the fewer (for instance, the abacus does not have many). Thus, special purpose systems provide a better opportunity for new technologies (such as optics) and underdeveloped paradigms (such as cellular automata).[4]

Midway between the extremes of special purpose and general purpose systems we can identify a class of systems which we may refer to as "quasi-general purpose" systems or "coprocessors." Such systems can perform a certain class of operations of general utility, such as math coprocessors or digital signal processing chips. Of course, the full picture is that there is a continuum of systems of varying degree of "general purposeness" between the two extremes of special purpose and general purpose systems.

---

[3] It does not take much to be able to do any task, if one allows for gross inefficiency.

[4] It is quite conceivable for a limited number of, say, image processing researchers to start from a description of the capabilities of an array of optical devices and devise algorithms and methods for performing tasks they are interested in. (Many researchers were interested when the systolic computation paradigm was introduced for VLSI systems, unveiling a new class of solvable but unsolved problems.) The key issue seems to be that it should be possible for a single group or working unit to be able to obtain fruitful results by themselves, since this will give them the incentive to attack the problem. On the other hand, the effort towards the general purpose system would require a much bigger effort, requiring strategic commitment by a larger institution.

Researchers and engineers make careers out of solving the problems associated with a certain level of abstraction in the mainstream system. They will not be willing to change their focus easily, since within the present system, the people working at the lower level are providing them with the technology to realize their stuff, and the people at the higher level want the stuff to realize whatever they are doing at their own level. No one will benefit from change unless everybody changes at once. This is a particularly severe kind of "chicken-and-egg problem," since it will not by itself change for the better once given a sufficient but small initial momentum. On the other hand, there are always people willing to work on special purpose systems, which due to their limited number of levels of abstraction, can be handled independently by a single person or group. Thus, successful exposition of the capabilities of optical technologies to the image processing and computer science communities may be rewarding.

An issue which perturbs these considerations is the fact that in areas of academia where theoretical achievements are valued, the interests of researchers may be independent of whether they can interface with upper or lower levels of abstraction.

The "programming" of a general purpose system can take place at various levels. A microprocessor is a custom designed chip which can be programmed at a fairly high level (assembly language). On the other hand, a system to do the same task can be programmed at the much lower hardware level, for instance, by customizing a gate array. Both the microprocessor and the gate array can be viewed as general purpose systems which can be programmed to perform special purposes; the difference is in the nature of the programming and the level at which it takes place.

Under the light of what has been discussed until now, it is no wonder most successful optical systems to date have been special purpose systems. Such systems can be designed to perform a certain task standing alone, or they might be designed as a self-complete component of a larger computing machine. There is nothing complicated with the former. As a very simple example, an array of optical devices might be used as an image amplifier in a medical imaging system. The intricacies involved in the latter case will be discussed later.

Optoelectronic systems such as memory with parallel access, state machines, matrix processors, neural networks, etc. may represent more realistic challenges as compared to general purpose systems, for short term development and validation of optoelectronic technologies. These systems do not involve too many levels of abstraction (which makes their conception possible), often involve regular patterns of information flow (which leads to simple physical architectures), and usually result in an interconnection bottlenecked system when implemented with purely electronic technologies. The major challenge with such a system is to either successfully interface it as a subsystem of a larger (possibly general purpose) system in a way that benefits from its high performance, or to find a special purpose application where it can directly exhibit its high performance.

An interesting case is that of subsystems with few inputs and outputs, which we will discuss in the remainder of this section.[5] In general one would expect a function with a small number of inputs and outputs to be implemented with a small number of components. However, in certain special cases it might be possible to come up with an efficient implementation involving redundant replication, outer product generation, etc. of the data, followed by some fairly simple or regular processing in parallel, followed by reduction to the desired answer. Such a subsystem would more easily fit into existing computing systems without requiring major architectural changes at the higher levels, since the parallelism of optics is exploited in an entirely transparent way, and does not lead to any interface problems.

[5] This case was pointed out by David A. B. Miller.

One example is the optical method of correlation where the input function is replicated in shifted versions, multiplied by a replicated mask, and then integrated. Digital optical implementation should be possible by performing the spreads and integrations in $\sim \log_2 N$ stages. Searching a large database is another example. Assume that we have a certain number of subject terms and we wish to retrieve the entries containing these subject terms. The input and output are small, but the search must take place through a large space. Yet another example is matrix-vector multiplication with a fixed matrix. It should be possible to increase the number of individual examples, but more important is to find a general class of such problems which have some central significance.

When an optical solution of this kind is found, we must immediately inquire whether more efficient electronic implementations exist, since the small number of inputs and outputs suggest that it may be possible to implement the desired system with a small number of components. For instance, systolic convolution or correlation on a linear array can be performed in the order of N time. If the input to the subsystem is arriving in serial manner, it will take this long to read it in anyway so that the optical method will not present any advantages. If, however, the input vector is available in parallel, there is a chance that the optical method might offer some advantages such as lower cost or greater speed.

5 INTRODUCING OPTICS INTO GENERAL PURPOSE SYSTEMS

Although we have seen that special purpose systems provide a conceptually simpler opportunity for optical technologies, we wish to explore how optics may be introduced into general purpose computing systems as well. Since we have seen that it is very difficult to come forward with an all-together novel system of levels of abstraction which would house optical technology in an efficient way, it is clear that general purpose computers will be mostly based on the mainstream system of levels of abstraction (which we might be able to modify to a limited degree).

First, let us consider modifying only the least abstract level of problem solving, the level of physical devices, wires, etc. In this approach, we start with the mainstream architectural and packaging paradigm and see whether it is possible to make a "better" machine by using optical components (interconnections and/or switches) instead of some of the electrical ones. Examples of this approach might be the introduction of optical backplanes or chip-to-chip modules instead of their electrical counterparts, while leaving the architectural conception and logical structure of the machine intact. This would change the job of the device physicist[6] and the person who designs the physical packaging, but would not affect people working at higher levels of abstraction, including those contemplating the logical and systemic architecture of the machine, as well as those providing the software.

This approach is appealing in that we do not have to worry about the development of new architectural concepts. However, there is no reason why the existing concepts should be particularly congenial to optical technology. In fact, they have historically developed to benefit from the strengths and accommodate the weaknesses of electrical technology, which are in some senses complementary to those of optics, so that this approach may not bring out the best of optical components. (VLSI architectures which try to minimize the length and number of chip to chip interconnections provide a good example.) Nevertheless, this may still be a valid and promising approach because it seems that replacing the longer wires with optical links does indeed result in a net advantage, even in existing systems.

If instead of the above simple approach, we wish to modify higher and higher levels of abstraction with the hope of better utilizing the particular optical technology at hand, we must face and overcome certain difficulties. For instance, we may attempt to replace a complete electronic combinatorial or sequential logic unit with an optical one which provides the same functionality, but in a "better" way. The interior structure and levels of abstraction of the optical unit may be entirely different, but it must interface with the system of levels of abstraction of the machine in which it is embedded at a certain level.

At relatively high levels of abstraction we might contemplate an optical microprocessor or digital signal processing coprocessor. At yet higher levels the physical and logical architecture of the machine will be altered significantly to suit the strengths of the optical technology. For instance, we might contemplate an optically interconnected parallel random access machine where the processor locations and algorithms are designed so as to match precisely the type of connection patterns that can be efficiently provided by optics.

Modifying the system at higher and higher levels of abstraction so as to better suit the optical technology becomes an increasingly difficult task as we move upwards because of the need to maintain continuity between the different lev-

[6] More precisely, it would create jobs for some device physicists while eliminating jobs for others.

els. If the central processing unit of a machine works on 32 bit wide words, its replacement must also work with 32 bit words. (It must be "plug compatible.") As another example, if we are to replace the existing electronic memory with an optical one, the input-output characteristics of the new memory must match those of what is being replaced. Notice that this requirement may sometimes resist improvements. A new optical memory, which provides much faster parallel access, may offer no system improvement, since the system in which it is embedded may not be able to utilize it. This makes it difficult to justify the optical technology, since the potential increase in performance offered by the optics cannot be utilized in this case, while its usually greater price will have to be paid. (On the other hand, perhaps an optical processor can be used to perform, say, some kind of parallel search on the data read from the optical memory, a feat which would be very expensive or slow with an electronic memory. The question now is whether the next higher level of the system can beneficially use the results of this fast parallel search. The answer would probably be yes if the search query as well as the result consist of small amounts of data.)

If we cannot succeed in getting a successful interface at one level, we might have to move up to a higher level and try our chance at that level. By modifying this higher level (which may or may not involve optical components), we might be able to exploit the higher parallelism or bandwidth offered by optics. If not, we might have to move another level up, until the intrinsic advantages of the optical technology seep through to the surface and translate directly into a user-level performance advantage (such as getting the job done in less time).

This discussion should also clarify what is meant by doing something "better." Doing some intermediate-level operation cheaper, faster, larger, etc. by introducing certain modifications at that level do not automatically result in user-level improvements. It may be necessary to make further modifications at higher levels, until the fastness, cheapness, etc. can seep through to the user level (which is the highest level).

Modifying the system at a certain level of abstraction might mean introducing an optical subsystem (such as an optical logic unit) into the machine, but this need not be the case. Remember that replacing all of the electronic switches and wires with optical ones does not alter the system of levels of abstraction at all, although we now have a computer consisting entirely of optical components. On the other hand, we may modify the architecture of the machine drastically, without introducing any optics at all. Despite the fact that the interconnection and packaging hierarchy often mirrors the levels of abstraction, the two concepts are distinct and must not be confused.

We now discuss a few examples to make the content of the last few paragraphs more concrete.

5.1 High-bandwidth "transparent" photonic switching

Research in guided-wave wideband switching networks has resulted in rather impressive switches whose various strengths and weaknesses are not exactly matched to the requirements of existing multiplexed switching networks, so that it seems they may find less application than originally hoped for. The weakness of these switches is that they have a limited number of spatial channels. Their strength is that they can route very high-bandwidth signals transparently. Efficient use of this bandwidth cannot be made if bitwise multiplexing is employed, since these systems cannot switch at a rate as high as their transmission bandwidth. However, if we make the higher-level modification of employing large-size block multiplexing instead of bitwise multiplexing, we can walk around this disadvantage. Now we must face the issue of whether the use of large-size block multiplexing is compatible with the next higher level of abstraction (which might be that of communications protocols and transmission standards). If not, we may try to push forward by suggesting modifications to the protocols and standards. If we do not arrive at a clear advantage within a few levels, we might have to give up.

5.2 Two-dimensional digital optical image processing

Optical technology will probably make it possible to construct image processing subsystems which can perform two-dimensional signal processing operations (such as the discrete Fourier transform (DFT), convolutions, etc.) in parallel at a very fast rate. Given the fact that digital electronic hardware is extremely strained to perform such operations of even moderate complexity, it initially seems that digital optical signal processing coprocessors would have much to offer. However, it is not immediately clear how such an optical coprocessor can be interfaced to the rest of the system. Setting up the two-dimensional input data serially from conventional electronic memory may largely nullify the potential advantages of such a system.[7] The limitation is actually that

[7] This has been referred to as the "fire hose problem" by David A. B. Miller.

of the electronic processing system as a whole, which cannot handle larger amounts of data in parallel, not of the optical coprocessor. But the bottom line is that it may not be possible to improve overall performance by simply replacing the coprocessor, because the rest of the system is not good enough to take advantage of the increased capacity and speed. (One should not exclude the possibility that in some cases the replacement might indeed prove beneficial, despite the bottleneck due to serial transfer at the interface, or there may be no interface problem because the data is already in optical form (coming from an optical memory or natural image). Nevertheless, it is likely that in most cases the capabilities of the optical subsystem will be largely underutilized due to this interface problem.)

Does this mean there is no possibility of employing optical technology short of contemplating an all optical system from scratch, which we argued was a very difficult task? Not necessarily. Some modification of the higher-level design of the system may enable a successful interface. The strength of optics in this case is that it can provide the interconnections necessary for the global flow of information, which electronics cannot. The strength of digital electronics is that it can provide complicated operations in a small space, which optics cannot. Thus, successful partitioning (or "factorization") of the overall problem to match the strengths of both technologies may lead to an architecture of the kind depicted in figure 1. The big block may represent an operation requiring regular global interconnections (implemented optically), whereas the smaller blocks represent digital electronic processing. The many smaller blocks on the left work on parts of the data independently in parallel and feed the optical subsystem in parallel, so that there is no serial bottleneck. After performing the necessary operations, the optical subsystem distributes the large array of data to the several digital processors on the right for subsequent processing.

In conclusion, we see that beneficial use of an optical subsystem may require integral redesign of the system architecture at one or more levels above that at which the optical subsystem is introduced.

5.3 Parallel memory access

Recent developments promise high-speed parallel access of huge amounts of data from silicon as well as optical memories. The former involves optical devices integrated with silicon, whereas the latter involves transmissive or reflective readout from optical storage media.

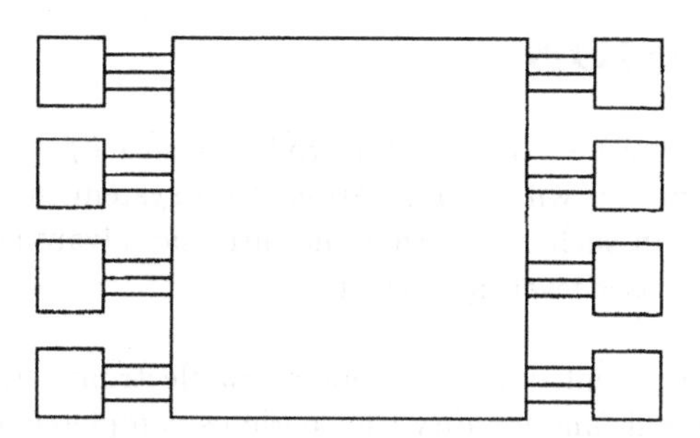

Figure 1 The large block in the middle represents global operations on the whole image. The smaller blocks on the left and right represent local operations on parts of the image.

The considerations here are similar to those in the preceding example. Although the large archival storage capacity of optical memory can be utilized without difficulty, its potential for fast parallel readout may not, unless system architectures are designed in a way that make this possible. Once again it seems that parallel architectures in the spirit of that suggested in the previous example might be useful. Alternatively, some form of optical processing may be used to distill the large amounts of data read from the optical memory, returning a small amount of information that can be handled by the electronics at the higher level. This might be useful in database applications.

5.4 Photonic digital (logical) switching

Digital photonic switches are essentially optical/optoelectronic logic circuits, often based on multi-stage switching network architectures, which enable a given pattern of connections to be established between a large number of incoming and outgoing communications lines [4].

Since the large number of incoming and outgoing lines coming from distinct locations are simply bundled together to form the "fire hose," the interface problem discussed in the image processing example does not arise in this case. This makes optical technology particularly suited to this application.

6 DISCUSSION

Optical technology can be beneficially introduced into a general purpose computer if we can come up with modifications to a system involving the use of optical components in such a way that the intrinsic advantages of the optical technology result in user-level improvements.

A particularly transparent case is to replace an electronic subsystem with an optical one of identical functionality (but perhaps completely different internal structure), in a situation where the user-level figure of merit is improved by the improvement in the external parameters of the subsystem (cost, speed, number of channels, etc.). In other words, one should identify a component or subsystem of a conventional electronic computer which constitutes a bottleneck, in the sense that replacing this component or subsystem with one that is faster (or cheaper or can handle larger amounts of data etc.), will result in the overall computer to be faster (or cheaper or able to handle larger amounts of data). (One might propose that every existing computer has a limiting component or subsystem, which is true. Nevertheless, situations in which replacement of a component or subsystem would result in substantial overall improvement may not be commonplace, since the historical evolution of design concepts and technology has a tendency to balance the various components and subsystems in a way that no component is "over-qualified" for the purpose it serves.)

Thus if such a component or subsystem is found, it would be beneficial to replace it with an optical one which exhibits improvements in the relevant characteristics. For instance, if the clock rate or power dissipation of a computing system is determined by the longest wires, and we can reduce the delay or dissipation along these wires by replacing them with optical channels, we can obtain a net improvement at the level of the overall computer. As another example, assuming that the speed of a computing system is solely determined by the memory access delay or the time it takes a coprocessor to invert a matrix, replacing these subsystems with optical ones may result in direct benefit.

The simplest special case of the above is to replace certain electrical switches or wires by optical ones without otherwise modifying the system. That is, we employ optics simply to wire up circuits designed under the conventional paradigm (low-level modification). Examples are optical backplanes, fixed free-space interconnections between circuit boards, etc. (In this case, no change is made to the existing system of levels of abstraction. A backplane or free-space interconnection system is not a subsystem in the sense of the previous paragraph.) This may not be the best way to utilize optics though, so that disappointments in this approach should not be used to judge the potential of optics.

However, it is very difficult to do something useful with an optical technology (despite its large speed and parallelism), if it cannot be interfaced to a certain level of abstraction of an established architecture in a seamless manner. Otherwise the user-level problem description is too many levels away to make optical device arrays or the like useful for general purpose applications. Thus if the conditions for successful application of the above approaches do not exist, either because the optical technology is not directly compatible with the higher levels, or because its intrinsic advantages are buried at that level, one must modify the system architecture a few levels upwards, as discussed in conjunction with the example on page 12.

On the other hand, special purpose applications where the product can be provided in a form which can be directly used without requiring any low-level development by the user or third parties are clearly promising. Examples might be integrated optical detection planes for image preprocessing, dedicated image processing functions, "smart" optical sensors, etc.

7 CONCLUSIONS

It is clear that large arrays of very fast and low-energy optical devices integrated with established electronic technology and interconnected with free-space optics has very large computational power in the raw sense, but realizing this potential may not be so easy. The difficulty stems from the fact that a whole system of paradigms and levels of abstraction has been constructed around the capabilities and limitations of purely electronic systems, and the dominance of this system of abstractions resists the introduction of a new technology with completely different capabilities and limitations. There does not seem to be much point in trying to build an optical microprocessor, and the user-level improvements obtained by replacing the longer wires in conventional systems may be limited. On the other hand, starting with an array of smart pixels, we are too many levels of abstraction away from being able to write a program that plays chess.

Since the construction of a totally new system of paradigms and platforms is an exceedingly difficult task, it is necessary to find ways in which a manageable degree of modification of the existing system would allow net benefits at the user level. The burden of doing this lies with those who want to promote their new technology, but it may also be possible to identify already existing computational paradigms and concepts which were until now only academic exercises but can now be implemented with optoelectronic technology.

One of the main features of the mentioned system of paradigms and levels of abstraction is that its various parts are more or less balanced in ability, in the sense that no part of the system is a bottleneck. (This is partly because more effort and resources are put into parts of the system that tend to create a bottleneck, and little effort and resources are put into those parts that are already too good compared to the rest of the system.) However, this state of affairs is dynamic; as applications change, new technologies evolve, and new ideas are introduced, it occurs that one part of the system appears as a bottleneck. Suddenly a flurry of activity begins to improve that part of the system, since any improvements in that part will automatically improve the overall machine.

A very important example that has been increasingly recognized in the past ten years is the interconnection bottleneck. Increasing use of memory, the ambition of processing large amounts of information such as with images and video, the advent of parallel computing, and purely geometrical and physical reasons are some of the factors that have contributed to the increasing importance of interconnections. The most widespread suggestion has been to replace the longer electrical interconnections with optical ones without otherwise modifying the logical architecture. Examples are optical backplanes, fixed free-space interconnections between circuit boards, etc. In this spirit, optical technology can be used to help wire up electronic circuits designed in the conventional way, by providing a large number of pinouts and high-performance long-distance connections. Although this approach certainly has a certain promise, it is not the one that we believe will bring the greatest rewards.

A more progressive approach is to replace an electronic subsystem with an optical one whose internal structure may be completely different from the electronic one it replaces. This is easier said than done, since the overall system that has been optimized with the low-performance electronic subsystem in mind may not be able to reflect the superior performance of the optical subsystem to the user level. User-level improvements would be observed only if that particular subsystem was already significantly bottlenecking the performance of the overall system, or if successful modification of the overall architecture can be made such that the optical subsystem is smoothly grafted to the overall system.

Special purpose applications in which only a few levels of abstraction are involved are excellent candidates for introducing optical technology in the short term since the architectural and systems issues that must be tackled are less severe than those associated with general purpose systems. The difficulty here is that many such applications do not require high performance, so that already existing technologies seem to suffice. "Smart image sensing" and image processing are two related special purpose applications which seem particularly promising. Optical switching networks seems to be another. There are certain characteristics that make these applications strong candidates: (i) they involve large volumes of data (both spatially and temporally) and require global flows of information so that they strain the limits of existing systems; (ii) the format of the data and the logical organization of the processing task map naturally onto optical architectures that we know can be efficiently implemented.

Acknowledgements

It is a pleasure to acknowledge the benefit of many discussions with David A. B. Miller of Stanford University, especially during the summer of 1994 at Bell Laboratories, Holmdel, New Jersey.

This work was first documented as part of [7], which was later released as [10]. We acknowledge the support of the institutions involved.

REFERENCES

[1] M. R. Feldman, S. C. Esener, C. C. Guest, and S. H. Lee. Comparison between optical and electrical interconnects based on power and speed considerations. *Applied Optics*, 27:1742–1751, 1988.

[2] M. R. Feldman, C. C. Guest, T. J. Drabik, and S. C. Esener. Comparison between optical and electrical interconnects for fine grain processor arrays based on interconnect density capabilities. *Applied Optics*, 28:3820–3829, 1989.

[3] J. W. Goodman, F. J. Leonberger, S.-Y. Kung, and R. Athale. Optical interconnections for VLSI systems. *Proceedings of the IEEE*, 72:850–866, 1984.

[4] H. S. Hinton. *Introduction to Photonic Switching Fabrics.* Plenum, New York, 1993.

[5] A. V. Krishnamoorthy, P. J. Marchand, F. E. Kiamilev, and S. C. Esener. Grain-size considerations for optoelectronic multistage interconnection networks. *Applied Optics*, 31:5480–5507, 1992.

[6] D. A. B. Miller. Optics for low-energy communication inside digital processors: quantum detectors, sources and modulators as efficient impedance converters. *Optics Letters*, 14:146–148, 1989.

[7] H. Ozaktas. Using planes of optoelectronic devices interconnected by free-space optics: systems issues and application opportunities. Technical Report BLO111680-940922-57TM, AT&T Bell Laboratories, Holmdel, New Jersey, Sept. 1994. AT&T Proprietary.

[8] H. M. Ozaktas. *A Physical Approach to Communication Limits in Computation.* PhD thesis, Stanford University, Stanford, California, June 1991.

[9] H. M. Ozaktas. Towards an optimal foundation architecture for optoelectronic computing. In *Third International Conference on Massively Parallel Processing Using Optical Interconnections (MPPOI'96)*, pages 8–15. IEEE Computer Society, 1996.

[10] H. M. Ozaktas. Using planes of optoelectronic devices interconnected by free-space optics: systems issues and application opportunities. Technical Report BU-CEIS-9606, Bilkent University, Department of Computer Engineering and Information Sciences, Bilkent, Ankara, May 1996. Formerly AT&T Proprietary Report released May 7, 1996.

[11] H. M. Ozaktas and J. W. Goodman. The limitations of interconnections in providing communication between an array of points. In S. Tewksbury, editor, *Frontiers of Computing Systems Research*, volume 2, pages 61–130. Plenum Press, New York, 1991.

[12] H. M. Ozaktas and J. W. Goodman. Implications of interconnection theory for optical digital computing. *Applied Optics*, 31:5559–5567, 1992.

[13] H. M. Ozaktas and J. W. Goodman. Comparison of local and global computation and its implications for the role of optical interconnections in future nanoelectronic systems. *Optics Communications*, 100:247–258, 1993.

[14] H. M. Ozaktas and J. W. Goodman. Elements of a hybrid interconnection theory. *Applied Optics*, 33:2968–2987, 1994.

2

SMART-PIXEL TECHNOLOGY CURRENT STATUS AND FUTURE TRENDS

Marc P.Y. Desmulliez and Brian S. Wherrett*

Department of Computing & Electrical Engineering,
Heriot-Watt University, Scotland, UK

**Department of Physics,*
Heriot-Watt University, Scotland, UK

ABSTRACT

Smart-pixel technology or hybrid-VLSI electronics is emerging as a serious candidate for the fabrication of massively parallel, large throughput bandwidth optoelectronics systems. This chapter attemps to review the current state of the art and the future trends in this field. Basic considerations show that the technology depends on few parameters in the optical and electronic domains. A system-specific calculation on the optimum throughput is carried out in the case of the bitonic sorter demontrator which is being built at Heriot-Watt University. The methodology developed emphasizes the need for more compact, high sensitivity, high gain and low power consumption photoreceiver amplifiers as well as the development of short pulse duration, high modulation rate, high energy laser sources.

1 INTRODUCTION

The escalation of chip performance forecast by the Semiconductor Industry Association (SIA) Roadmap places stringent requirements on the clock frequency, gate density, power dissipation and pin-out of future electronic processors [36]. Although all-electronic solutions are currently being investigated to attain the Roadmap specifications, an emerging technology, called "smart-pixel technology" or optoelectronics-Very-Large-Scale Integration (VLSI) circuit technology, allows the construction of demonstrators which exhibit today the same aggregate bandwidth as that foreseen by electronics for the year 2007 [6, 16, 28, 34].

P. Berthomé and A. Ferreira (eds.), Optical Interconnections and Parallel Processing: Trends at the Interface, 19-48.

This new technology exploits the strengths of free-space optics and electronic processing whose combination has been made technically possible by the recent integration of micron-size optoelectronic transceivers with VLSI circuits [12, 29]. Massive parallelism, low i/o driving energy and synchronous processing of hundreds of channels of input data enable smart-pixel-based systems to provide potentially very large on/off chip communication rates. For example, the bitonic sorter being built at Heriot-Watt University shows a theoretical maximum aggregate throughput of the order of $5\ 10^{11}$ pin-Hz [7, 41], which is at least one order of magnitude higher than presently achievable by electronics alone [40].

Over the last five years, the implementation of this promising technology has provoked various studies on the issues involved in determining the degree of complexity, that is, the "smartness" of the processing elements, such as to optimize the overall system performance [6, 7, 11, 22, 28, 32]. With the notable exception of references [11, 23, 32] and a chapter of this book, such studies, however, have been system specific: a bitonic sorter [6, 7], a banyan network [28], a photonic buffer [22], an optical backplane [37]. All smart-pixel arrays, albeit based on different electronic logic families and optoelectronic transceivers, receive and modulate or emit information in the optical domain and process it in the electrical domain. A generic performance study should therefore be possible, encompassing these common features, whilst accounting for the differences in circuit and transceiver designs. Having in mind this rationale, the purpose of this chapter is fivefold:

1. To provide a brief introduction to the concept of smart-pixel technology and to discuss the motivation behind the worldwide drive towards this technology.

2. To show that, irrespective of the electronic logic family and of the optoelectronic transceivers used, the performance of smart-pixel based systems relies on a very small set of basic parameters which characterize the processing element in the optical and electrical domains. The optimum system performance is shown, as a result, to lie in a narrow niche of the resulting parameter space.

3. To describe a semi-empirical method aimed at determining the performance metrics of a given system. This method relies on analytical and simulation-based calculations of the system optical and electrical power budgets.

4. To indicate consequently which areas of device development should be pursued in order to exploit the full aggregate bandwidth of such systems.

5. Finally, to present a potential and unexpected novel application for which smart-pixel technology could be useful.

The plan of the chapter is as follows: a definition of smart-pixel technology is presented in Section 2. The benefits of this technology are also described in this section. Although the term "smart-pixel technology" encompasses different techniques which are being researched currently worldwide, fundamental constraints limit the performance metrics of this technology. These technological constraints which belong to the optical and electrical domains, are described in Section 3. For a more detailed performance calculation, system-specific considerations need to be carried out. Section 4 provides a detailed example of the methodology used in the case of the bitonic sorter. This semi-empirical method combines analytical and simulation results in order to calculate the throughput rate of the system at all limits. From these results, it can be inferred what aspect of the technology needs to be developed in order to reach the maximum theoretical throughput rate. These future trends are presented in Section 5. They include the development of novel photoreceivers and high-modulation rate lasers. Finally, Section 7 describes a potential novel application of smart-pixel technology which could contribute to the development of low-power electronic systems.

2 SMART-PIXEL TECHNOLOGY

Whereas optical fibre communications have long found industrial applications, it is only recently that space-multiplexed arrays of optoelectronic components have been utilized in computer hardware, such as in the propagation of optical clock signals in multiprocessor systems [21], or the fabrication of optical backplanes [37]. Further applications that exploit the strengths of optical interconnections and electronic logic processing are currently being sought. Hence, in this section the advantages and disadvantages of smart-pixel technology over other conventional technologies are reviewed and the basic properties which characterize smart pixels are summarised.

2.1 Definition and Rationale

The generic layout of a smart-pixel is shown in Figure 1.

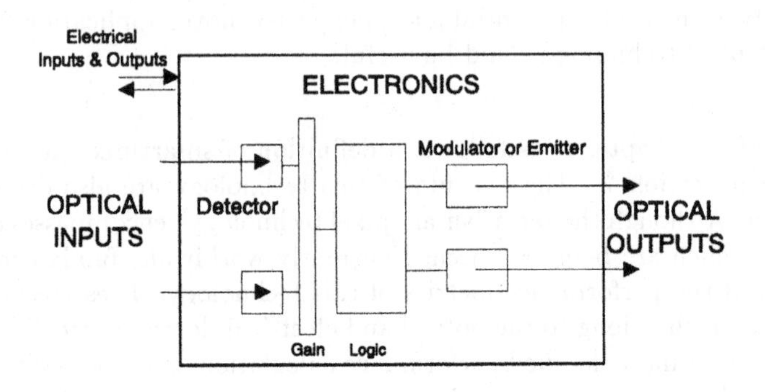

Figure 1 Generic picture of a smart pixel

Optical information pulses are input onto one or several photodetectors, the converted signals are then amplified and processed electronically, the resulting outputs drive modulators or emissive devices. The exampled modulators are electro-absorptive devices based either on bulk material and exploiting the Franz-Keldysh effect [13, 19] or based on multiple-quantum-well structure and exploiting the Quantum-Confined Stark effect [30]. The emissive devices can be Light Emitting Diodes (LEDs) or Vertical Cavity Surface Emitting Lasers (VCSELs). Arrays of perhaps 10^4 (usually) identical pixels occupying around 1 cm^2 chip area are interconnected to other chips using free-space optics whereas logic and local on-chip interconnection is electrical. In this context, smart pixel arrays which could also be defined as dense space-multiplexed arrays of optoelectronic components, benefit from the strengths of optics and electronics. The well-known advantages of optical interconnects, combined with the processing power of electronics, allow the design and construction of high aggregate bandwidth systems. Whereas the clock frequency and parallelism of a smart-pixel array is determined by the detection time and by the electronics, the performance-gain over all-electronics solutions is in the aggregate bandwidth and flexibility of the optical interconnection. Smart-pixel technology addresses the following issues:

1. Networking and processing devices are putting strong demands on high pin-out (750 to 1100) and high operating frequency (250 MHz) electronic packages. Electronic companies such as LSI have recently addressed these needs in part by the introduction of an 35 mm, 1089-leads ceramic ball grid array (CBGA) package [27]. The move to higher pin-out count would require the use of quite complex multilayer technology [35]. Smart-pixel

technology, by the optical addressing of the optoelectronic transceivers in the third dimension, provides inherently a higher connection density at a lower technological cost. In addition, data addressing to other hardware in the computer, and data acquisition from other chips, other sensory devices and in the future from optical disc storage, may be in parallel.

2. Non-local and space-variant interconnect patterns may be implemented in a single cycle. Interconnect reconfiguration on single-cycle timescale is also becoming possible. A major rethink in the computer science community on how to benefit from these two properties would be welcome.

3. Power consumption for off-chip interconnection is lower than for terminated wire [10, 14]. For example, up to a third of the total on-chip power could be saved in electronic processing chips if optical clock distribution were implemented [18].

4. By the transmission of information through optics, lower clock-skew occurs in the electronics and higher frequency of operation may be possible.

5. The cost of the shielding required for the integrity of the electronic signals is much reduced in smart-pixel technology since the information is transmitted optically.

2.2 Smart-Pixel Technologies at present

Table 1, shows a small fraction of the combinations of optoelectronic components and electronic circuitry that are being researched worldwide; this list is by no means exhaustive.

To the best of the authors' knowledge, the first smart pixel arrays were monolithic GaAs-based arrays in which the Field Effect Transistors (FET) GaAs electronic circuits. These were laid out on the same substrate as the multiple-quantum-well based optoelectronic components [42]. The development of this technology was hampered by the unavailability of enhancement-mode transistors and by the unpredictability of the device threshold voltage. Level-shifting diodes were needed between each logic stage (Buffered FET Logic or BFL), resulting in a large power consumption. In this new FET-SEED technology, the required pixel areas, even for modest logic functionality, proved to be well beyond 100x100 μm^2.

As early as September 1993, Heriot-Watt University decided to implement the technique of flip-chip solder-bump bonding [15] in order to interface an InGaAs-based optoelectronic device array (the opto-chip) onto Silicon-based CMOS logic (the electronic chip). It is worth noting that this technique, invented by IBM more than 30 years ago under the name "Controlled-Collapse Connection Chip" or C4, was intended, at that time, to overcome the pin-out problem of electronic chips. In the smart-pixel technology, by fabricating arrays of solder-wettable pads on chips, a high one-to-one interconnect density can be achieved between the opto-chip and the electronic chip. In the case of InGaAs-based MQW devices fabricated on a GaAs substrate, the opto-chip transceivers can be addressed through the substrate [41]. The signals, detected as photocurrents, pass via the solder bumps onto the CMOS chip where they are amplified and processed. The signals are then sent back through a separate solder-bump array to the InGaAs modulators. This hybrid technology allows the smart-pixel designer to build from any improvement of the Silicon electronics.

2.3 Electronic and Optical Limits

There exist numerous issues involved in determining the logic complexity of smart-pixel arrays, such as to optimize the overall system performance. For example, pixels with large electronic intelligence demand large real-estate of silicon, so that few can actually be accomodated per chip. The resulting low on-chip parallelism may not take benefit of the established advantages of non-local or space-variant optical interconnects. Alternatively, the sheer cost of fabricating diffraction-limited multi-lens optical components may prohibit either the layout of vast numbers of simple pixels or the optical addressing of complex pixels over a large chip area. The power dissipated on chip by the pre-amplifiers at the front-end interface and the logic circuitry downstream may also limit the pixel density. For example, for a 32-by-32 pixel array, the power dissipation at the electronic front-end, of 1024 transimpedance amplifiers, each dissipating 1.5 mW, would contribute significantly to the total on-chip power dissipation. Moreover insufficient heat-removal capabilities induce thermal non-uniformity across the array, which affects particularly the opto-electronic components, thereby degrading the system performance. Finally, if the processing array is to be operating at a given frequency, the limited amount of laser-source power available to transmit optical information on and off-chip, imposes, for some systems, a maximum pixel density. These limiting factors need be accounted for during the optimization process.

Group	Input	Logic	Output
AT&T	GaAs-SEED	GaAs-FET	GaAs Modulator
Huazhong University,China	GaAs-SEED	GaAs-FET	GaAs Modulator
Zuerich	GaAs	GaAs-MESFET	GaAs LED
NTT, Opticomp	MSM GaAs	GaAs MESFET	GaAs VCSEL
Honeywell	MSM-GaAs	GaAs-MESFET	GaAs VCSEL
UC San Diego	Si	MOS	PLZT Modulator
Edinburgh, Colorado	Si	CMOS	LC Modulator
Lucent Technologies	GaAs	Si-CMOS	GaAs Modulator
Heriot-Watt University	InGaAs	Si-CMOS	GaAs Modulator
McGill University	GaAs	Si-CMOS	GaAs Modulator
Georgia Tech	ELO InGaAsP	Si-CMOS	ELO InGaAsP LED
Erlangen,Siemens, Colorado	Si PIN	Si-CMOS	GaAs VCSEL

Table 1 Smart-pixel technologies and smart-pixel groups worlwide.

Electronic domain				
Electronics switching energy, E_e (pJ)	0.1			
Maximum heat dissipation, Q (Wcm^{-2})	10			
Electronics frequency, F_e(MHz)	100			
Number of gates per cm^{-2}	10^6			
Gate-switches per chip (gates-Hz)	10^{14}			
Optical domain				
Optical conversion energy, E_o(pJ)	0.1			
Source average power, P_o (Watt)	1			
Source-to-detector efficiency, η	0.1			
Communication rate (pin-Hz)	10^{12}			
Number of (optical) input pins	10^2	10^3	10^4	10^5
Potential optical data-rate per pin (MHz)	10^4	10^3	100	10
Potential electronic frequency (MHz)	100	100	100	100
Number of gates per pin	10^4	10^3	100	10
Gate-switches per chip (gate-Hz)	10^{14}	10^{14}	10^{14}	10^{13}
Communication rate (GHz)	10	100	1000	1000

Table 2 Bird's-eye view of smart-pixel technology.

3 BASIC CONSIDERATIONS

Due to the success of smart-pixel technology, various studies have appeared on the desirable degree of smartness and optical pinout per pixel, of the array such as to optimize the overall aggregate bandwidth of the system. Most of these studies have been specific to the architecture and the technology used for both the opto-chip and the electronic chip. It is in fact possible to demonstrate that, irrespective of the architecture and technologies implemented, a small set of parameters belonging to the optical and electronic domains characterize the performance of the hybrid processing element as shown in Table 2.

As a result, it is shown in this section that, however crude the present model, the optimum performance lies in a narrow window of the resulting parameter space.

3.1 Electronic Domain

Most smart-pixel arrays designed so far share two common characteristics : the pixels are regularly distributed along both dimensions and they are activated at the same time because of the inherent parallelism of the array. The logic circuitry associated with each pixel lies in the proximity of its transceivers (detectors, emitters, modulators) rendering the whole array highly partitioned. As a result, the gate density, albeit dependent on the technology (CMOS, ECL, BiCMOS) and scaling geometry, is likely to be less than that of an equivalent all-electronic chip. Moreover, the occupation ratio or duty factor of the electronic gates, k, defined as the ratio of the average number of active gates over the total gate number, is also expected to be close to unity. This ratio depends mainly on the electronic depth of each individual pixel. Assume a frequency of operation F_e(Hz) for the electronics, an energy of E_e(J) necessary to switch the gate and charge the inter-gate electrical connection, a gate density G_e, and a maximum heat dissipation of $Q(\mathrm{Wcm}^{-2})$. The on-chip performance, defined as the number of gate-switches per second, is then : the gate density G_e is then

$$G_eF_e \leq Q/(kE_e) . \tag{2.1}$$

It could be that the electronic power deliverable onto a chip, P_e, is insufficient to drive the gates at their highest frequencies. A second limitation, in principle, is therefore

$$G_eF_e \leq P_e/(kE_e) . \tag{2.2}$$

E_e defined here should be regarded as the sum of two energies : the energy which is dissipated by the switching operation of the gate and the energy which is needed to charge the metal interconnection between gates. With the continuously decreasing minimum feature size of VLSI circuits, densely populated chips have been known for a long time to be wire-limited [20]. As a result, it is most likely that the interconnection energy constitutes an increasing fraction of E_e. Moreover, in the above equation, the partitioning of the chip into independent electronic islands, which is a recognizable feature of smart-pixel arrays, has been neglected at first approximation. Although the gate density is therefore likely to be less than that given by the equation above, we are only concerned here in quantities within an order of magnitude. Two factors, however, are likely to reduce the gate density. A peak current of several Amperes needs to be provided for the whole array since the pixels are activated at the same time. The design of multiple wide power lines running across the chip is thus certainly required and is in fact a feature of all smart-pixel chips. This is also true for the provision of clock signals, if no optical clock distribution scheme is implemented on chip [5], or of any electrical signal needed to control the pixels. As a result, silicon areas on the smart-pixel chip are required to provide the power and signal (clock) tracks, which reduces the real-estate devoted to logic processing. More importantly, the derivation of the gate-switches number assumes that only dynamic power is dissipated. This is not the case for smart-pixel arrays which include photoreceiver amplifiers of large quiescent power at their front-end interface, such as the transimpedance amplifiers. Indeed new transimpedance amplifiers have been designed which are switched off whenever not needed for amplification in order to reduce the total power consumption [43].

3.2 Optical Domain

For a laser source of average power P_o(Watt), some fraction, η, is incident on the smart-pixel array. η is determined by the most lossy path from the source to the detectors and by the insertion loss of the modulators (if used). Defining the optical switching energy, E_o(J), as the optical energy needed to convert an optical signal into a valid electronic signal, the expected chip-to-chip communication rate, expressed in pin-Hz, is $\eta P_o/E_o$. The off-chip performance is therefore limited by,

$$G_o F_o \leq \eta P_o / E_o \,, \tag{2.3}$$

where G_o is the number of optical channels and F_o is the optical data rate. In principle the optical power delivered onto the chip must be accounted for in the heat dissipation argument (equation (1) should be modified). In practive however, $P_o << P_e$ and the thermal budget is dominated by the dissipation of the electronics so that Q is roughly equal to P_e. This is a significant factor in determining the nature of smart-pixels, and one that is often overlooked, that it is far easier to deliver high electrical power on-chip than optical power.

The strength of the smart-pixel technology lies in the magnitude of the off-chip performance. For η=0.1, P_o=1 Watt and E_o=0.1 pJ, say, the on/off chip communication rate is of the order of 10^{12} pin-Hz and exceeds by approximately two orders of magnitude the capabilities of electronic communication. In order to fully exploit this throughput rate the optical data-rate per channel should equal the electronic processing rate of the chip providing $G_o = P_o/(E_o F_e)$ optical pins which corresponds to 10^4 optical channels. A higher number of channels would decrease the optical energy deliverable per pulse so that the system would not run at the optimal electronics frequency to benefit from the raw electronic processing power. At a lower number of channels the smart-pixel array would not fully exploit the offered optical bandwidth, unless a measure of multiplexing/demultiplexing were introduced in the electronics/optics interface. The pixel smartness (S) is well defined by the number of electronic gates per optical channel,

$$S = G_e/G_o = P_e\, E_o\, F_o/(\eta k\, P_o\, E_e\, F_e)\ . \tag{2.4}$$

If 10 Watts of electronic power is provided onto the chip, the pixel smartness $S=100\ E_o/E_e$, depends only on the switching energy ratio. With this figure-of-merit, the optimum configuration for the smart-pixel technology (10^4 pins per chip) provides 100 logic gates per channel (assuming $E_o=E_e=0.1$ pJ), 10 to 20 of which are used for the receiver amplification stages.

The above methodology, as expressed in table 2, allows one to determine how technological changes would alter the architecture of smart-pixels. For example, were only 1 pJ detectors available then the potential optical data-rates would decrease by one order of magnitude; the 10^4 pin-out system could then only be operated at 100 GHz communication rate and at 10^{13} gate-Hz but the 10^3 pin-out system performance would be unaltered. In contrast, reduction in the electronic switching energy towards 0.01 pJ would allow the 10^3 pin-out chip to operate at 10^{15} gate-Hz and 1 THz communication bandwidth.

3.3 Conclusion

The technological constraints: 1 watt of laser source power, 100 MHz electronic frequency and 0.1 pJ (0.1pJ) electronic (optical) switching energy, are representative of the smart-pixel technology. We also take $\eta=0.1$ and $k=1$. Some of these figures can easily be increased or decreased by one order of magnitude. The main results of this section however prevail : the niche for smart-pixel technology is rather narrow and relies only on a small set of criteria which characterise the pixel in the electrical and optical domains. Within our assumptions, the niche lies between the regions of 100 to 1000 gates per channel and 10^4 to 10^3 pixels per chip, as indicated in table 2. As crude as the above ballpark numbers may appear, they provide a surprisingly good estimate of the characteristics of present demonstrators [6, 7, 22]. Most smart-pixel arrays have between 1000 and 5000 pixels, each of which encompasses between 50 to 300 gates. In particular, the maximum on/off chip communication rate for the SCIOS bitonic sorter discussed below is about 5 10^{11} pin-Hz [7], which is close to the number calculated above.

4 THE BITONIC SORTER

More detailed calculations of the maximum on/off chip performances for specific systems involve careful studies that include the optical and electrical power budgets. These budgets depend on the architecture and the algorithm being implemented. To illustrate this statement, the case of the bitonic sorter demonstrator, which is being built at Heriot-Watt University, is developed in this section.

4.1 Architecture and Algorithm

Beyond the provision of dense photonic connectivity, data routing, data sorting and discrete-fast-Fourier-transformation are three classic tasks which exemplify the benefits of smart-pixel technology. These problems require high interconnectivity, modest logic between communication stages and non-local space-variant interconnect patterns. For high parallelism, electronics takes several clock periods to achieve the connectivity that can be provided by a single optical interconnect, whether it is a perfect shuffle, a banyan network or a radix-2 interconnect. The demonstrator currently being constructed at Heriot-Watt uses the Extended-Interconnect Cellular Logic Image Processor (EX-CLIP) architecture for data sorting, shown in Figure 2.

An array of 2^m L-bit long words to be sorted are input optically in bit-slice form from a spatial light modulator (SLM). The L bit-planes enter the processing loop at a 2-D smart-pixel array acting as a buffer or memory unit. The bit-planes from the memory unit pass through the interconnection pattern to be processed (routed) by the sorting node array. The processed data are then fed back onto the memory unit for further processing. Several iterative loops are necessary to sort the whole data set. Full details of the implementation of the Batcher bitonic sorting algorithm [2] are given in reference [6]. The routing of the data is carried out within a single chip which accomodates the whole logic complexity of the sort. The demonstrator chip, based on the CMOS-SEED technology, has 512 2-input, 2-output sorting nodes, each of which treats a pair of input data words. At each processing cycle prior to the arrival of the data planes, the sorting node array is reconfigured by an optically-embedded control data plane, called the control mask, which sets up the functionality of each sorting node within the chip. Defining the clock period as the time between optical data planes, the sorting of 2^m L-bit words requires

$$N = [m^2 - m + 1]\,(L + F)\,, \qquad (2.5)$$

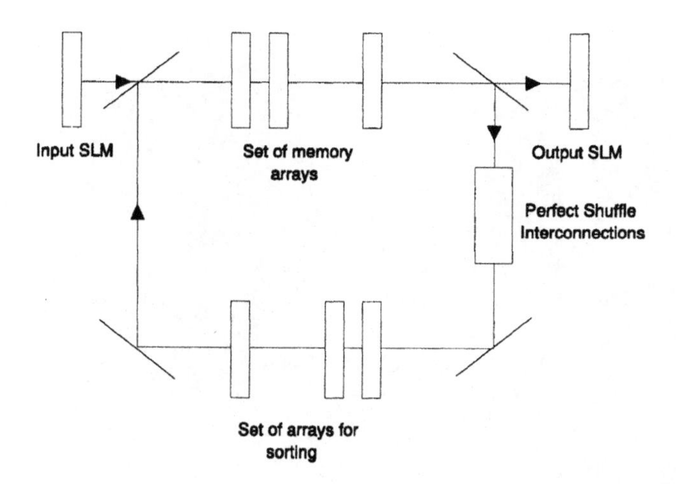

Figure 2 Functional schematic of the EX-CLIP architecture used for data sorting. The unsorted 2-D array of input data circulates a fixed number of times around the processing loop before being output to the output SLM. In the optoelectronics-VSLI circuit option, there is only one memory array and one array for sorting.

clock periods, where F is the number of clock signals required to load a control mask [7]. For our implementation m=10, L=8, F=2, therefore N=910. The CMOS-SEED processing array has been designed and fabricated under the European Europratice Programme. Using the 1μm CMOS process, each node has around 200 gates and an area of 200-by-400μm^2. The assembled demonstrator is shown in Figure 3.

4.2 Optical Power Budget

The optical power budget at the system level must take into account three factors: (i) The type and mode of operation of the laser source, (ii) The optical response of the modulators and detectors and (iii) The losses incurred by the optical hardware. Each of these factors is reviewed here.

Figure 3 Photograph of the bitonic sorter demonstrator. The sorter chip and the memory chip can be seen on the left and right-hand side.

The Laser Source

The minimum sensitivity required of the photodetector at a given operating frequency and the losses of the optical path (taking into account any optical fan-out) allows one to calculate the total output power needed for the demonstrator. The on-chip optically-induced dissipated power can then be calculated by the power incident onto the detectors and the modulators as well as by the mode of operation of the laser sources. The laser sources for smart-pixel circuitry can be operated cw or at high modulation rate quasi-cw. Typical cw power levels of 1 Watt are now available commercially for semiconductor diode-laser bars. In the quasi-cw regime, a laser source such as the master-oscillator power amplifier [45] has been successfully operated at 0.6-W time-averaged power for a modulation rate of 1 Gbit/s. Quasi-cw, forced Q-switched AlGaAs bow-tie lasers have also shown a modulation rate of 2 Gbit/s with pulse durations of around 25 ps [39]. It can therefore be assumed that, in all laser modes of operation, a maximum (peak)power of 1 W is available.

Photodiode and Modulator Optical Responses

The optoelectronics components used to convert the optical signals into current or voltage variations are MQW p-i-n photodetectors of unit internal quantum efficiency and responsivity S. In the case of smart-pixel devices, the electronic front-end interfaces are best operated either at a fixed voltage level (in the case of current-mode amplifiers) or at small voltage swing ΔV_{in} (in the case of voltage gain amplifiers). In both cases, the replacement of the responsivity function S by its mean value S_{avg} at the voltage level or about the threshold voltage is justified. The same type of photodiodes are also utilized as electro-absorption modulators to convert the voltage output from the electronic circuitry into low (R_m) or high (R_p) reflectivity levels. The modulators are operated at wavelengths longer than that of the excitonic peak wavelength in order to achieve a good output contrast ratio. The precise wavelength used for the hybrid pixels is usually determined by the simultaeous optimization of the photoreceivers and the modulators. At this wavelength, saturation effects of the photodiodes can safely be neglected for the optical powers discussed here. Whilst the absorption increases with the increased field, the resulting increase of the photogenerated carriers does not bleach the excitonic peak because the carriers are rapidly swept away by the strong field.

Optical Losses

Systems which use smart-pixel technology based on MQW detectors and modulators flip-chip bonded onto CMOS have been shown to be laser-power limited [7]. The throughput rate depends ultimately on the amount of optical power incident onto the photodetectors. The most lossy optical path from a laser source to the detectors of a smart-pixel array determines the overall system frequency. The efficiency of this path, η is

$$\eta = \eta_{doe}\eta_{bulk}\eta_{inter} , \tag{2.6}$$

where η_{doe} is the efficiency of the diffractive optical element, η_{bulk} the efficiency of the bulk optics and η_{inter} the optical interconnection. Typical efficiency values are shown to be 0.8 for the DOE, 0.5 for the bulk optics which includes relay lenses and beam splitters, 0.7 for the shuffle interconnects. The resultant efficiency is therefore η=0.28. The total laser source power P_{tot} is divided by 2^{m+1} beams in order to sort 2^m words since the bitonic sorter uses dual-rail logic encoding [17]. The photodetectors therefore receive the powers:

$$\begin{aligned} P_{high} &= 2^{-m-1}\eta\, R_m\, P_{tot} \\ P_{low} &= 2^{-m-1}\eta\, R_p\, P_{tot} \end{aligned} \tag{2.7}$$

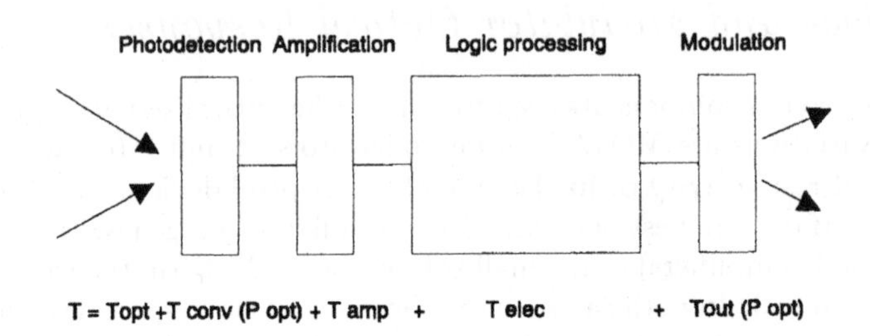

Figure 4 Functional schematic of the smart pixel with latency times

The difference in these levels will create either a current variation or a voltage variation which will need to be amplified for logic processing. The time needed for conversion of these optical signals is then a determining factor in the calculation of the system throughput rate

$$G_o F_o = 2^{-m-1} \eta \, P_{tot} \, (R_m - R_p)/E_o \, . \tag{2.8}$$

4.3 Electrical Power Budget

The laser source power and optical losses determine the maximum power incident on the photodetectors. In this sub-section, it is shown that some of the latency times occuring within a smart-pixel depend on the optical power available. Additionally an electrical power budget has to be calculated in order to predict whether heat dissipation acts also as a constraint in the throughput rate of the system. It is explained below that, in the case of hybrid-VLSI circuits, a quasi-CW laser mode of operation helps reduce the on-chip power dissipation.

The Clock Period

Knowledge of the clock period is required to determine the throughput rate. A schematic diagram of a generic smart pixel decomposed into functional blocks with their respective latency times is shown in Figure 4. The optical-to-electronic conversion time, T_{conv} depends on the optical power incident on the detectors. In order to reduce this power, smart-pixels have one or several amplification stages. T_{amp} is the time needed to amplify the optically induced current or voltage swing for logic processing. T_{conv} and T_{amp} overlap each other. For example, in the case of a voltage gain stage, amplification of the electronic signals will take place as soon as the input voltage has reached the transistor threshold voltage. T_{elec} is the propagation time of the electronic signals from the output of the amplification stage to the input of the last electronic stage. Although approximate calculations could be performed by looking at the logical depth of the pixel and the nature of the gates at each electronic stage, T_{elec} is usually determined by simulation computer programs such as SPICE. The switching time of the output electronic stage stage, T_{out}, may depend on the laser source power if the laser is operated cw as explained in the next subsection. Finally, the time of flight of the optical signals between arrays, T_{opt}, of the order of a nanosecond per foot, has also to be included. The processing time, T, of a smart-pixel is the sum of the different times described above. The clock period must be as long as the time T.

Timing of the Hybrid Chip

In the quasi-cw mode, the read beams of one array are switched off as soon as the full input voltage or current swing has been reached by the receivers of the next array. For simplicity the laser source is assumed to generate square pulses of duration w and peak power P_{peak}. A single front-end voltage gain with zero output conductance is considered here. Because of the dual-rail encoding of the signals, the differential input optical energy must satisfy the relation :

$$w(P_{high} - P_{low}) \geq \Delta E_{min} = C_{in} \frac{\Delta V_{in}}{S_{avg}} , \tag{2.9}$$

where C_{in} is the total input capacitance formed by the detectors, the bonding pad, and the input transistor gate. Typical values for this hybrid technology are given in table 3. For an amplifier of constant transconductance g_m, which sets up at its output a logic-level voltage swing ΔV_{log}, the conversion and amplification times are [26, 44] :

$$T_{conv} + T_{amp} = \frac{3}{4} C_{in} \frac{\Delta V_{in}}{S_{avg}} \frac{1}{(P_{high} - P_{low})} + 2C_{out} \frac{\Delta V_{log}}{g_m \Delta V_{in}} , \tag{2.10}$$

where C_{out} is the output capacitance of the amplification stage. T_{out} depends also on the mode of operation of the laser beam. In the quasi-cw mode, the modulator states are prepared prior to the reading of the modulators, so that we have

$$T_{out} = T_{out}(P = 0) = 2 \frac{C_{ex} V_0}{\Delta I_{trans}} , \tag{2.11}$$

where V_0 is the voltage to be set at the modulator, C_{ex} is the output capacitance of the smart pixel and ΔI_{trans} is the difference of the transistor currents flowing

Table 3 Definitions and selected values of the S-SEED technology device parameters.

DEFINITION	SYMBOL	VALUE
Applied voltage	V_0	8 V
Mean responsivity	S_{avg}	0.5 A/W
CMOS-SEED window area	A	10x10 μm^2
Maximum input voltage swing	ΔV_{in}	0.25 V
Logic voltage swing	ΔV_{log}	1 V
Word length	L	8
Transconductance	g_m	10^{-3}S
Input gate capacitance	C_{inp}	60 fF
Bond pad capacitance	C_{pad}	200 fF
First electronic stage output capacitance	C_{out}	120 fF
Detector capacitance per area	C_{cap}	200 aF/μm^2

to the output capacitance. In the cw mode of operation, the modulators are continuously read and the electronic output drivers must be properly scaled to sink the photocurrents generated by the modulators. The time required to set up the modulators at the voltage level V_0 is [25]:

$$T_{out}(P) = T_{out}(0)(f/2)\ln(\frac{f+1}{f-1}) , \qquad (2.12)$$

where $f=\Delta I_{trans}/\Delta I_{ph}$ is the ratio of the difference of the transistor currents to the difference of the modulator photocurrents. $T_{out}(0)$ is of the order of tens of picosecond, so that the output time can generally be neglected.

The processing time of the hybrid chip is then :

$$T_{hybrid} = \frac{3}{4}C_{in}\frac{\Delta V_{in}}{S_{avg}}\frac{1}{(P_{high} - P_{low})} + 2C_{out}\frac{\Delta V_{log}}{g_m \Delta V_{in}} + T_{elec} + T_{opt} , \qquad (2.13)$$

where T_{elec} is determined generally by SPICE simulations. In the case of the bitonic sorter, T_{elec} is of the order of 10 ns. Using the values of Table 3, the second term on the right-hand side of the above equation contributes about 1 ns. Because of the optical power dependence of the first term on the right-hand side, its value depends strongly on the array size. For a 32-by-32 array and 1 Watt of laser source power, the conversion time is of the order of 3 ns.

Thermal Limitations

For the hybrid-VLSI technology, the heat created by the power absorbed at the transceivers and the power dissipated by the electronic circuitry can be a serious constraint since the modulator output response is particularly sensitive to thermal effects (0.3 nm/degree). Heat-removal capabilities need be included in the benchmarking of smart-pixel arrays. In the following a maximum power dissipation of 10 W/cm^2 is assumed. It is shown below that the heat dissipated depends on the mode of operation of the laser source.

In the quasi-cw mode of operation, the total average power dissipated is:

$$P_{quasi-cw} = \frac{(P_{sw} + P_{out})T_{conv} + P_{elec}T_{elec}}{T} , \tag{2.14}$$

where P_{elec} is the power dissipated by the electronic circuitry during the time T_{elec}. P_{sw} is the power dissipated at the photodetector during the conversion time and P_{out} is the power dissipated by the modulator during the reading phase. P_{elec} is usually calculated by the use of SPICE simulation package. The detailed derivations of P_{sw}

$$P_{sw} = 2^{-m}\eta\ \eta_{doe}P_{tot}[R_m + R_p]S_{avg}\Delta V_{in} \tag{2.15}$$

and P_{out}

$$P_{out} = 2^{-m}\eta_{doe}P_{tot}S_p\ V_0 , \tag{2.16}$$

are given in reference [7].

In the cw mode of operation, more power is dissipated by the smart-pixel since the laser beams are ON during the whole processing time at the transceivers. The total average power dissipated is in this case [7] :

$$P_{cw} = \frac{(P_{sw}/2 + P_{out})[T + T_{conv}] + T_{elec}P_{elec}}{T} . \tag{2.17}$$

Figure 5 shows the average power dissipated per node and the resulting laser-source power required for the bitonic sorter under quasi-cw operation.

As anticipated, the frequency of operation and average power dissipated increase as the input power increases. The bitonic sorter, which encompasses 32-by-64 optical channels, can be operated at a maximum of 70 MHz operating frequency (point A, figure 5). At this frequency, about 1 mW/node of laser source power is available and 2 mW/node of electrical plus optical is dissipated. In the CW regime, only 53 MHz is achievable for the bitonic sorter. At any

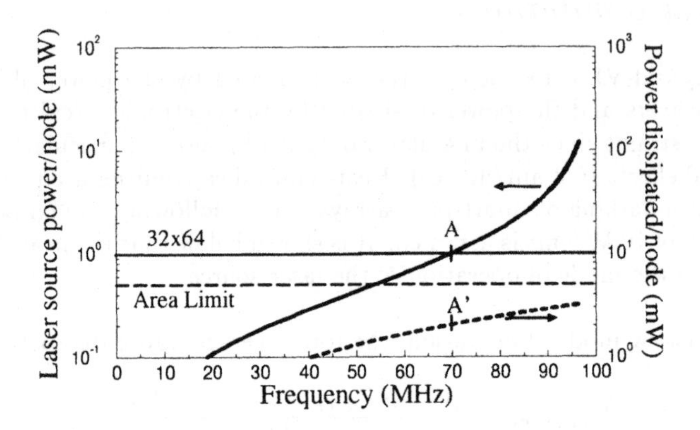

Figure 5 Average power dissipated and laser-source power required for the bitonic sorter node plotted as a function of frequency of operation.

frequency, the quasi-cw mode of operation is the more energy efficient because the optical beams are applied only when needed. In both laser regimes, the bitonic sorter is laser-source-power limited.

4.4 Maximum On/Off Chip Communication Rates

High chip operating frequency or/and complex smart-pixels do not necessarily provide the largest system throughput rate. The interplay of algorithmic, architectural and technological constraints imposes different regimes of pixel-smartness, parallelism and operating frequency. It is desirable, however, to define a figure of merit for smart-pixel chips which is system independent and relies solely on the technological constraints defined in the previous sections. A suitable performance metric is the number of bits transmitted per second on and off/chip, defined as the aggregate communication rate G which is expressed in effect pin-Hz. In the limit of large fan-out, the amount of optical power incident onto the photoreceivers is small and the conversion time is the dominant term in the expression for the processing time. The clock period becomes inversely proportional to the laser-source power per channel so that

G is independent of the array size. For the bitonic sorter

$$G_{\text{hybrid}} = 2\,\eta\,\frac{S_{\text{avg}}(R_m - R_p)}{C_{\text{in}}\,\Delta V_{\text{in}}}\,. \tag{2.18}$$

This limit provides a theoretical maximum communication rate. Using the values shown in table 3, 5.2 x 10^{11} pin-Hz achievable for 1 W of source power. The practical value, 3 x 10^{11} pin-Hz/W, shown in Table 4, differs from the maximum value because of the electronic timing which is still significant for the bitonic sorter operated at 100 MHz. For the sake of completeness, table 4 shows also the performance metrics of similar sorters based on other technologies [7].

5 FUTURE TRENDS

The sorting module design for which the performance analysis was undertaken operates close to limits. 1 watt of laser source power was assumed, the chip area used for the sorting module was taken as 1 cm^2 and no more than 10 Watts per cm^2 of thermal dissipation were allowed. Relaxation of more than one of these limits is needed in order to allow the sorter to go beyond the 200 Gbit/s communication rate (1024 channels operating at 100 MHz in and out of the chip). Various strategies can be implemented to increase the performance of smart-pixel arrays, some of which are being pursued by research groups worldwide.

5.1 High modulation rate and/or high-power lasers

The importance of modulated lasers is best explained in terms of pulse duration. In the calculation of the optical data-rate in section 3, the duration of the beam was assumed to be inversely proportional to this frequency; that is, the laser is on all the time. For the same time-average power of 1 Watt, it is in fact more judicious to provide the necessary optical switching energy in the shortest pulse duration so that the logic processing can be implemented as quickly as possible.

	S-SEED	CMOS-SEED	FET-SEED
THROUGHPUT (Mops)	3.4	140	4.8
Clock period (ns)	46	14.3	3.3
Optical pulse duration (ns)	23	4.3	0.13
Channel number	32x64	32x64	8x8
Active chip area (cm^2)	0.02	0.8	0.6
Number of logic chips required	8	1	1
Number of clock periods	13244	1000	64480
Time to sort a single 32^2-data set (μs)	609	14.3	213
Laser source power per chip(W)	1	1	1
Maximum power dissipation on-chip (W)	0.16	2.5	5.4
COMMUNICATION RATE (pin-Hz)	$4.5\ 10^{10}$	$2.9\ 10^{11}$	$3.9\ 10^{10}$
Data rate (MHz)	22	70	303
Optical pin number (i/o)	2048	4096	128
On-chip gate/switches per second	$4.5\ 10^{10}$	$2.4\ 10^{13}$	$2.3\ 10^{12}$

Table 4 Performance characteristics of S-SEEDs and hybrid-SEED-based smart-pixels for data sorting. The sorting of 32-by-32 8-bit words is chosen in order to compare the different technologies.

The mark-to-space ratio of the laser should therefore be as low as possible; this advocates for a pulse mode of operation of the laser source. The need for a high modulation laser source obviously becomes less important for a low number of optical pins since the required pulse duration is a tiny fraction of the electronic processing time. In all cases, however, the use of a modulated laser is beneficial since the laser beams would be incident on the smart pixel array only when needed, thereby decreasing the optically induced power dissipation compared to the CW case as shown in Section 4.

If more laser source power were available for the bitonic sorter, 10^{12} pin-Hz could be achieved for the same channel number. In effect, the optical pulse duration required at the detectors is reduced, which allows the data to be propagated at a higher clock rate. Alternatively, either the number of pixels could be increased at the expense of less gates per pixel. The benefits specified above can only be realized if either:

1. The heat removal capability is increased to permit faster electronic clock rate.
2. Or the chip size is increased, thereby allowing a larger number of pixels without decreasing their complexity. This strategy can only be implemented in the limit of sensible field-of-view for the optical hardware.

The design of new laser sources emphasized in Section 4.2 (page 32) addresses precisely the needs for high-power, high-modulation rate, short-pulse duration lasers.

5.2 The design of new photoreceiver circuits

The case for high sensitivity, low power dissipation photoreceivers can now clearly be formulated. Highly sensitive receivers reduce the required optical switching energy. This allows the system to be operated either at a lower laser source power (constant communication rate) and/or to increase the optical data-rate per pin (constant number of gates per pins). If such receivers have also low power dissipation the raw electronic processing power of up 10^{14} gate-switches per chip can be fully utilized for the sole use of logic processing. An appealing solution is the design of Current Controlled Voltage Sources or Current Controlled Current Sources [33] (CCCS or CCVS), which reduces the effective input capacitance seen by the optically converted signal. Several of such circuits, such as low-power transimpedance amplifiers [43], current conveyors [38], charge-sensitive amplifiers [8] and current-sense amplifiers [43] have been proposed to avoid seeing the front-end capacitance.

The reduction in optical and amplifier energy required per detector has the advantages of increasing the number of pixels that could be driven (if the chip area was allowed to increase) and of reducing the power dissipated by the electronics, which can therefore be clocked faster. Alternatively it will allow the design of demonstrators with off-the-shelf compact laser sources which would offer better packaging solutions. It should be noted that the design of amplifiers for smart-pixel technology differs significantly from the design of photoreceiver amplifiers used in the telecommunication industry in three aspects. The lack of real-estate on the silicon prevents, on the smart-pixel array, the design of Retiming, Reshaping and Regenerating circuitry (R^3 systems) which is usually encountered in high performance telecommunication systems. Albeit at the moment not noise-limited, the smart-pixel photoreceivers must be designed so that the Bit Error Rate (BER), for a given bit-rate, must lower than that of the telecommunication receivers. In addition, they must dissipate less energy and be more compact because of their sheer number on 1 cm^2 area chip.

Finally, the last strategy consists of reducing the electronics feature size and the voltage requirement. This will happen in all circumstances because of the progress of conventional electronics. The pixel complexity or the clock rate can be then increased. Moreover, the gain of the amplifiers at the front-end interface can then be reduced, thereby increasing the potential clock rate.

6 POTENTIAL APPLICATION

Beyond the tremendous value of the on/off chip communication rate, the hybridization of conventional electronic logic with optical i/o offers a possible unexpected application in the field of low power electronics.

6.1 Low Power Electronics

Low power strategies in VLSI electronics are becoming increasingly important in the development of new portable electronic information systems for high throughput applications. Portable computers has become the fastest growing segment in the computer industry and future personnal communications services (PCS) are expected to encompasses full-motion digital video and control via speech recognition. Unfortunately these high-throughput applications are unlikely to benefit from the classical power-down schemes encountered in today's laptops [24]. This situation is all the more exacerbated by the unabated increase in clock frequency, gate density and power consumption of the VLSI chips as predicted by the Semiconductor Industry Association Roadmap [36]. In addition, the severe restrictions on power imposed by portable computers are unlikely to be eased in the near feature because of the small improvements forecast in battery performance [9] and low-cost thermal management.

To address this issue, the dynamic power consumption on a CMOS chip can be reduced through a decrease of the clock frequency, the power supply and /or the device and interconnect capacitance. The most promising strategy, that is the scaling down of the voltage, provides a quadratic improvement in the power-delay product of a logic family at the expense however of a linear increase in the gate delay. As the main motivation is to maintain the aggregate bandwidth, architectural and algorithmic strategies need to be implemented to compensate for the increased delays occuring at low voltages [1]. Several solutions have been proposed which range from the use of precomputation logic to reduce the duty factor or occupation ratio in the following clock cycle [1] to the design of self-timed circuits and adaptive scaling of the supply voltage [31]. The most promising approach seems to rely on a combination of parallelisation and pipelining of the system architecture in order to reduce the circuit logic depth and the speed requirement [3]. Whereas the reduced logic depth diminishes the power induced by the dynamic hazards, the decreased clock frequency allows the circuit to be biased at a lower voltage supply. Such an architecture therefore shows substantial decrease in power consumption through these two mechanisms.

As an example, Chandrakasan and co-workers have demonstrated that the implementation of an 8-bit long adder-comparator circuit dissipates 5 times less power than its straightforward implementation [3]. The amount of parallelism needed, however, must be traded-off against the resulting silicon area, power dissipated and pin-out overheads which result in the calculation of an optimal voltage. Unfortunately, the applications targeted for such schemes (video-compression, speech recognition) use time-multiplexed architectures which prevent the speed reduction of the logic circuitry, and therefore do not allow the reduction in the power supply voltage.

We believe that smart-pixel technology is able to answer effectively all these drawbacks. The massive parallelism provided by free-space optics alleviates the need for time-multiplexing architectures necessary in conventional electronics for these types of applications. The required pin-out overheads can be tackled effectively by the optical addressing of the chip in the third dimension. On the issue of electrical power consumption, optical interconnects have long been known to dissipate less power than electrical connections above a certain frequency and communication length. Optical clock distribution schemes, for example, can provide up to 30% decrease of overall power dissipated on-chip [4, 18]. In analogy with low-power electronics, the parallelisation of smart-pixels clocked at lower frequency exhibit also a net decrease in power dissipation. Research work has started to demonstrate that smart-pixel technology can provide the possibility of achieving large throughput, low-power systems for computationally intensive applications needed in future portable environments.

7 CONCLUSION

The goal of the chapter was to present basic and system-specific considerations about smart-pixel technology and to outline possible future trends of this technology. Although this field of activity is rapidly changing, smart-pixel technology is fundamentally constrained by the chip area, the laser source power and the heat dissipated by the system. These three constraints which belong to the optical and electrical domains were shown to determine within an order of magnitude the smartness and granularity of the pixel. For a more refined calculation of the performance metrics of this technology, it is necessary to revert to the system under study. In that respect, a semi-empirical method was described which combines analytical and simulation results related to the optical and electrical power budget. With this method, the aggregate throughput rate can not only be calculated but also be extrapolated when some of the constraints are relaxed. Small-area, low-power-dissipation and high-sensitivity photoreceivers as well as high-modulation rate, high-power laser sources have been identified as the two main fields which need to be improved upon if the full theoretical throughput rate offered by smart-pixel technology is to be harvested. Finally, new research at Heriot-Watt and McGill University has started to evaluate the potential of smart-pixel technology in the field of low-power electronics. The pipelining-parallelisation of novel low-power systems could benefit from the high-communication rate feature of this promising technology.

Acknowledgements

The Scottish Collaborative Initiative in Optoelectronic Sciences (SCIOS) involves research groups from several Scottish Universities : Heriot-Watt University, Edinburgh University, the University of St Andrews, and the University of Glasgow. SCIOS is funded by the Engineering and Physical Sciences Research Council (EPSRC) of the United Kingdom. Contributions from member of SCIOS and prior members of the Heriot-Watt Optical Computing Group are gratefully acknowledged.

Studies on the potential application of smart-pixel technology in low power electronics are made possible thanks to the financial support of NATO through the NATO grant referenced CRG 961061 held by Dr. Desmulliez and Prof. F. Tooley, from McGill University, Montreal, Canada. Studies on the optical and electrical limits of smart-pixel technology have been made possible through the Esprit program SPOEC (Smart-Pixel Opto-Electronic Connections) funded by the European Commission. One of the author (M. Desmulliez) acknowledges stimulating discussions with A.Z. Shang from McGill University.

REFERENCES

[1] M. Alidina, S. Devadas, A. Ghosh, and M. Papaefthymiou. Precomputation-based sequential logic optimization for low power. *IEEE Transactions on VLSI Systems*, 2:426–435, 1994.

[2] K. E. Batcher. Sorting networks and their applications. In *Spring Joint Computer Conference*, pages 307– 314, 1968.

[3] A. P. Chandrakasan, S. Sheng, and R. W. Brodersen. Low power CMOS digital design. *IEEE Journal of Solid-State Circuits*, 27:473–484, 1992.

[4] M. P. Y. Desmulliez, P. W. Foulk, and M. R. Taghizadeh. Optical clock distribution for multichip module. Accepted for Optical Review, 1997.

[5] M. P. Y. Desmulliez, P. W. Foulk, and B. S. Wherrett. Hybrid technology for optoelectronic parallel processing. basic considerations. Accepted to OSA Topical Meeting in Optics in Computing 97, Lake Tahoe, Nevada, USA, 1997.

[6] M. P. Y. Desmulliez, F. A. P. Tooley, J. A. B. Dines, N. L. Grant, D. J. Goodwill, D. Baillie, B. S. Wherrett, P. W. Foulk, S. Ashcroft, and P. Black. Perfect-shuffle interconnected bitonic sorter: optoelectronic design. *Applied Optics*, 34:5077–5090, 1995.

[7] M. P. Y. Desmulliez, B. S. Wherrett, A. J. Waddie, J. F. Snowdon, and J. A. B. Dines. Performance analysis of SEED-based smart-pixel arrays used in data sorting. *Applied Optics*, 35(32):6397-6416, 1996.

[8] J. A. B. Dines. Smart-pixel optoelectronic receiver based on a charge sensitive amplifier design. *IEEE Journal on Selected Topics in Quantum Electronics*, 2:117–120, 1996.

[9] Eager. Advances in rechargeable batteries pace portable computer growth. In *Silicon Valley Personal Computer Conference*, pages 693–697, 1991.

[10] M. R. Feldman, S. C. Esener, C. C. Guest, and S. H. Lee. Comparison between optical and electrical interconnects based on power and speed considerations. *Applied Optics*, 27:1742–1751, 1988.

[11] D. Fey. Characterization of massively parallel smart-pixels systems for the example of a binary associative memory. In *Second International Workshop on Massively Parallel Processing using Optical Interconnections*, pages 76–83, San Antonio (USA), Oct. 1995. IEEE Computer Society Press.

[12] S. R. Forrest and H. S. Hinton. Special issue on smart pixels. IEEE Journal of Quantum Electronics, Vol. 29, 1993.

[13] W. Franz. Z. Naturforsch. Teil. A **13**, 1958.

[14] J. W. Goodman. *Optical Processing and Computing*, chapter Optics as an interconnect technology, pages 1–32. Academic Press, San Diego, 1989. H.H. Arsenault, T. Szoplik, and B. Macukow, Eds.

[15] M. Goodwin, A. Moseley, M. Kearley, R. Morris, C. Kirby, J. Thompson, R. Goodfellow, and I. Bennion. Opto-electronic component array for optical interconnection of circuits and subsystems. *Journal of Lightwave Technology*, 9:1639–1644, 1991.

[16] S. H. Hinterlong and H. M. Hall. Bringing photonics to broadband switching. *AT&T Technical Journal*, pages 71–80, 1994.

[17] H. S. Hinton, T. J. Cloonan, F. B. McCormick, A. L. Lentine, and F. A. P. Tooley. Free-space digital optical systems. *Proceedings of the IEEE*, 82:1632–1649, 1994.

[18] A. Iwata. Optical interconnections for ULSI technology innovation. *Optoelectronic Devices and Technology*, 9:778–783, 1994.

[19] L. V. Keldysh. Zh. Eksp. Teor. Fiz. **34**, 1958. [Sov. Phys. - JETP **7**, pp. 788 (1958)].

[20] R. W. Keyes. The wire-limited logic chip. *IEEE Journal of Solid-State Circuits*, 17:1232–1233, 1982.

[21] D. R. Kiefer and V. W. Swanson. Implementation of optical clock distribution in a supercomputer. In *Optical Computing*, number 10, pages 261–263, 1995. OSA Technical Digest Series (Optical Society of America, Washington DC, 1995).

[22] A. V. Krishnamoorthy, J. E. Ford, K. W. Goossen, J. A. Walker, A. L. Lentine, S. P. Hui, B. Tseng, L. M. F. Chirovsky, R. Leibenguth, D. Kossives, D. Dahringer, L. A. D'Asaro, F. E. Kiamilev, G. F. Aplin, R. G. Rozier, and D. A. B. Miller. Photonic page buffer based on GaAs multiple-quantum-well modulators bonded directly over active silicon CMOS ciruits. *Applied Optics*, 35:2439–2448, 1996.

[23] A. V. Krishnamoorthy and D. A. B. Miller. Scaling optoelectronic-VLSI circuits into the 21st century: a technology roadmap. *IEEE Journal on Selected Topics in Quantum Electronics*, 2:55–76, 1996.

[24] Z. L. Lemnios. Manufacturing technology challenges for low power electronics (LPE). DARPA Project, 1996. http://eto.sysplan.com.

[25] A. L. Lentine, L. M. F. Chirovsky, L. A. D'Asaro, E. Laskowski, S. Pei, M. Focht, J. Freund, G. Guth, R. Leibenguth, L. Smith, and T. K. Woodward. Field-effect transistor self electrooptic effect (FET-SEED) electrically addressed differential modulator array. *Applied Optics*, 33:2849–2855, 1994.

[26] A. L. Lentine, L. M. F. Chirovsky, and T. K. Woodward. Optical energy considerations for diode-clamped smart-pixel optical receivers. *IEEE Journal of Quantum Electronics*, 30:1167–1174, 1994.

[27] Ceramic ball grid array package. Semiconductor International, Nov. 1996. page 64.

[28] D. T. Lu, H. Ozguz, P. J. Marchand, A. V. Krishnamoorthy, F. Kiamilev, R. Paturi, S. H. Lee, and S. C. Esener. Design trade-offs in optoelectronics parallel processing systems using smart-SLMs. *Optical and Quantum Electronics*, 24:S379–S403, 1992.

[29] D. A. B. Miller. Hybrid SEED - massively parallel optical interconnections for silicon ICs. In *Second International Workshop on Massively Parallel Processing using Optical Interconnections*, pages 2–7, San Antonio (USA), Oct. 1995. IEEE Computer Society Press.

[30] D. A. B. Miller, D. S. Chemla, T. C. Damen, A. C. Gossard, W. Wiegmann, T. H. Wood, and C. A. Burrus. Band-edge electroabsorption in quantum well structures - the quantum confined Stark effect. *Physics Review Letters*, 53:2173, 1984.

[31] L. S. Nielsen, C. Niessen, J. Sparso, and K. Van Berkel. Low-power operation self-timed circuits and adaptive scaling of the supply voltage. *IEEE Transactions on VLSI Systems*, 2:391–397, 1994.

[32] H. Ozaktas and J. W. Goodman. *Frontiers of Computing Systems Research 2*, chapter The limitations of interconnections in providing communication between an array of points, pages 61–130. Plenum Press, New York, 1991.

[33] A. Payne and C. Toumazou. Analog amplifiers : classification and generalization. *IEEE Tansactions on Circuits and Systems-I*, 43:43–50, 1995.

[34] D. V. Plant, B. Robertson, H. S. Hinton, W. M. Robertson, G. C. Boisset, N. H. Kim, Y. S. Liu, M. R. Otazo, D. R. Rolston, and A. Z. Shang. An optical backplane demonstrator system based on FET-SEED smart pixel arrays and diffractive lenslet arrays. *IEEE Photonics Technology Letters*, 7:1057–1069, 1995.

[35] A. Schiltz. A review of planar techniques for multichip modules. *IEEE Transactions on Components, Packaging, and Manufacturing Technology*, 15:236–244, 1992.

[36] Semiconductor Industry Association. The national technology roadmap for semiconductors, 1994. San Jose, California.

[37] T. H. Szymanski and H. S. Hinton. A reconfigurable intelligent optical backplane for parallel computing and communications. submitted to Applied Optics, 1997.

[38] N. Tan and S. Eriksson. Low-power chip-to-chip communication circuits. *Electronic Letters*, 30:1732–1733, 1994.

[39] P. P. Vasilev, I. H. White, D. Burns, and W. Sibbett. High-power, low-jitter encoded picosecond pulse genration using an RF-locked, Q-switched multi-contact GaAs/GaAlAs diode laser. *Electronic Letters*, 29:1593, 1993.

[40] Vitesse VSC864A-2. Gallium arsenide 64x64 crosspoint switch. Preliminary data sheet, 1993.

[41] A. C. Walker, M. P. Y. Desmulliez, F. A. P. Tooley, D. T. Neilson, J. A. B. Dines, D. A. Baillie, S. M. Prince, L. C. Wilkinson, M. R. Taghizadeh, P. Blair, J. F. Snowdon, B. S. Wherrett, C. Stanley, F. Pottier, I. Underwood, D. G. Vass, W. Sibbett, and M. H. Dunn. Construction of demonstration parallel optical processors based on CMOs/InGaAs smart pixel technology. In *Second International Workshop on Massively Parallel Processing using Optical Interconnections*, pages 180–187, San Antonio (USA), Oct. 1995. IEEE Computer Society Press.

[42] T. K. Woodward, L. M. F. Chirovsky, A. L. Lentine, L. A. d'Asaro, E. Laskowski, M. Focht, G. Guth, S. Pei, F. Ren, G. Przybylek, L. Smith, R. Leibenguth, M. Asom, R. Kopf, J. Fuo, and M. Feuer. Operation of a fully integrated GaAs-$Al_xGa_{1-x}As$ FET-SEED - a basic optically addressed integrated circuit. *IEEE Photonic Letters*, 4:616–618, 1992.

[43] T. K. Woodward, A. V. Krishnamoorthy, A. L. Lentine, and L. M. F. Chirovsky. Optical receivers for optoelectronic VLSI. *IEEE Journal on Selected Topics in Quantom Electronics*, 2:106–116, 1996.

[44] T. K. Woodward, A. L. Lentine, and L. M. F. Chirovsky. Experimental sensitivity studies of diode clamped FET-SEED smart pixels optical receivers. *IEEE Journal of Quantum Electronics*, 30:2319–2324, 1994.

[45] A. Yu, M. Krainak, and G. Unger. 1047-nm laser diode master ocillato Nd:YLF power amplifier laser system. *Electronic Letters*, 29:678–679, 1993.

3

ENERGY REQUIREMENT AND SPEED ANALYSIS OF ELECTRICAL AND FREE-SPACE OPTICAL DIGITAL INTERCONNECTIONS

Gökçe I. Yayla
Philippe J. Marchand
and Sadik C. Esener

University of California San Diego
ECE Department
La Jolla, CA

1 INTRODUCTION

Scaling of VLSI technology has been dramatically increasing microelectronic device densities and speeds. However, the interconnection technology between devices does not advance proportionally. Limited available interconnect materials compatible with VLSI and packaging technologies, increased wire resistance as a result of scaling, residual wire capacitance due to fringing fields and fields between interconnect wires are among the factors that prohibit drastic improvement of the electrical interconnect performance. As a result, the performance of VLSI systems become increasingly more dominated by the performance of long interconnects. To overcome this limitation, free-space optical interconnects have been suggested, where long electrical interconnects are replaced by an optical transmitter, a photodetector, and interconnection optics between them [2, 10, 13–15, 19, 22, 27]. This scheme, although devoid of electrical interconnection parasitics, has its own difficulties. Unavailability of monolithically integrated optical transmitters on silicon imposes hybrid integration schemes with larger parasitic capacitance and increased cost. Transformation of information from electrical to optical domain and vice versa introduces severe inefficiencies into the energy budget. There are voltage incompatibility issues between some transmitter technologies and VLSI technology as well.

It is therefore essential to clarify the regions of superiority between electrical and optical interconnects, and identify some of the important trade-offs. This would help system designers choose the proper interconnect technology for the

P. Berthomé and A. Ferreira (eds.), Optical Interconnections and Parallel Processing: Trends at the Interface, 49-128.

application in question. In this chapter, we compare the speed performance and energy cost of a class of electrical and optical interconnections for their use in large scale computing systems. A similar analysis was presented by Feldman et al. [7] about ten years ago. This study expands that analysis in many ways. In addition to on-chip interconnects, this study considers off-chip electrical interconnects as well. The analysis presented here involves the design of the interconnections to maximize the speed. Interconnects are compared after they are designed in detail. The effect of light modulator and emitter drivers as well as electrical line buffers are included. Multiple-Quantum-Well (MQW) modulator saturation phenomena, the dependence of laser output power on speed, the dependence of laser threshold on output power, the effect of higher transmitter supply voltage than VLSI supply voltage, effect of parasitics due to the hybrid integration of silicon devices with optical transmitters are all included. In the case of off-chip interconnects, the effects of both series and parallel termination schemes are discussed. Superbuffers are designed to minimize the propagation delay between minimum logic and off-chip line drivers. Optimum repeater design is adopted to maximize the speed of distributed wafer-scale interconnections. Both return-to-zero (RZ) and non-return-to-zero (NRZ) transmission schemes are used where appropriate in order to minimize energy dissipation. The comparison is extended to clock distribution application for both multichip module (MCM) and wafer-scale integration (WSI) cases.

In Section 2, we present the basic assumptions and limitations of the study. In Section 3, we define the interconnection in the scope of this analysis, and introduce fundamental equations of energy calculation which are referred to in the subsequent sections. The design and analysis of off-chip and on-chip electrical interconnects for data transmission as well as for clock distribution are presented in Sections 4 and 5 respectively. Section 6 is devoted to the analysis of the optical interconnects, separately for the cases of PLZT (lead lanthanum zirconium titanate) modulators, MQW modulators and Vertical Cavity Surface Emitting Lasers (VCSEL's) as optical transmitters. In each case, the performance of the optical interconnect is compared to on-chip and off-chip electrical interconnects, for one-to-one and fan-out connections. The suitability of different optical transmitter technologies for large-scale clock distribution is also considered, and compared to their off-chip and on-chip electrical counterparts. In Section 7, we discuss some of the effects of technology scaling on the performance of electrical and free-space optical interconnects. Finally, some conclusions are presented in Section 8.

2 ASSUMPTIONS

In this chapter, we use the term "interconnection" to refer to the physical medium designed for digital communication between electronic sub-systems. The following are the assumptions under which we analyze such interconnections:

A1 We do not consider a certain type of computing algorithm or architecture to calculate the required interconnection line lengths or fan-out in a particular system. We take interconnection length (L_{int}) and fan-out (N) as independent variables.

A2 The application that we are considering is: "*large-scale digital computing systems using dense interconnections*". This leads to the assumption of silicon CMOS VLSI technology for monolithic integration and Multi-Chip Module (MCM) technology for packaging of silicon chips. This restricts our analysis to a certain class of on-chip and off-chip digital interconnections. Off-chip interconnections are those that interconnect chips on an MCM substrate, whereas on-chip connections start and end within a chip. Since interconnection line length is an independent parameter in our analysis, increasing on-chip line length automatically extends the analysis to the wafer scale integration (WSI) domain.

A3 We analyze on-chip and off-chip electrical interconnections separately. Performance of systems using both type of interconnections in the same channel can be estimated by combining the results of the independent studies.

A4 We consider both one-to-one and fan-out type of connections in our analysis. Analog fan-in is not considered due to the assumption of digital communication.

A5 To ease the communication protocol, we assume synchronous communication where data is forced to the transmitter end of the interconnection by the rising edge of a global clock signal, and is sensed at the receiver end of the interconnection by the falling edge of the clock. Therefore, assuming that overall system speed is limited by the speed of interconnections, the sum of the delays of components through the channel as well as the maximum clock skew determine the minimum clock period, or maximum frequency of synchronous operation.

A6 We assume static CMOS logic design with rail-to-rail voltage swings. However, same analysis methodology can be applied to dynamic or reduced voltage swing logic designs by proper adjustment of voltage swings and currents in the analysis.

A7 In our study, we do not include the scaling analysis of VLSI technology. We use 0.5 micron CMOS technology parameters for numerical illustrations. However, in Section 7, we discuss the first order effects of technology scaling on both electrical and optical interconnect performance.

A8 We consider free-space optical interconnects for optical interconnection. However, we do not involve with the design of the optical routing sub-system, rather, we model it with an optical time-of-flight delay depending on the optical interconnection path length, and an optical power transfer efficiency depending on whether light modulators or light emitters are used as optical transmitters. The power efficiency is assumed to be independent of the interconnection length.

A9 For optical interconnection, we consider flip-chip bonded light transmitters (PLZT and MQW light modulators, and VCSELs) to silicon, and monolithically integrated reverse-biased silicon p-n junctions as photodiodes.

A10 For the series terminated off-chip electrical interconnection as well as for the modulator-based optical interconnections, we assume " non-return to zero " type of communication where the channel logic level is not altered unless a new data bit to be transmitted is different than the previously transmitted data bit. For the parallel terminated electrical interconnection and the VCSEL-based optical interconnection cases, we assume return to zero transmission scheme, because the DC power consumption due to the parallel termination resistor or the laser current is much higher than the power consumption of the interconnection due to switching.

A11 In the optical interconnection channel, we will assume that required bit-error-rate can be achieved by requiring a certain voltage swing at the photo-diode output which, in turn, requires a certain input optical power. In our calculations, we will assume a photo-diode output voltage swing of 330mV, which is approximately equal to the transition width of the CMOS inverter transfer characteristic for 3.3V power supply voltage.

A12 In the off-chip interconnection case, we assume that the interconnection conductor is lossless. This approximation holds within the limits of the independent parameters used.

3 DEFINITION OF INTERCONNECTION AND ESTIMATION OF ENERGY

Based on Assumption A2, we define the interconnection in our scope as follows: an interconnection is the physical implementation of a 1-bit wide digital communication channel within or between digital VLSI subsystems (chips, wafers), involving parasitics of the medium as well as active and passive design components used to force, restore, enhance, route(optical) and sense the data in the channel. This definition is illustrated in Figure 1. Note that, we require the interconnection to connect minimum geometry gates due to the large-scale integration requirement.

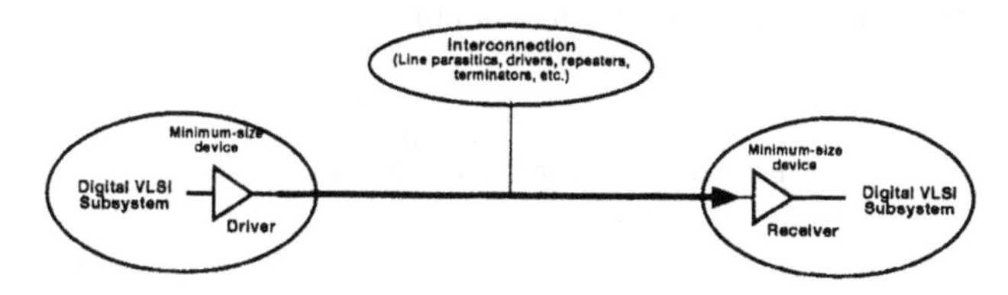

Figure 1 Definition of interconnection in the context of this chapter

Let us now consider the average energy requirement of 1 bit data transmission through the interconnection. Since in CMOS design, any logic gate (or a combination of gates) can be represented electrically with an equivalent inverter, we will consider the simple inverter circuit shown in Figure 2.

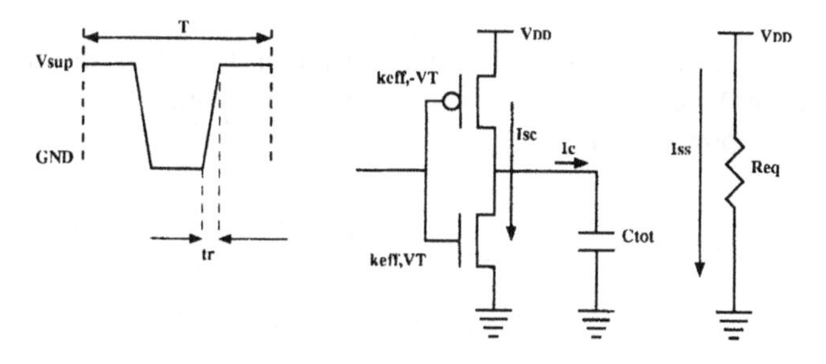

Figure 2 Electrical model of interconnection for the calculation of energy requirement. An inverter models all the logic devices in the interconnection, and the resistor models all the devices that require steady-state power. C_{tot} is the total interconnection capacitance that is switched during transmission of a digital bit.

This inverter represents all the logic devices in the interconnection, whereas the capacitance C_{tot} represents the total capacitance switched during data transmission. As the input to the inverter switches from V_{sup} to ground, a current flows from power supply through the PMOS transistor. Part of the average value of this current is used to charge C_{tot} (capacitive component), while the remaining part flows to ground through the NMOS transistor, which is partly "ON" during switching (short-circuit component). The instantaneous capacitive current is given by:

$$i_c = C_{tot} \cdot \frac{dv}{dt}, \tag{3.1}$$

and the supply power associated with this current is:

$$p_c = i_c \cdot V_{sup} = C_{tot} \cdot \frac{dv}{dt} \cdot V_{sup}. \tag{3.2}$$

The energy is the integral of the power over the period of time that the power is dissipated:

$$E = \int_t p \cdot dt, \tag{3.3}$$

using Equation 3.2 in Equation 3.3 and the integration over the full voltage range (based on Assumption A6) gives the capacitive component of the energy drawn from the power supply:

$$E_c = C_{tot} \cdot V_{sup}^2. \tag{3.4}$$

From the electrostatic theory, we know that half of this energy is stored on C_{tot}; the other half therefore, must have been dissipated as heat over the PMOS transistor's resistance during charging of C_{tot}. Note that during the switching of the input from 0 to V_{sup}, the inverter does not require any capacitive energy from the supply, since the capacitive discharge current originates from the energy stored on C_{tot} and not from the power supply. Therefore, Equation 3.4

represents the total capacitive energy requirement from the power supply, in a period of the input signal involving two opposite transitions.

On the other hand, the average short-circuit current of a CMOS inverter (due to two switchings in one period and assuming equal rise and fall times) is given as [25]:

$$I_{SC} = \frac{1}{12} \cdot k_{eff} \cdot \frac{(V_{sup} - 2V_T)^3}{V_{sup}} \cdot \frac{t_r}{T}, \tag{3.5}$$

where k_{eff} is the effective transistor transconductance parameter (modeling the equivalent transconductance of all the gates in the interconnection), V_T is the transistor threshold voltage, and t_r and T are the rise time and the period of the input signal respectively. Multiplying Equation 3.5 with V_{sup} to calculate the power, and applying Equation 3.3 over the period gives the short-circuit component of the energy drawn from the power supply as a result of two opposite switchings of the input signal:

$$E_{SC} = \frac{1}{12} \cdot k_{eff} \cdot t_r \cdot (V_{sup} - 2V_T)^3. \tag{3.6}$$

So far we have considered energy due to capacitive and short-circuit currents. Some circuitry in the interconnection may also consume considerable steady-state current due to termination, biasing or high leakage. Covering such cases, we can express the total energy (per period of the input signal) as:

$$E_{/T} = E_C + E_{SC} + E_{SS}, \tag{3.7}$$

where E_{SS} represents the energy consumed due to the steady-state currents:

$$E_{SS} + V_{sup}(I_H T_H + I_L T_L). \tag{3.8}$$

In Equation 3.8, I_H and I_L are the average steady-state currents from supply to ground during steady-state high and low level of the input signal, and T_H and T_L are the durations of high and low logic levels.

Let us now consider Figure 3, which shows the average scenario of 4 bits of serial data transmission based on Assumptions A5 and A10. From Figure 3(a), we observe that, in non-return to zero case, on the average, the channel experiences only one pair of opposite switching (low-to-high, high-to-low) per 4 bits of data transmission, and T_H and T_L are each equal to 2 clock periods. Average energy per bit can be expressed as:

$$E_{nrz/bit} = E_C/4 + E_{SC}/4 + E_{SSnrz}/4, \quad (3.9)$$

where:

$$E_{SSnrz} = 2V_{sup}T(I_H + I_L). \quad (3.10)$$

In the return-to-zero case (Figure 3(b)), the channel performs two pairs of switching per 4 bits of data transmission, and T_H and T_L are equal to 1 and 3 clock periods respectively. In this case, the average energy per bit, is then:

$$E_{rz/bit} = E_C/2 + E_{SC}/2 + E_{SSrz}/4, \quad (3.11)$$

where:

$$E_{SSrz} = V_{sup}T(I_H + 3I_L). \quad (3.12)$$

Note that, in the case of parallel terminated electrical interconnect where the parallel termination impedance is at the end of the line, one must take into account the finite time-of-flight delay (t_f) of the electrical waveform (from source to the end of line) in the energy calculation of the termination. Specifically, t_f should be subtracted from the high-level duration of data T_H at the driver chip output, since only after t_f delay of the waveform that the signal reaches the termination and causes a current through it. When the channel logic level switches back to "0" at the transmitter output, parallel termination resistor continues to conduct current, but the source of this energy does not come from the supply (since the transmitter driver's output is disconnected from the power supply), but it comes from the energy stored in the transmission line.

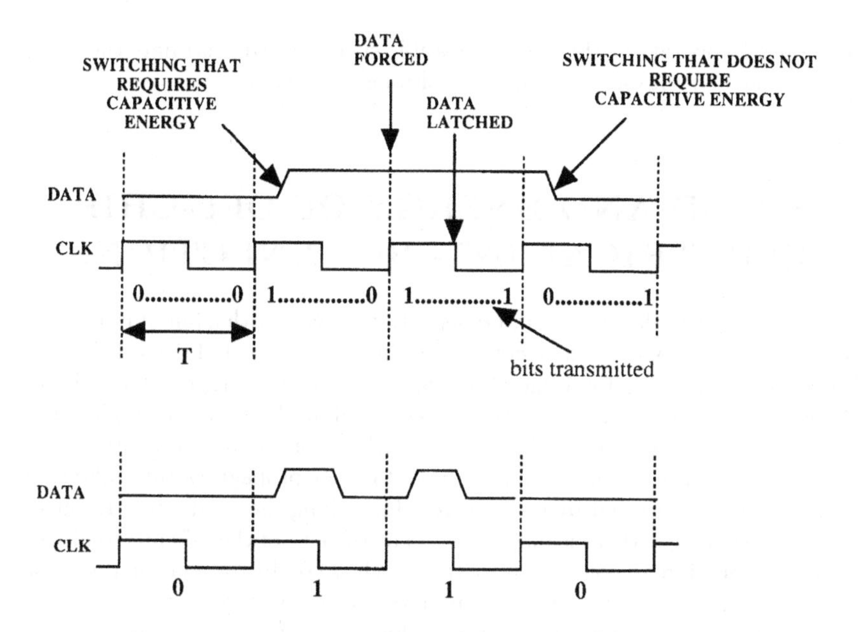

Figure 3 Average 4 bit transmission through the channel, a) non-return to zero, b) return to zero transmission.

Equations 3.4, 3.6, 3.9 and 3.11 are necessary and sufficient to calculate the energy requirement of the interconnection. Beside the technology parameters V_{sup} and V_T, evaluation of these equations require the following five parameters be known:

1. C_{tot}, total capacitance switched during transmission,
2. k_{eff}, effective transconductance of all active devices in the interconnection,
3. I_H, I_L, high and low level steady-state currents during data transmission,
4. t_r, rise time of the signal in the interconnection,
5. T, period of the synchronous system clock.

In order to calculate the above five parameters, we will apply the well-established circuit techniques used to minimize the propagation delay through interconnections, such as the use of superbuffers, optimum repeaters or transmission line terminations. Once the interconnection is designed using these techniques to operate at the highest possible speed, we will estimate the energy requirement of 1 bit data transmission through the channel.

4 SPEED AND ENERGY OF OFF-CHIP ELECTRICAL INTERCONNECTIONS

Figure 4 illustrates the off-chip interconnection scheme with a fan-out of N. The transmitter chip involves an electronic sub-block composed of dense minimum-size devices necessary for large-scale computation or storage. This block is followed by a superbuffer to drive a large size line driver where a global off-chip interconnection is needed. The line driver forces the data into the off-chip conductor via an output pin. The data then propagates throughout the conductor and is received by the receiver chips along the conductor. In one-to-one connection case, there is only one chip at the end of the off-chip conductor. Each receiver chip is connected to the off-chip conductor via an input pin, which is then connected to a minimum-size inverter to receive and restore the data for use in the following electronic block. If parallel termination scheme is used, then the off-chip conductor is terminated with a parallel termination resistor R_T. If series termination is used, then the line driver output resistance is matched to the off-chip conductor's impedance.

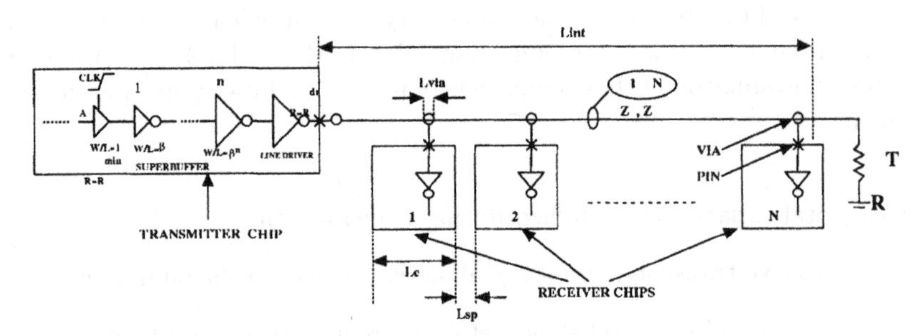

Figure 4 Model of off-chip electrical connection with fan-out.

Because of the large width and thickness of the off-chip conductors as well as the low resistivity of the materials used, such off-chip connections offer very low unit length resistance (< 1 ohm/cm), and thus for practical purposes, they can be considered lossless. For short interconnection lengths, transmission line behavior can be neglected, and the line acts as a lumped capacitor. In this regime, the optimum superbuffer (to minimize the propagation delay) can be designed based on the total load capacitance of the driver chip [17]. The interconnection can be treated as a lumped capacitor when [1, 9]:

$$t_r > 5t_f, \tag{3.13}$$

where t_r is the rise time of the data signal in the interconnection and t_f is the time-of-flight delay of the signal propagation through the interconnection conductor. For long interconnection lengths, such that $t_r < 2.5t_f$, the transmission line phenomena becomes dominant [9]. To enable the continuity of the analysis, we will use a stricter criteria in this study, and assume that the transmission line phenomena needs to be considered when:

$$t_r < 5t_f. \tag{3.14}$$

4.1 Interconnection in the lumped capacitor regime ($t_r > 5t_f$)

In this regime, a superbuffer is necessary and sufficient to drive the total load capacitance of the driver chip with minimum delay [9, 17]. The load capacitance of the driver chip can be estimated (Figure 4) as:

$$C_L = C_{dr} + L_{\text{int}} C_{\text{int}off} + N C_{rc}. \tag{3.15}$$

In Equation 3.15, L_{int} and $C_{\text{int}off}$ are the off-chip interconnection length and interconnection capacitance per unit length, and C_{dr} and C_{rc} are the driver chip output and receiver chip input capacitances:

$$C_{dr} = C_{pin} + C_{sb,o}, \tag{3.16}$$

$$C_{rc} = C_{pin} + C_{\min,i} \tag{3.17}$$

where C_{pin} is the chip package pin capacitance given by the packaging technology, $C_{\min,i}$ is the minimum inverter input capacitance given by the on-chip integration technology, and $C_{sb,o}$ is the last superbuffer stage output capacitance given by Equation 3.101 in Appendix A. Using Equation 3.15 in Equation 3.99, solving n, number of superbuffer stages, from Equation 3.101 and equating to Equation 3.99 provides C_L as a function of solely technology parameters and independent variables:

$$C_L = \frac{C_{pin} + L_{\text{int}} C_{\text{int}off} + N C_{rc}}{1 - \frac{C_{\min,o}}{\beta C_{\min,o}}}, \tag{3.18}$$

where $C_{\min,o}$ is the minimum inverter output capacitance and is the superbuffer tapering factor (see Appendix A for details). From Figure 4, we see that the interconnection length as a function of the number of chips (N) is:

$$L_{\text{int}} = L_{eff}(N+1), \tag{3.19}$$

where:

$$L_{eff} = L_c + L_{sp}. \tag{3.20}$$

In Equation 3.20, L_c is the side length of a chip and L_{sp} is the spacing between subsequent chips. Using Equation 3.19 in Equation 3.18 gives the total load capacitance of the driver chip for the fan-out connection:

$$C_L^N = \frac{C_{pin} + L_{eff} C_{\text{int}off} + N(L_{eff} C_{\text{int}off} + C_{rc})}{1 - \frac{C_{\min,o}}{\beta C_{\min,o}}}. \tag{3.21}$$

For one-to-one connection, $N = 1$, and Equation 3.18 reduces to:

$$C_L^1 = \frac{C_{pin} + L_{int} C_{\text{int}off} + C_{rc}}{1 - \frac{C_{\min,o}}{\beta C_{\min,o}}}. \tag{3.22}$$

Substituting Equation 3.21 for load capacitance in Equation 3.99 gives the required number $n_{1,N}$ of stages in the superbuffer. Using this in Equation 3.102 provides the total parasitic superbuffer capacitance C_{sb}. The total capacitance of the interconnection is then:

$$C_{tot}^{1,N} = C_{sb}^{1,N} + C_{pin} + L_{\text{int}} C_{\text{int}off} + NC_{rc}. \tag{3.23}$$

The effective transconductance k_{eff} of the interconnection is equal to the sum of the transconductances of the superbuffer stages, and is calculated by substituting $n_{1,N}$ in Equation 3.103 Because there is no biasing or termination resistor, there is no steady-state current consumption in this regime of operation:

$$I_H = I_L = 0. \tag{3.24}$$

The rise time the of the signals in the superbuffer is estimated by Equation 3.100. The superbuffer propagation delay is found by substituting $n_{1,N}$ in Equation 3.97. Finally, based on the synchronous operation assumption, the minimum clock period T_{CLK} is required to be as long as the superbuffer propagation delay:

$$T_{CLK} = t_{sb,p}. \tag{3.25}$$

This concludes the calculation of the five parameters of the interconnection (listed at the end of Section 3) necessary to estimate the speed performance and energy requirement of the interconnection in the lumped capacitor regime. In the next section, we extend the analysis to the case of long off-chip interconnections which behave as transmission lines.

4.2 Interconnection in the transmission line regime ($t_r < 5t_f$)

In this regime, a series or parallel termination scheme should be used to minimize reflections and spurious transitions. Figure 4 illustrates these termination schemes. In series termination, driver output resistance is matched to the impedance of the interconnection conductor. Due to the equal impedance of the conductor and driver, the initial voltage transfer to the line is only half the supply level. To achieve full supply level across the conductor, the signal has to bounce from the receiver end, and propagate back to the driver site, thus requiring a round-trip propagation of the signal. In the parallel termination scheme, the receiver end is terminated with an impedance equal to the line impedance. This way, no reflections occur from the receiver end, but the driver should be strong enough to provide initially a high (or low) enough logic voltage to the line. Because only a one way trip of the signal is needed, parallel termination provides faster data transmission than the series termination, but it also requires more energy due to a steady-state current through the parallel termination resistor. Since either termination scheme requires much larger than minimum geometry drivers, a superbuffer is needed to connect minimum size logic to the line drivers in order to minimize the propagation delay.

In the case of one-to-one connection, the interconnection line is unloaded, and the characteristic line impedance can be calculated by [9]:

$$Z_1 = \frac{1}{\nu \cdot C_{\mathrm{int}off}}, \tag{3.26}$$

where ν is the speed of wave propagation in the medium and $C_{\mathrm{int}off}$ is the parasitic off-chip line capacitance per unit length. In the case of fan-out throughout the conductor, the line can be treated as distributed if the spacing between the receivers is short. Under this assumption the loaded line impedance can be calculated as [9]:

$$Z_N = \frac{Z_1}{\sqrt{1 + \frac{C_N}{C_{\mathrm{int}off}}}}, \tag{3.27}$$

where C_N is the fan-out capacitance per unit length, which, from Figure 4, is seen to be:

$$C_N = C_{rc}/L_{eff}. \tag{3.28}$$

Similarly, time-of-flight delays for one-to-one and fan-out connections can be estimated as:

$$t_f^1 = \frac{L_{int}}{\nu}; \tag{3.29}$$

$$t_f^N = \frac{L_{int}}{\nu}\sqrt{1 + \frac{C_{rc}}{L_{eff}C_{\text{int}off}}}. \tag{3.30}$$

Note that in Equations 3.27 and 3.30, we neglected the increasingly smaller effect of the driver chip output capacitance as interconnection gets longer. Now, we are in a position to estimate the boundary of the lumped capacitor and transmission line regimes. Using Equation 3.30 in Equation 3.14, and solving for L_{int} provides the region of the transmission line operation:

$$L_{\text{int},1} > \frac{t_r \cdot \nu}{2.5}; \tag{3.31}$$

$$L_{\text{int},N} > \frac{t_r \cdot \nu}{2.5\sqrt{1 + \frac{C_{rc}}{L_{eff}C_{\text{int}off}}}}. \tag{3.32}$$

In the case of series source-end termination, the output resistance of the last superbuffer stage in the driver chip should match the characteristic line impedance (Z_1 in one-to-one and Z_N in the fan-out case):

$$Rdr_{1,N}^{ser} = Z_{1,N}. \tag{3.33}$$

In the parallel termination case, while the buffer impedance does not have to match the line impedance, it should be low enough to provide the necessary high level voltage across the parallel termination resistor:

$$Rdr_{1,N}^{par} = R_T\left(\frac{V_{DD}}{V_H} - 1\right) = Z_{1,N}\left(\frac{V_{DD}}{V_H} - 1\right). \tag{3.34}$$

where R_T is the termination resistor whose value is matched to the line impedance, and V_H is the minimum acceptable logic high level voltage.

After the resistance of the last superbuffer stage is calculated by Equation 3.34, the entire superbuffer can be designed: the size S of the nth (the biggest) buffer is: $S = \beta^{n-1} = R_{\min}/R_{dr}$ where $R_{\min}$ is the minimum inverter output resistance. This allows the calculation of the number of stages in the superbuffer:

$$n_{1,N}^{ser,par} = \frac{1}{\ln \beta} \ln \left(\frac{R_{\min}}{Rdr_{1,N}^{ser,par}} \right), \tag{3.35}$$

and, from biggest to the smallest, the stages diminish in size by a factor of β. Generally, a 3-stage buffer is sufficient for this application. After the number of stages is determined, the effective transconductance k_{eff} of the line is calculated from Equation 3.103. The total capacitance of the superbuffer (C_{sb}) is obtained by substituting Equation 3.35 for n in Equation 3.102. As in the lumped capacitor regime, the total interconnection capacitance is calculated as:

$$C_{tot_{ser,par}^{1,N}} = C_{tot_{ser,par}^{1,N}} + C_{pin} + L_{\text{int}} C_{\text{int}off} + N C_{rc}. \tag{3.36}$$

Note that in the series termination case there is no steady state current consumption, but in the parallel termination case, there is a steady state current as long as the logic level of the interconnection is high:

$$I_{H,par}^{1,N} = \frac{V_H}{Z_{1,N}}; I_{H,ser}^{1,N} = 0 \tag{3.37}$$

$$I_{L,par}^{1,N} = 0; I_{L,ser}^{1,N} = 0. \tag{3.38}$$

Since the interconnection is driven by the superbuffer, the rise time of the data signals in the interconnection is calculated by Equation 3.100.

The minimum clock period T_{CLK} is equal to the sum of the superbuffer propagation delay given by Equation 3.97, and one-way time-of-flight delay for parallel, or, round-trip time-of-flight delay for series termination:

$$T_{CLK} = t_{sb,p}^{1,N} + m \cdot t_f^{1,N}, \tag{3.39}$$

where $m = 1$ for parallel, and $m = 2$ for series termination. In the above equation, we assumed for simplicity that the last superbuffer stage propagation delay is equal to the propagation delay of a previous stage, although its size is determined by the matching condition rather than the output capacitance. In Equation 3.39 we also neglected the rise time of the signal at the receiver end, which is quite small due to the small flip-chip bond and receiver input capacitance. Note that in Equation 3.39, the superbuffer propagation delay is different for series and parallel termination cases due to Equation 3.35. This difference however, is small, because the minimum high level voltage V_H is generally around half supply level, requiring a buffer size as big as a buffer designed for series termination.

For short interconnection lengths, estimations of the interconnection speed with the lumped capacitor or transmission line models provide close results. For this reason we will only plot the speed performance estimated by the transmission line theory for all interconnection lengths. The technology constants used in the numerical illustrations are presented in Table 1. Figure 5(a) shows the maximum clock speed and energy per bit transmitted as a function of interconnection length for one-to-one connection. Figure 5(b) shows the same characteristic for the fan-out connection. As a result of Equations 3.29 and 3.30, the speed decreases with interconnection length, from above 1Ghz to around 300MHz in the slowest case. The speed of parallel terminated lines is higher due to the one-way trip delay of the signal, as stated by Equation 3.39. The speed of fan-out connections is slightly lower due to two facts. First, the impedance of the interconnection conductor is reduced as a result of periodic loading, which demands a bigger driver inside the driver chip, thus increasing the propagation delay of the signal between minimum logic and the line driver. Second, the periodic loading increases the unit length capacitance of the line, which in turn increases the time-of-flight delay of the signal through the interconnection conductor.

In Figure 5, the speed difference between one-to-one and fan-out connections is very small due to the small flip-chip bond capacitance. Note that, no matter how low the line impedance gets (as a result of loading), the signal rise time within the superbuffer stays constant because lower output impedance requirement of the buffer is met by increasing the number of stages, while keeping the driving strength/load capacitance ratio of the individual stages constant. The propagation delay of the superbuffer, on the other hand, increases as the line impedance gets lower, due to the increased number of stages. This dependence,

Symbol	Description	Value
V_{DD}	VLSI power supply voltage level	3.3 V
V_T	Transistor threshold voltage	0.5 V
V_H	Minimum acceptable logic high-level voltage	$V_{DD}/2$
R_{min}	Minimum-size transistor average resistance	8700 ohm
RC_{min}	Minimum-size inverter internal propagation delay	100 ps
k_{min}	Minimum-size transistor transconductance parameter	80 μA/V^2
$C_{\text{min},i}$	Minimum-size inverter input capacitance	6 fF
$C_{\text{min},o}$	Minimum-size inverter output capacitance Optimum superbuffer tapering factor	6 fF 5
C_{pin},	Flip-chip bond capacitance	20 fF
C_{bond}	(also used as pin capacitance in the text)	
L_c	Side length of a chip in MCM packaging	1 cm
L_{sp}	Spacing between MCM chips	0.2 cm
v	Speed of wave propagation on MCM substrate	15x109 cm/s
$C_{\text{int}off}$	Off-chip interconnection capacitance per unit length	1 pF/cm
$R_{\text{int}off}$	Off-chip interconnection resistance per unit length	0.8 ohm/cm
$C_{\text{int}on}$	On-chip interconnection capacitance per unit length	1.4 pF/cm
$R_{\text{int}on}$	On-chip interconnection resistance per unit length	90 ohm/cm
$C_{minN,i}$	Minimum NMOS transistor input capacitance	3 fF
$C_{minN,o}$	Minimum NMOS transistor output capacitance	2 fF
W_{min}	Off-chip conductor minimum width	25μm
C_{ff}	Off-chip conductor fringing field capacitance per cm	1pF/cm
d_{on}	Side length of an electronic block (on a wafer) to which clock is routed	1 cm
d_{off}	Side length of a chip (on an MCM) to which clock is routed	1 cm

Table 1 VLSI and electrical packaging constants.

however is logarithmic as observed in Equation 3.35. The strongest dependence of the delay on the interconnection length comes from the time-of-flight delay, which increases linearly with line length. Since the overall speed is controlled by all three of the above delay contributions, the speed decreases less than linearly with interconnection length in all cases in Figure 5. For short interconnections, as suggested by Equations 3.31 and 3.32, the line is in the lumped capacitor regime. In this regime, energy increases linearly with interconnection length due to linear dependence of both C_{tot} and k_{eff} on L_{int}. As interconnection gets longer, transmission line phenomena becomes dominant, resulting in a sudden increase in energy at the boundary of this regime. This increase is small in the series termination case, since this scheme requires nothing more than a slight increase of the line driver size. In the parallel termination case, however, the jump is drastic since this scheme requires a parallel termination resistor which consumes high steady-state power. Note that the boundary between lumped-capacitor and transmission line regimes is not precisely defined due to the approximate separation of the two regions by Equations 3.13 and 3.14.

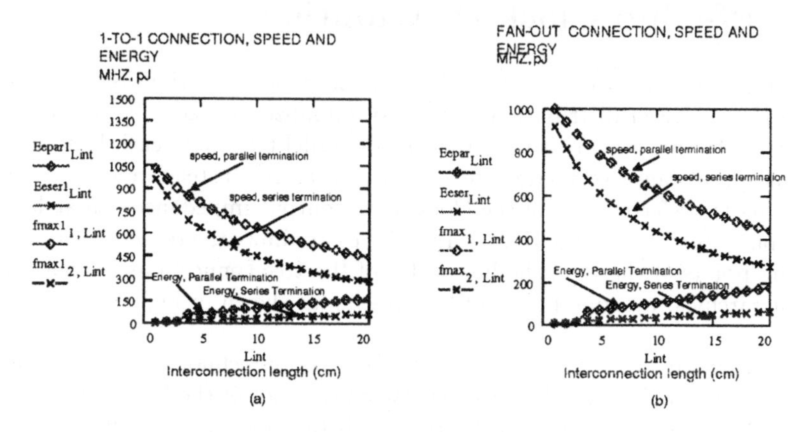

Figure 5 Speed performance and energy requirements of off-chip electrical interconnects as a function of interconnection length for serial and parallel terminated lines. a) 1-to-1 connection, b) fan-out connection.

For short interconnections, the biggest contribution to the total energy comes from the parallel line terminator. As line length gets longer, the capacitive component becomes quickly the dominant component, constituting about 70% of the total energy. For short interconnections, the short-circuit component of the energy is about 20% of the other components. As line length gets longer, its effect reduces to about 10%.

Because the line impedance is independent of line length, once the line impedance is matched (in the series termination case), the superbuffer size remains constant as line length gets longer. However, as the line length gets longer, the total interconnection capacitance increases linearly, resulting in an overall linear increase of energy with interconnection line length. In the fan-out connection case, the energy is higher due to the increased buffer size and interconnection capacitance. The slope of increase is also slightly higher in this case, as a result of the increased unit length capacitance of the line due to periodic loading. In the parallel termination case, the slope of the energy increase is higher than in the series termination case. This is due to the fact that, in addition to the linear increase of the interconnection capacitance with line length, the energy requirement of the parallel termination resistor also increases with line length, because, longer interconnection slows down the transmission speed of an electronic bit, which results in the dissipation of power on the termination resistor over a longer period of time.

4.3 Off-Chip Clock Distribution

A special case in off-chip connections is high speed clock distribution. In this case, an H-tree based distribution is used to minimize the skew [9], as shown in Figure 6. In this case, it is impractical to parallel terminate each leaf unit, since this would cause very large steady-state currents. Instead, a source-end termination is preferred, where the clock driver output impedance is matched to the input impedance of the H-tree. In order to minimize the reflections at the branching points of the tree, the line width of the tree connections is doubled at the hierarchical merging points of the tree [9], as shown in Figure 6.

Since minimum line width is limited by the fabrication technology, this scheme results in increasingly wider line widths toward the root of the H-tree, thereby reducing the input impedance to the H-tree. If the clock is to be distributed over $M = N^2$ processors on a plane (see Figure 6) using an H-tree, the tree involves $\log_2 M$ levels of connection hierarchy and $\log_2 M$ branching points. If the lowest level hierarchy (leaf) connections are implemented physically with minimum width lines (W_{min}), then the width of the highest hierarchy connection (root wire) is $W_R = MW_{min}$. The unit length capacitance of the root wire is calculated as:

$$C_R = C_{pp} \cdot W_{min} M + C_{ff}, \quad (3.40)$$

where C_{pp} is the parallel plate capacitance per unit area, and C_{ff} is the fringing field capacitance per unit length. Since the characteristic impedance of the H-tree is equal to that of the root segment:

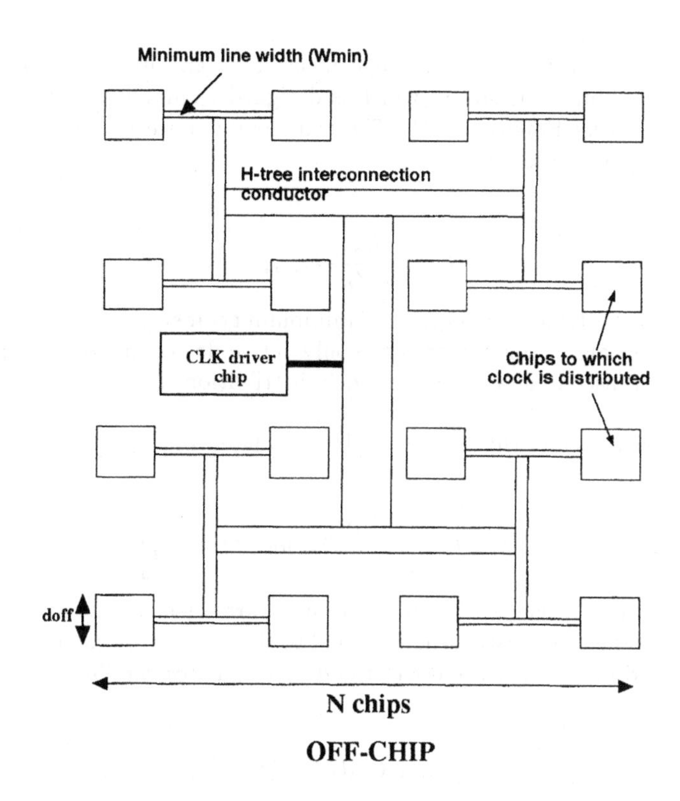

Figure 6 Off-chip electrical clock distribution scheme using an H-tree. The line width of the tree is doubled at each hierarchical merging point of the tree to allow a homogeneous line impedance throughout the tree and reduce reflections. Wider interconductors exhibit a lower characteristic impedance, thus requiring a larger impedance-matched driver.

$$Z_H = \frac{1}{\nu \cdot C_R}. \tag{3.41}$$

Based on the characteristic impedance, an impedance-matched driver can be designed as before. Although in the off-chip interconnection case, a separate chip can be devoted to clock generation, which could possibly be built in a different technology than CMOS, this study assumes a CMOS chip for clock generation as well. Because the clock chip does not need to be integrated at a very large scale, we do not need to assume a minimum geometry clock generator circuitry, which would necessitate a superbuffer to drive the output load. Rather, we will assume a single inverter as an output buffer (in the clock generator chip), the size of which is designed to match the H-tree input impedance given by Equation 3.41. The transconductance k_{dr} of this inverter should be designed to satisfy:

$$k_{dr} = \frac{R_{min}}{Z_H} k_{min}, \tag{3.42}$$

where k_{min} is the transconductance of minimum geometry inverter. Since we assumed no repeaters, this buffer is the only active device in the network, and the effective transconductance of the clock distribution tree is $k_{eff} = k_{dr}$.

The wire capacitance of the H-tree can be calculated as:

$$C_H = M \cdot d_{off} \left[C_{pp} \frac{W_{min}}{2} \log_2 M + C_{ff} \right] \tag{3.43}$$

where d_{off} is the side length of a processor in the array (see Figure 6). The total capacitance of the clock distribution is equal to the H-tree wire capacitance plus the output and input capacitances of the driver and receiver chips:

$$C_{tot\,H} = C_H + C_{rc}(M+1) + C_{buf}, \tag{3.44}$$

where C_{buf} is the sum of the driver chip buffer input and output capacitances estimated by:

$$C_{buf} = \frac{R_{\text{min}}}{Z_H} (C_{\text{min},i} + C_{\text{min},o}). \tag{3.45}$$

Since series termination is used, there is no steady-state current:

$$I_H = I_L = 0. \tag{3.46}$$

The propagation delay from the buffer input to the leaf units of the H-tree can be estimated as:

$$t_{H,p} = t_{buf,p} + tf_H, \tag{3.47}$$

where $t_{buf,p}$ is the buffer propagation delay which can be approximated as:

$$t_{buf,p} = Z_H C_{pin} + R_{\min} C_{\min,o}, \tag{3.48}$$

and tf_H is the time-of-flight delay (from root to the leaf of the H-tree) which is calculated by:

$$tf_H = \sqrt{M} \frac{d_{off}}{\nu}. \tag{3.49}$$

For acceptable clock waveform, we will assume that $t_r = 0.2T_{CLK}$, where T_{CLK} is the minimum clock period estimated as:

$$T_{CLK} = t_{H,p}. \tag{3.50}$$

In Figure 7, we show the maximum speed and energy requirement (at the maximum speed) of the H-tree clock distribution as a function of the number of processors on a side of the H-tree.

It can be observed from Figure 7 that as N increase, the clock speed falls sharply initially and then levels off around 500MHz. This is due to the square root dependence of the root-to-leaf line length of the H-tree on the total number of processors ($M = N^2$) (see Equation 3.49). The energy increases approximately linearly with M due to Equation 3.44. Note that this energy figure is much larger than in the data transmission case. This is due to the extremely low input impedance of the H-tree, which may decrease to as low as a few ohms when the clock distribution area gets larger. This is observed from Equations 3.40 and 3.41.

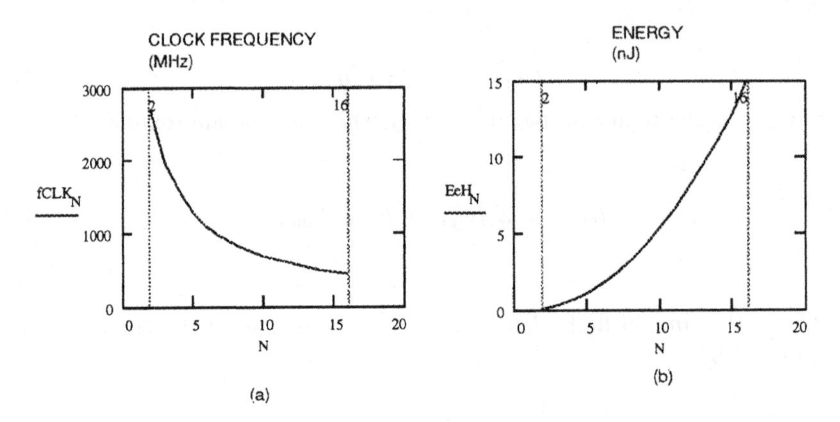

Figure 7 a) speed, b) energy of off-chip clock distribution as a function of the number of processors an a side of the H-tree.

5 SPEED AND ENERGY OF ON-CHIP ELECTRICAL INTERCONNECTIONS

Figure 8 illustrates a typical on-chip (wafer) interconnection configuration. Since an on-chip conductor line is lossy (R_{int} 100 ohm/cm), distributed line behavior is dominant, as opposed to the transmission line. In this case, for short interconnection lengths, a superbuffer is sufficient to drive the interconnection with small delays.

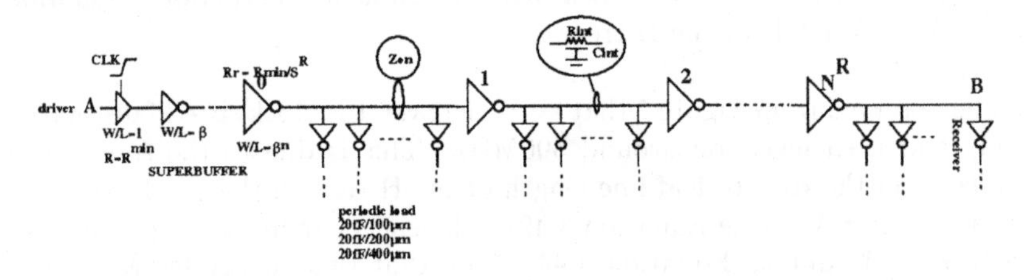

Figure 8 Model of on-chip interconnection with fan-out.

As interconnection length gets longer however, line resistance becomes non-negligible, and the propagation delay through the interconnection can only be minimized through the use of optimally-sized and spaced repeaters [9]. Since a repeater is large in size, a superbuffer is still needed to drive the first repeater. We consider 3 different fan-out loading conditions: a minimum inverter input load every 100, 200, or 400 μm. Again, assuming distributed load behavior, extra loading due to fan-out can be absorbed in the parasitic line capacitance.

Given the line parasitics per unit length and the minimum inverter parameters, optimum number (N_R) and size (S_R) of the repeaters can be calculated as [9]:

$$N_R = \sqrt{\frac{0.4R_{\text{int}on}(C_{\text{int}on} + C_N)}{0.7R_{\text{min}}C_{\text{min},o}}}; \tag{3.51}$$

$$S_R = \sqrt{\frac{R_{\text{min}}(C_{\text{int}on} + C_N)}{R_{\text{int}on}C_{\text{min},o}}}, \tag{3.52}$$

where $R_{\text{int}on}$ and $C_{\text{int}on}$ are the on-chip interconnection resistance and capacitance per unit length, R_{min} and $C_{\text{min},o}$ are the output resistance and input capacitance of minimum size inverter, and C_N is the extra capacitance per unit length due to fan-out. The resulting propagation delay through the interconnection (including the delay of the repeaters) is found as [9]:

$$t_{p,RP} = 2.5\sqrt{R_{\text{min}}(C_{\text{int}on} + C_N)R_{\text{int}on}C_{\text{min},o}}. \tag{3.53}$$

Based on typical 0.5 μm technology parameters, there has to be one repeater, 150 times larger than minimum geometry inverter, approximately every centimeter of on-chip interconnection length in order to minimize the propagation delay (in the one-to-one connection case). Since the repeater size is much larger than minimum, the first repeater is to be driven by a superbuffer. The repeater input capacitance is calculated as:

$$C_{R,in} = S_R C_{\text{min},i}. \tag{3.54}$$

The superbuffer load capacitance is equal to the sum of the repeater input capacitance given by Equation 3.54 and output capacitance of the last superbuffer stage:

$$C_{sb,L} = C_{R,in} + C_{sb,o}, \tag{3.55}$$

substituting Equation 3.60 in Equation 3.99 for the load capacitance and solving for n with the help of Equation 3.101 gives the number of superbuffer stages

$n_{sb,on}$. The total superbuffer capacitance $C_{sb,on}$ and propagation delay $t_{sbp,on}$ are calculated by Equations 3.102 and 3.97. Generally, a two-stage superbuffer is sufficient to drive the repeater. The superbuffer effective transconductance is calculated by Equation 3.103. The effective transconductance of the interconnection is then estimated as:

$$k_{eff} = k_{eff}^{sb} + N_R S_R k_{min}, \tag{3.56}$$

and the total capacitance of the interconnection is:

$$C_{tot} = \frac{k_{eff}}{k_{min}}(C_{\min,i} + C_{\min,o}) + L_{\text{int}}(C_{\text{int}on} + C_N). \tag{3.57}$$

Since there is no termination or biasing requirement, there is no steady state current consumption, and $I_H = I_L = 0$.

Because of the use of a superbuffer cascaded with repeaters in the interconnection, the rise time of the signals varies slightly throughout the interconnection. Since the biggest contribution to the energy for long connections comes from the repeaters, for simplicity, we will use the signal rise time at the input of a typical repeater for the entire interconnection. This rise time can be estimated by weighing the distributed RC terms (between two successive repeaters-about a centimeter) by 1 and lumped ones by 2.3 [9]:

$$t_r = R_{\text{int}on} C_{\text{int}on} + 2.3\left\{\frac{R_{\min}}{S_R}(C_{\text{int}on} + C_N + C_{R,in}) + R_{\text{int}on} C_{R,in}\right\}. \tag{3.58}$$

Finally, minimum clock period is calculated to be:

$$T_{CLK} = t_{sbp,on} + T_{p,RP}. \tag{3.59}$$

Note that, unlike in the off-chip interconnection case, in transmission line regime, in the on-chip interconnection case, k_{eff} is a function of the interconnection length, since longer interconnections involve more repeaters. This results in a quicker increase of energy as a function of Lint. Figure 9 illustrates the results of the analysis.

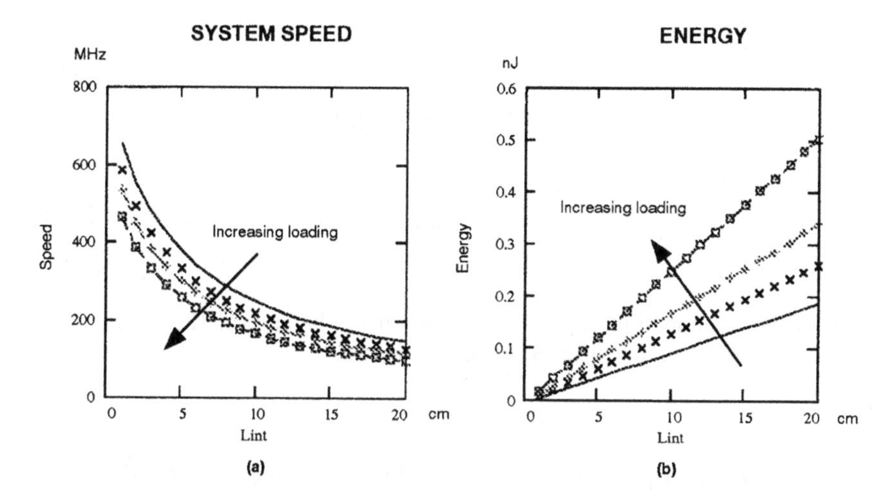

Figure 9 Speed performance and energy requirement of on-chip electrical interconnects as a function of interconnection length for different loading conditions (no load, 50fF/mm, 100fF/mm, and 200fF/mm), a) speed, b) energy.

The effect of loading to the speed seems to be small. The energy increases linearly due to the linear dependence of the number of repeaters as well as the line capacitance on L_{int}. The biggest contribution to the energy comes from the capacitive component, while the short circuit component constitutes about 25% of the overall energy. The energy requirement of the on-chip interconnection is comparable to that of off-chip interconnection. While there is no need for terminations, on-chip interconnects require periodic repeaters as a result of the lossy nature of the interconnection conductor. For the same reason, on-chip interconnections are also slower than their off-chip counterparts.

5.1 On-Chip Clock Distribution

A special case in on-chip interconnections is the H-tree signal distribution. As shown in Figure 10, a common practice to reduce the propagation delay, is to use a repeater at every branching point of the tree, the output resistance of which is matched to the characteristic line impedance [9].

Therefore including the buffer at the output of the clock generator circuitry, there are M buffers. Effective transconductance of the interconnection is:

$$k_{eff} = k_{min} \cdot M \cdot S_{buf}, \tag{3.60}$$

where S_{buf} is the buffer size designed to match the wire impedance:

$$S_{buf} = \frac{R_{\min}}{Z_{on}} = R_{\min} \cdot \nu \cdot C_{\text{inton}}. \tag{3.61}$$

The total interconnection capacitance is equal to the sum of the wire capacitance and input/output capacitances of the buffers:

$$C_{tot,H} = 2M \cdot S_{buf}(C_{\min,i} + C_{\min,o}) + M \cdot d_{on}\left(C_{pp}\frac{W_{\min}}{2}\log_2 M + C_{ff}\right). \tag{3.62}$$

Since there is no termination, steady-state currents are zero: $I_H = I_L = 0$.

Assuming distributed line behavior, the rise time of the signals in the tree gets shorter toward the leaves of the tree, as the length of the line segments between the repeaters decreases in this direction, and the inverters with larger than unity gains improve the rise time. Although calculation of an average rise time throughout the tree is possible, for simplicity, we will consider a worst-case and use the longest rise time in the tree for all the repeaters (since the short-circuit current increases as the rise time increases).

The longest rise time is observed at the input of the first hierarchy repeaters, as shown in Figure 10. The rise time can be estimated using the distributed line approximation by weighing the RC product of distributed terms by 1 and the others by 2.3 [9]:

$$t_r = R_r \cdot C_r + 2.3(Z_{on} \cdot C_r + Z_{on} \cdot 2S_{buf} \cdot C_{\min,i} + R_r \cdot 2S_{buf} \cdot C_{\min,i}), \tag{3.63}$$

where we used Z_{on} for repeater output impedance according to Equation 3.61. In Equation 3.63, C_r and R_r are the capacitance and resistance of the root wire which can be calculated as:

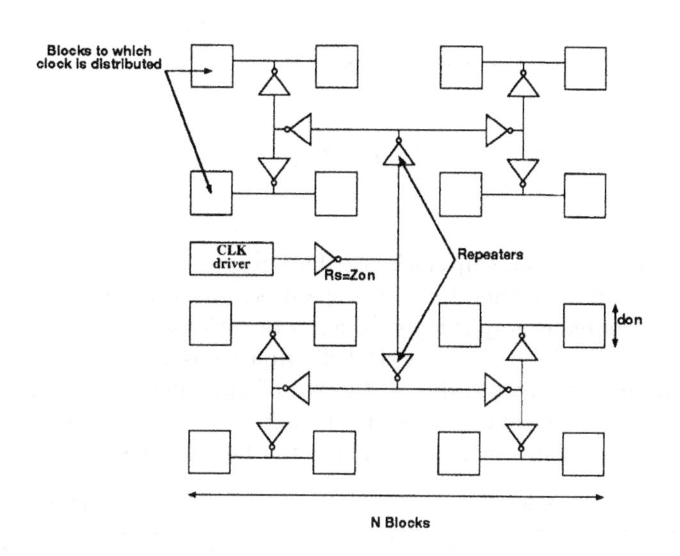

Figure 10 On-chip (wafer-scale) electrical clock distribution scheme using an H-tree. A common practice is to use an impedance-matched repeater at the splitting points of the tree to reduce propagation delay and sharpen the clock transitions.

$$C_r = \frac{\sqrt{M}}{2} d_{on} \cdot C_{\text{int}on} ; \tag{3.64}$$

$$R_r = \frac{\sqrt{M}}{2} d_{on} \cdot R_{\text{int}on} . \tag{3.65}$$

The propagation delay from the root of the tree to a leaf unit can be calculated, again using the distributed line approximation, by weighing the distributed terms by 0.4 and others by 0.7 [9]. Since the propagation delay is different for each segment, a summation over the hierarchy levels of the H-tree is needed, yielding to:

$$\begin{aligned} t_{H,p} = {} & 0.07 R_{\text{int}on} C_{\text{int}on} d_{on}^2 (4^{\log_2 M+1} - 1) \\ & + \left(2^{\frac{\log_2 M}{2}+1} - 1\right) d_{on} (0.7 Z_{on} C_{\text{int}on} + 2 S_{buf} C_{\min,i} R_{\text{int}on}) \\ & \log_2 M \cdot Z_{on} 2 S_{buf} C_{\min,i} . \end{aligned} \tag{3.66}$$

Finally, the clock period is estimated as:

$$T_{CLK} = t_{H,p}. \tag{3.67}$$

As shown in Figure 11, on-chip clock distribution agrees with the general behavior of the on-chip interconnections; the speed is lower (for long interconnects) and yet the required energy is higher than in the off-chip interconnection case. The required energy for wafer-scale clock distribution is immense, reaching hundreds of nanojoules. This is due to the use of large number of repeaters (of large size) at the branching points of the H-tree to reduce clock propagation delay as well as to sharpen the clock rising and falling edges.

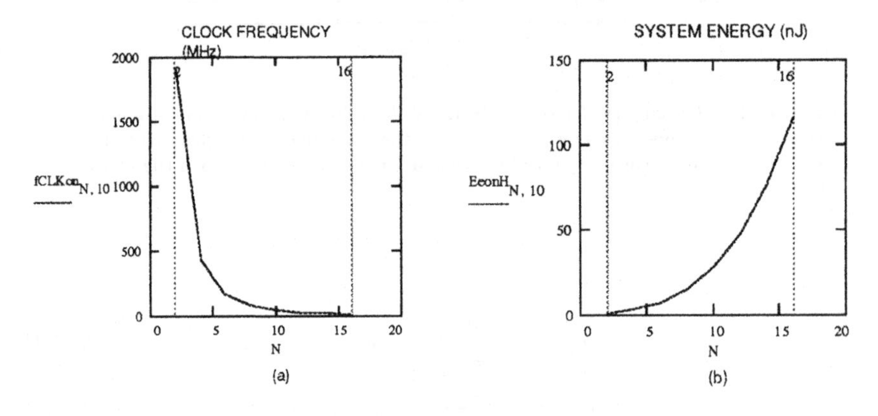

Figure 11 a) speed, b) energy of wafer-scale clock distribution as a function of the number of processors on a side of the H-tree.

6 OPTICAL INTERCONNECTIONS

Figure 12 shows the optical interconnection model considered in this study. As in the electrical interconnection case, a superbuffer amplifies the minimum logic driving capability to drive the optical transmitter driver, which, in turn, switches the optical transmitter device. The transmitter is modeled by a current source and a capacitor. For a VCSEL, the current source models the laser current necessary to produce the required laser output power. For an MQW modulator, the current models the modulator current due to absorption. For a non-absorptive modulator such as PLZT modulator, this current is zero.

The transmitter capacitance absorbs the transmitter driver output capacitance, the transmitter device capacitance and flip-chip bond capacitance (based on Assumption A9).

Minimum logic that drives the interconnection uses the logic-level power supply voltage V_{DD}. Due to transmitter device requirements, on the other hand, the transmitter, transmitter driver, and the superbuffer all use a separate supply level V_{TR}, which is generally larger than V_{DD}. In this study, we assume that V_{TR} is less than the limiting breakdown voltage on the chip, such that no special circuitry is needed in these stages. Furthermore, if V_{TR} is only a couple times larger than V_{DD}, then an inverter in the logic is capable to drive the properly-biased, higher voltage inverter (first inverter in the superbuffer), since this inverter would provide the necessary gain. Note that, the first inverter of the superbuffer never turns off completely due to the limited voltage swing at its input, resulting in steady-state power dissipation. Alternatively, the superbuffer could be operated at VLSI supply level V_{DD}, and the amplification could be performed by the transmitter driver. In this "late amplification" scheme, the superbuffer consumes less energy due to the reduced supply level, but the transmitter driver consumes more energy because the transmitter driver does not switch off completely in a static design. Since the transmitter driver is generally larger than minimum, the energy requirement is higher than that of the first superbuffer stage. In the "early amplification" case where the amplification is performed by the first inverter of the superbuffer, superbuffer consumes more and the transmitter driver consumes less energy. A detailed analysis shows that, for transmitter voltages less than 10V, early amplification scheme is more energy efficient. In a dynamic design, or in cases where transmitter driver does not have to be large "late amplification" would be more beneficial.

In some cases, the transmitter may need to be biased at a certain voltage for optimum operation; a separate supply voltage V_b is used for this purpose (see Figure 12). If a modulator is used as transmitter, then optical power needs to be routed to each modulator from a system optical power supply, as also depicted in Figure 12.

The free-space optical interconnects, which route the transmitter output to N receivers, is modeled by a time-of-flight delay and a power transfer efficiency. At the receiver site, a current source models the absorbed photo-current in a reverse-biased p-n junction. The current source IL models the photodiode load current. Two clamping diodes are used to limit the photodiode output voltage swing to about 330mV (based on Assumption A11), for fast switching. A minimum size inverter is used to threshold the photodiode output and restore the logic levels.

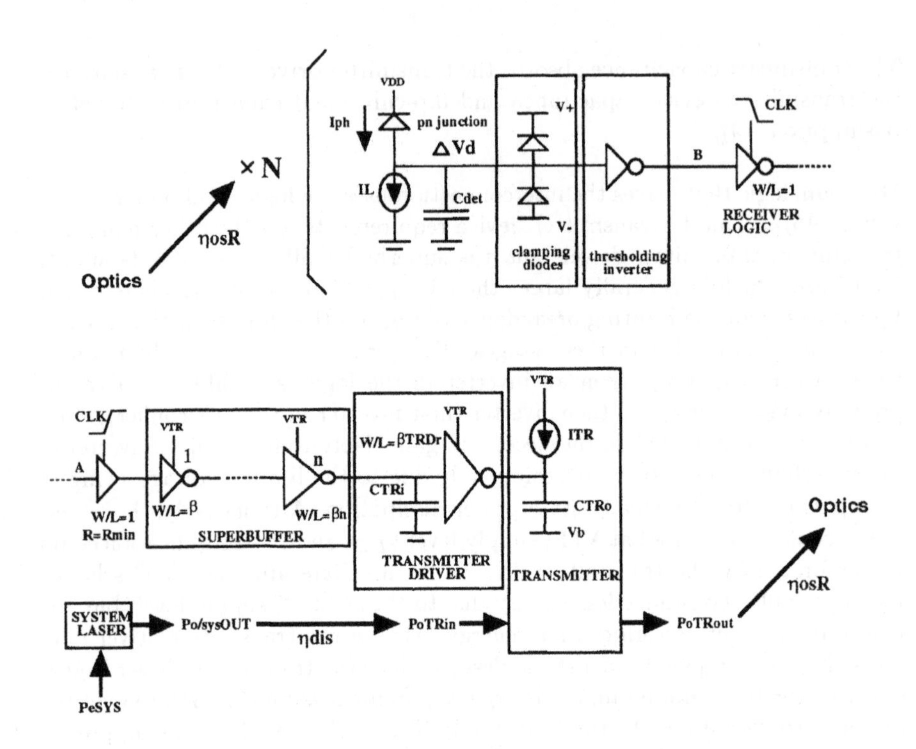

Figure 12 Model of interconnection using free-space optical interconnects. On the transmitter site, the interconnection involves a transmitter, a transmitter driver, and a superbuffer in case the transmitter driver is too large to be driven by minimum logic. The receiver site involves a photodiode together with a thresholding current source, clamping diodes to limit the voltage swing, and a minimum-size inverter to amplify the photodiode output signal and to restore the logic.

Independent of what type of transmitters are used, the design of the interconnection always starts with the estimation of the required photocurrent dynamic range at the receivers, which depends on the technology parameters, operation speed and the photodiode output voltage swing. After the detector photocurrent dynamic range is determined, the transmitter optical output power dynamic range can be estimated based on detector responsivity, fan-out (N), and optical link power transfer efficiency. This is followed by the design of the particular transmitter used.

Symbol	Description	Value
ηOSRlas	VCSEL-based optical interconnection power routing efficiency	0.7
ηOSRmod	Modulator-based optical interconnection power routing efficiency	0.5
ηdis	Optical power distribution efficiency	0.9
ηL,sys	System laser, current-to-optical power conversion efficiency	0.3 W/A

Table 2 Optical routing/power supply constants.

6.1 PLZT modulator as transmitter

PLZT (Lead Lanthanum Zirconium Titanate) is an electro-optic ceramic material with high transmission for wavelengths of 400 nm to 6 μm [11]. Depending on the ratio of lead to lanthanum, the refractive index of the ceramic material can be either a linear or a quadratic function of an externally applied electric field. PLZT 9/65/35 exhibits a large quadratic electro-optic coefficient and offers good potential for achieving a large modulation depth with a relatively low external voltage swing. The modulator frequency response is typically dominated by the slew rate of the driver amplifier and can be rather high if the required voltage swing is small. The use of such modulators have also been demonstrated in free-space optoelectronic systems [18]. Consequently, PLZT 9/65/35 is considered for modulator applications in this section.

Analysis of PLZT modulator Interconnect

For a given speed and technology, the required photocurrent dynamic range at the receivers is expressed in Appendix B. Based on this, the necessary optical power dynamic range at the PLZT modulator output can be estimated as:

$$DR_{opt} \equiv P_{opt,H} - P_{opt,L} = \frac{DR_I}{\eta_{os,R} \cdot R_{ph}} \tag{3.68}$$

where $P_{opt,H}$ and $P_{opt,L}$ are the high and low level optical output powers from the modulator, is the optical link efficiency from transmitter output to the receiver inputs, R_{ph} is the photodiode responsivity at the wavelength used, including surface reflections, and DR_I is the photocurrent dynamic range given by Equation 3.108. After the required PLZT modulator output power dynamic range is determined, the modulator input power dynamic range can be estimated based on the modulator transfer characteristic:

$$P_{opt,in} = \eta_{PLZT}^{-1} \cdot DR_{opt}, \tag{3.69}$$

where the PLZT modulator power efficiency is calculated by:

$$\eta_{PLZT} = T(V_b + V_{TR}) - T(V_b), \tag{3.70}$$

V_b is an experimentally determined optimum bias voltage, V_{TR} is the modulation voltage, and the transmission function $T(V)$ is given as:

$$T(V) = \sin^2 \left[\frac{\Pi}{2} \left(\frac{V}{V_\alpha} \right)^2 + \frac{\Pi}{2} \left(\frac{V}{V_\beta} \right) \right] \tag{3.71}$$

where V_α and V_β are the modified halfwave voltages. A large modulation voltage results in a higher modulator optical power transfer efficiency, thus requiring less input optical power at a certain output power requirement, but consumes more on-chip electrical power due to high supply voltage. Our study shows that optimum operating point, where optical power is equal to the electrical power, is around 10V for 1-to-1 connection. As fan-out increases, the optimum voltage increases too. In this analysis, we will use 10V PLZT modulation voltage for all cases in order to keep on-chip voltages below breakdown levels.

Once the input optical power requirement is determined by Equation 3.69, the high and low level optical output powers from the modulator are determined as follows:

$$P_{opt,H} = P_{opt,in} \cdot T(V_b + V_{TR}) \tag{3.72}$$
$$P_{opt,L} = P_{opt,in} \cdot T(V_b) \tag{3.73}$$

Given the low-level and high-level optical powers, the low and high level photodiode currents as well as the necessary thresholding load current are calculated as:

$$I_{ph,H} = DR_I \cdot \frac{T(V_b + V_{TR})}{T(V_b + V_{TR}) - T(V_b)}; \tag{3.74}$$

$$I_{ph,L} = DR_I \cdot \frac{T(V_b}{T(V_b + V_{TR}) - T(V_b)}; \tag{3.75}$$

$$I_{Load} = \frac{I_{ph,H} + I_{ph,L}}{2} \tag{3.76}$$

Because the PLZT modulator is non-absorptive and practically does not saturate, the modulator driver design is independent of the input/output optical power. The modulator driver is designed to drive the modulator capacitance at the required speed. The modulator is sized optimally based on the modulation voltage used and the optical beam dimensions. The load capacitance of the modulator driver can be expressed as:

$$C_{L,PLZT} = C_{sb,o} + C_{bond} + C_{PLZT}, \tag{3.77}$$

where $C_{sb,o}$ is the last buffer stage output capacitance, C_{bond} is the flip-chip bond capacitance, and C_{PLZT} is the modulator device capacitance. Using Equation 3.77 in Equation 3.99 and solving simultaneously with Equation 3.101 gives the load capacitance as a function of solely technology parameters and independent variables:

$$C_L = \frac{C_{bond} + C_{PLZT}}{1 - \frac{C_{\min,o}}{\beta C_{\min,i}}}. \tag{3.78}$$

Using Equation 3.78 in Equation 3.99 gives the number n_{PLZT} of the superbuffer stages, and using Equation 3.99 in Equation 3.102 provides the total parasitic superbuffer capacitance C_{sb}. Plugging n_{PLZT} in Equation 3.103 gives the superbuffer effective transconductance, which in this case constitutes the effective line transconductance k_{eff} (note that the thresholding inverters at the receiver sites behave as analog amplifiers rather than logic gates, and their contributions to energy –due to their conductance– are calculated separately in Appendix C.

The total effective interconnection capacitance is estimated as:

$$C_{tot} = (C_{sb} + C_{bond} + C_{PLZT})A_{TR} + NC_{rc}, \tag{3.79}$$

where the receiver capacitance is:

$$C_{rc} = (C_{cs} + C_{ph} + 2C_{clamp} + C_{\min,i})A_{rc} + C_{\min,o} + C_{\min,i}. \tag{3.80}$$

C_{cs} is the load current source output capacitance, C_{ph} is the photodiode capacitance and C_{clamp} is the clamping diode capacitance. Note that in Equation 3.79, the transmitter capacitance terms are switched between ground and transmitter supply level (which is larger than VLSI supply level) whereas the photodiode output capacitance terms in Equation 3.80 switches only by an amount ΔV_d around mid-supply level. To account for these facts, we included two coefficients in Equations 3.79 and 3.80 defined as:

$$A_{TR} = \left(\frac{V_{TR}}{V_{DD}}\right)^2 \; ; \; A_{rc} = \left(\frac{\Delta V_d}{V_{DD}}\right)^2 . \tag{3.81}$$

Both the transmitter and receiver sites consume steady-state current. The transmitter site current results from the amplification of the VLSI supply level V_{DD} to the transmitter voltage level V_{TR} by the first inverter in the superbuffer. The average value of this current, I_{TR}, is given in Appendix C. The receiver site steady state currents result from the clamping of the thresholding inverter input voltage and use of a thresholding current source load. The inverter steady state current, I_{RC} is also given in Appendix C. The high and low state currents in the interconnection are therefore calculated as:

$$I_H = I_{TR} + I_{Load} + I_{RC}; \tag{3.82}$$

$$I_L = I_{TR} + I_{ph,L} + I_{RC}. \tag{3.83}$$

As before, the superbuffer propagation delay $t_{sb,p}$ is found by substituting n_{PLZT} in Equation 3.97. The minimum clock period is estimated as:

$$T = t_{sb,p} + t_{p,PLZT} + t_{fopt} + t_{p,ph} + t_{p,det} + RC_{\min}, \tag{3.84}$$

where $t_{p,PLZT}$ is the PLZT modulator internal propagation delay, t_{fopt} is the optical time-of-flight delay estimated in Appendix D, $t_{p,ph}$ is the photodiode internal propagation delay, and $t_{p,det}$ is the photodetector propagation delay which is a function of the input optical power as suggested by Equations 3.108 and 3.109. The minimum meaningful design value for $t_{p,det}$ is a minimum inverter propagation delay $RC_{\min}$. Below this value, very little is gained in speed at a very high energy cost. Increasing $t_{p,det}$ above $RC_{\min}$ results in savings in energy while slowing down the system. In this analysis, we will observe this speed/energy trade-off by varying $t_{p,det}$.

Finally, the rise time of signals in the interconnection is approximated by the rise time of signals in a superbuffer given by Equation 3.100.

In all case of optical interconnections using light modulators, an additional energy source is needed to create the optical power modulated by the modulators. This power is generated by a system light source, most commonly a laser, and then distributed optically to the modulators via an optical power distribution network. The calculation of the modulator input power requirement by Equation 3.69 allows the estimation of the energy requirement of the system laser due to a single modulator:

$$P_{e,sys} = \frac{P_{opt,in}}{\eta_{dis} \cdot \eta_{L,sys}}, \tag{3.85}$$

where $P_{e,sys}$ is the system laser electrical power requirement per modulator, and the terms in the denominator represent the optical distribution efficiency of power from system laser to the modulators, and the system laser's electrical to optical power conversion efficiency. Note that this power is not dissipated in the transmitter or receiver chips and should be excluded from the calculation of the on-chip power dissipation.

This concludes our analysis of the PLZT-based optical interconnects. In the next sections, we compare the speed performance and energy cost of the PLZT-based optical and electrical interconnect technologies. The PLZT interconnection parameters used for numerical illustrations are given in Table 3 and Table 4 while the on-chip and off-chip interconnection parameters are identical to those used in the previous sections.

Comparison To Off-Chip Electrical Interconnects

1-to-1 connection

In Figure 13(a), we show the speed performance of the PLZT-based 1-to-1 optical interconnection and a comparison to off-chip electrical interconnections. The graph illustrates the speed performance of the optical interconnection as a function of the planar (electrical) interconnection length for different photodetection speeds (see Appendix D for the relationship between planar interconnection length and actual optical interconnection path length). The optical

Symbol	Description	Value
ΔV_d	Photodiode output voltage swing	330 mV
A_{ph}	Photodiode area	50 μm^2
A_{diode}	Clamping diode area	10 μm^2
C_{ph}	Photodiode device capacitance per unit area	0.2 fF/μm^2
C_{clamp}	Clamping diode capacitance per unit area	0.2 fF/μm^2
R_{ph}	Photodiode responsivity	0.3 A/W
$t_{p,ph}$	Photodiode internal propagation delay	100 ps

Table 3 Photodetector Constants

Symbol	Description	Value
VTRplzt	PLZT modulator supply voltage	10 V
V_b	PLZT modulator optimum bias voltage	85 V
V_α	Empirical modified half wave voltage	241 V
V_β	Empirical modified half wave voltage	225 V
C_{plzt}	PLZT modulator device capacitance	700 fF
$t_{p,PLZT}$	PLZT modulator internal propagation delay	100 ps

Table 4 PLZT Modulator Constants

interconnection speed is controlled by varying the speed of light detection. In the fastest case, photodetection delay is equal to the minimum inverter propagation delay $RC_{\min}$. By letting this delay increase, the energy requirement of optical interconnection is reduced at the expense of speed. This is illustrated in the figures with different curves; the second subscript in the legend of a curve represents the photodetection delay in multiples of $RC_{\min}$.

In Figure 13(a), we also plotted the speed performance of the off-chip 1-to-1 electrical interconnect for series and parallel termination. The plot shows that, in the fastest photodetection case, optical interconnect offers higher speed than the fastest, parallel terminated electrical interconnection for short connection lengths. Note that for longer interconnections, this difference reduces. The speed advantage of the PLZT-based optical interconnect is mainly due to the smaller superbuffer requirement of the optical transmitter compared to the electrical line driver. The shorter time-of-flight delay advantage of the optical interconnect is neutralized by the longer travel of the optical path. For

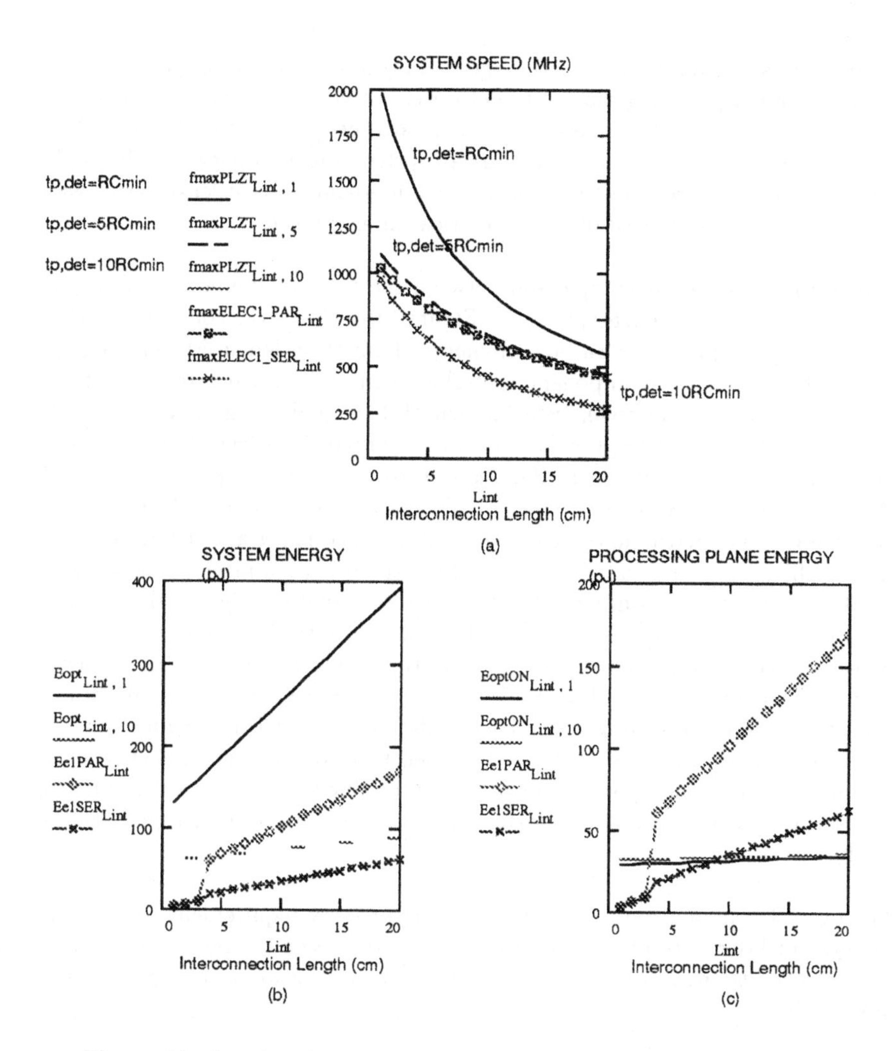

Figure 13 Speed and energy comparison between off-chip electrical and PLZT-based optical interconnects for 1-to-1 connection, a) speed, b) total system energy requirement, c) processing plane energy requirement. The higher energy and speed curves are those representing parallel terminated lines. Performance of optical interconnects is illustrated at different detection speeds (tp,det) to allow comparison to electrical interconnects running at the same speed.

short 1-to-1 interconnections, the speed advantage of the optical interconnects is more pronounced, because the superbuffer propagation delay constitutes a bigger percentage of the overall transmission delay. Compared to the energy saving series termination case, the speed advantage is more pronounced, due to the round-trip propagation of the electrical wave. In this case, optical path length is approximately equal to the electrical wave propagation length, which emphasizes the shorter time-of-flight delay of the optical propagation. Roughly, for all practical interconnect lengths, optical interconnect is at least twice faster than the series terminated off-chip electrical interconnect.

Figure 13(b) shows the comparison between the two interconnect technologies for the overall system energy, whereas Figure 13(c) shows the comparison when only processing plane (or on-board) energy dissipation is considered (i.e., the system laser's energy component is excluded in the optical interconnect energy calculation). The break-even line length for equal energy is around 6 cm for processing plane energy considerations, compared to fastest, parallel terminated electrical connection. This is due to the very sharp increase of the electrical energy requirement when parallel termination is needed. For 20 cm interconnection length, an optical 1-to-1 interconnection is about 6 times more energy efficient than parallel terminated electrical interconnection, and yet, it is 50% faster. When compared to the series termination, the break-even line length is around 7 cm. For long lengths, optical interconnect is twice faster than the series terminated electrical connections, but also 2 times more energy efficient. Therefore, PLZT-based 1-to-1 optical interconnect offer a simultaneous speed/energy advantage over 1-to-1 off-chip electrical interconnects when processing plane energy is considered. In terms of overall system energy, optical interconnect's energy requirement is comparable to the electrical one in the fastest case. This is due to the external optical power required to supply the modulators.

The biggest contribution to the overall system energy requirement comes from the system laser current consumption. While the second biggest contribution is due to the transmitter steady-state current consumption resulting from amplification.

Fan-out connection

Figure 14(a) shows the comparison between optical PLZT and electrical interconnects for fan-out connections. The characteristic is similar to the 1-to-1 connection case, but speed advantage of optical connection becomes more significant. PLZT modulators are non-absorptive and practically do not saturate, fan-out does not affect the design of the modulator driver. The speed performance of the optical interconnection is therefore the same for 1-to-1 or fan-out connections. For the electrical connection, however, extra capacitive loading due to fan-out slows down the transmission speed as a result of increased time-of-flight delay and superbuffer propagation delay. For long interconnections, it is observed from the graphs that optical interconnection offers more than twice the transmission speed than series terminated electrical interconnect: while the series terminated electrical interconnect operates at 250 MHz for 20 cm line lengths, optical interconnect provides an operation speed of almost 600 MHz. Compared to the parallel termination, the optical transmission speed is 50% higher for long interconnections.

Figure 14(b) shows the comparison for overall system energy, whereas Figure 14(c) shows the comparison when only processing plane energy dissipation is considered. The much higher energy requirement of the optical fan-out interconnection, compared to one-to-one connection, is due to the fact that the energy dissipation of a free-space optical interconnection of fan-out N is equal to the sum of the energy dissipations of N 1-to-1 connections. In other words, each fan-out receiver requires its own energy resource in the transmitter. In the electrical interconnections, however, distributed fan-out loads require only a square root growth of the line driver on fan-out. From the processing plane energy requirement point of view, there is a break-even line length around 5 centimeters above which optical interconnect is more energy efficient than electrical interconnection. For long interconnections, PLZT-based optical interconnect offers 4 times better on-chip energy efficiency and yet 150 MHz faster transmission speed than the parallel terminated electrical interconnection.

Note that, in the fan-out connection case, the biggest contribution to the processing plane energy requirement comes from the steady-state currents of the receivers.

Clock distribution connection

Figure 15 shows the results of the speed and energy comparison between the two interconnect technologies for clock distribution. Figure 15 suggests that the speed performance of the two technologies is similar, while optical interconnects are up to 30 times more energy efficient than their electrical counterparts when on-chip consumption is of concern. In terms of overall system energy, however, electrical interconnect are more efficient.

Comparison To On-Chip Electrical Interconnects

1-to-1 connection

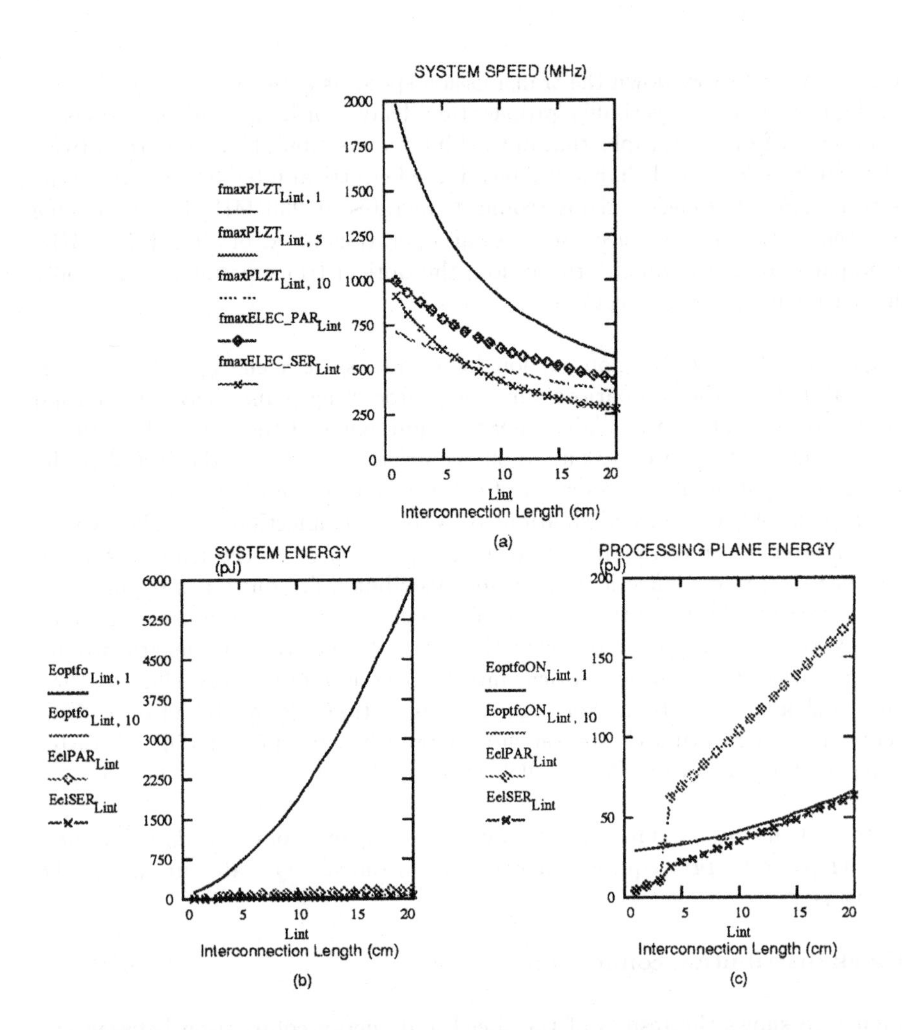

Figure 14 Speed and energy comparison between off-chip electrical and PLZT-based optical interconnects for fan-out connection, a) speed, b) total system energy requirement, c) processing plane energy requirement. Curves related to electrical interconnect are illustrated with symbols. Performance of optical interconnects is illustrated at 3 different detection speeds; as fast as minimum logic, 5 times slower, and 10 times slower.

Figure 16(a) illustrates the speed comparison between PLZT-based optical and on-chip electrical 1-to-1 interconnections. Compared to on-chip connections of up to a few centimeters length, optical interconnects provide much better speed performance. In the wafer-scale domain, the speed of on-chip connections

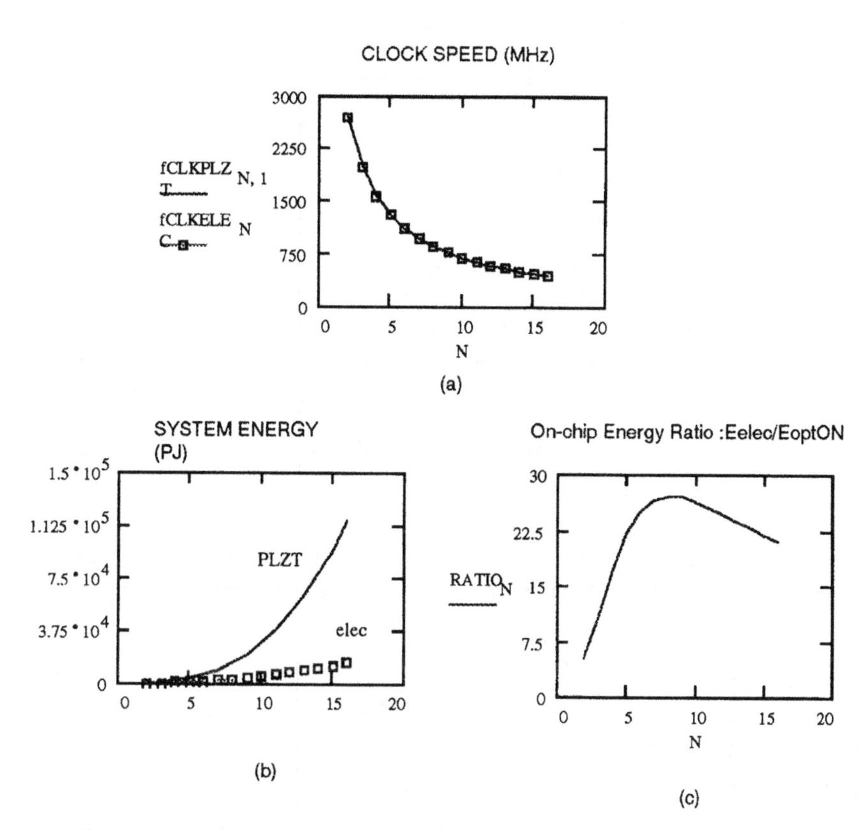

Figure 15 Speed and energy comparison between off-chip electrical and PLZT-based optical interconnects for clock distribution, a) speed, b) system energy c) electrical-to-optical ratio of processing plane energy requirements.

decreases sharply, falling to 120 MHz for 20 cm interconnects. At this point, optical interconnects provide 4 times better (about 500 MHz) speed performance. Figures 16(b) and 16(c) illustrate the energy comparison between the two type of interconnect technologies. The break-even line length for equal on-chip energy dissipation is around 4 cm. For long wafer-scale interconnections, optical interconnect offers about 5 times smaller energy cost simultaneously with 4 times better speed performance. Thus, simultaneous speed/energy advantage of optical interconnect reach a factor of 20.

Fan-out connection

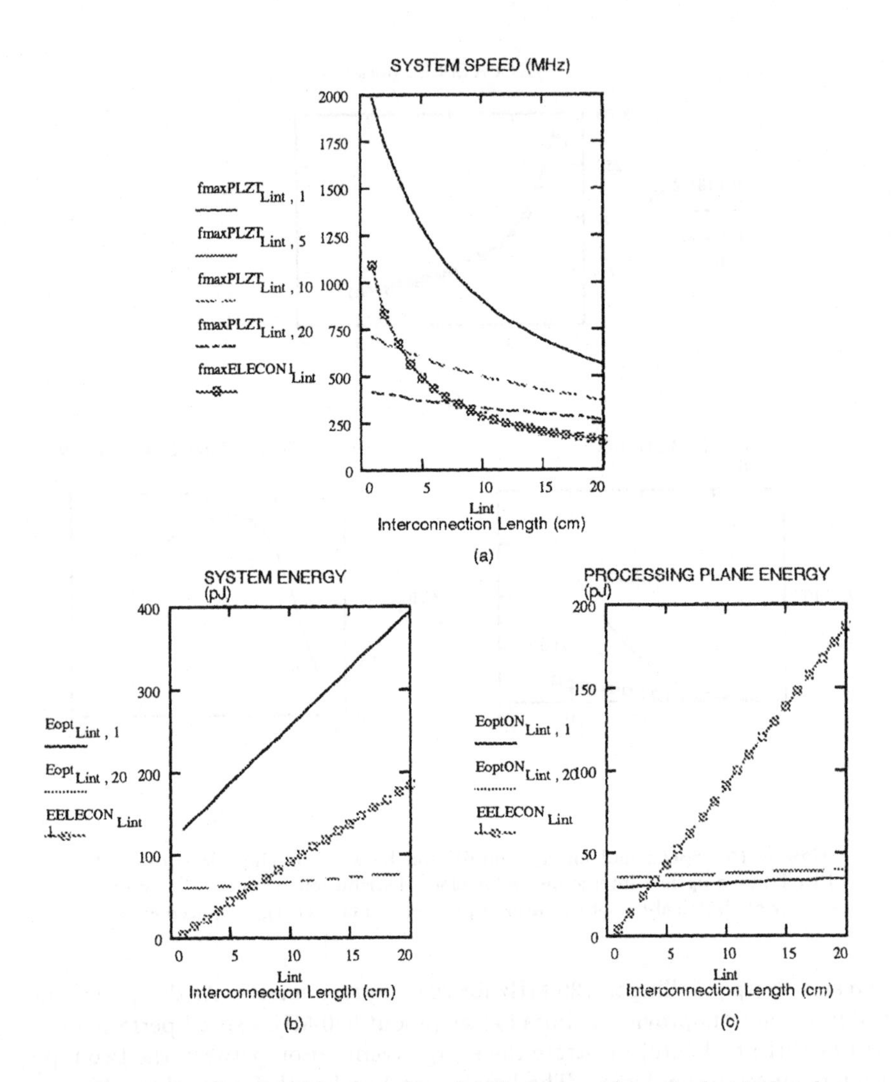

Figure 16 Speed and energy comparison between wafer-scale electrical and PLZT-based optical interconnects for 1-to-1 connection, a) speed, b) total system energy, c) processing plane energy. Curves related to electrical interconnect are illustrated with symbols.

For the fan-out connection (Figure 17(a)), the speed difference between the two interconnect technologies increases slightly. In terms of processing plane energy requirement, the break-even line length is around 3 cm. For long interconnects, optical on-chip energy consumption is 4 times lower, yet it is about 5 times faster. Again, the simultaneous speed/energy advantage of optical interconnections over on-chip electrical connections is around 20.

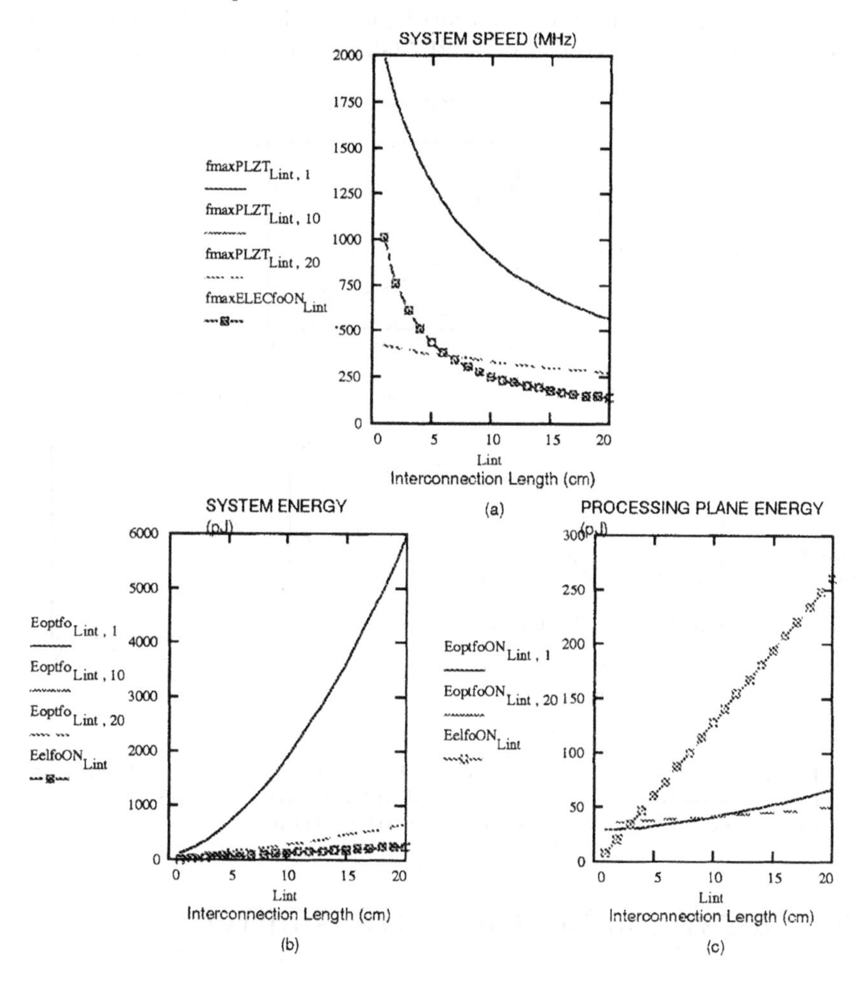

Figure 17 Speed and energy comparison between wafer-scale electrical and PLZT-based optical interconnects for fan-out connection, a) speed, b) total system energy, c) processing plane energy.

From overall system energy point of view, it is possible to operate the detectors 20 times slower than maximum, and achieve twice faster speed than maximum electrical interconnect speed, but at a cost of twice the energy.

Clock distribution connection

Figure 18 shows that optical clock distribution based on PLZT modulator technology is considerably faster than electrical wafer-scale clock distribution, but the energy price is extremely high: 3 times higher on-chip, and 300 times more overall system energy is needed by the PLZT-based optical clock distribution.

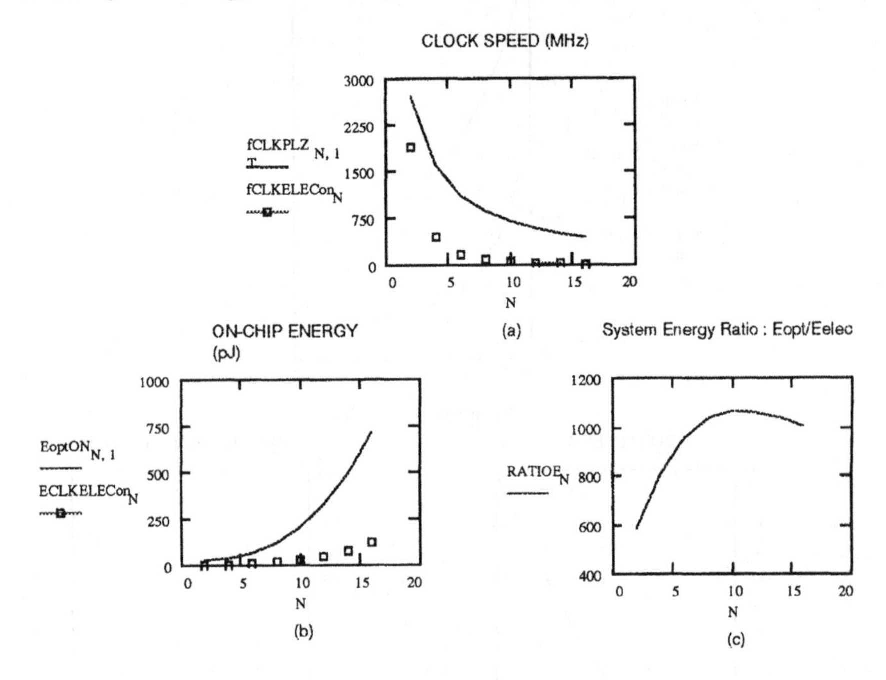

Figure 18 Speed and energy comparison between wafer-scale electrical and PLZT-based optical interconnects for clock distribution, a) speed, b) processing plane energy, c) optical-to-electrical ratio of system energy.

6.2 MQW modulator as transmitter

For optical interconnection applications, optical intensity modulation can be directly achieved in a Multiple Quantum Well structure through electrically modulating the excitonic absorption. This effect is commonly referred to as the quantum-confined Stark effect [3,6]. The MQW modulators are, potentially, very high speed devices and their use in free-space optoelectronic interconnection systems has been demonstrated [16]. The modulator contrast ratio, insertion loss, and absorption saturation are important issues which constrain the applications of these modulators [20]. It has been observed experimentally that the contrast ratio and the insertion loss of the MQW modulators saturate at high optical intensities [21], and this saturation has been attributed to carrier screening in the material and analyzed in the past [26]. In this section, we will neglect the saturation effect by sizing the devices appropriately as a function of optical power.

Analysis of MQW modulator interconnect

In the case of optical interconnection using MQW modulators, the design of the interconnection is similar to the PLZT case, with the exception that the modulator driver should be designed to take into account the absorbed modulator current. In addition, for MQWs, the modulator size is a function of the input optical power to avoid the saturation phenomena. Higher speed or larger fan-out requires more modulator power, thus resulting in a larger modulator with bigger capacitance and current, affecting the design of the driver. Unlike in the PLZT case, the transmitter itself also contributes to the total steady-state current due to absorption. Appendix E presents the details of the MQW modulator driver design. Based on the input capacitance of the driver given in Equation 3.125, a superbuffer can be designed as before if needed. The total capacitance C_{tot} is then:

$$C_{tot} = (C_{sb} + C_{TR,i} + C_{TR})A_{TR} + NC_{rc}, \tag{3.86}$$

where C_{sb} is the superbuffer total parasitic capacitance, $C_{TR,i}$ and C_{TR} are the transmitter driver input and transmitter capacitances given by Equations 3.125 and 3.123 respectively, and C_{rc} is the receiver capacitance given by Equation 3.80. The effective interconnection transconductance is:

$$k_{eff} = k_{eff}^{sb} + k_{dr}, \tag{3.87}$$

where k_{dr} is the transmitter driver transconductance given in Appendix E. The high and low level steady-state currents are calculated as:

$$I_H = I_{TR} + I_{MQW,H} + I_{Load} + I_{RC} \; ; \tag{3.88}$$

$$I_L = I_{TR} + I_{ph,L} + I_{RC}, \tag{3.89}$$

where I_{TR} and I_{RC} are the transmitter and receiver site steady-state currents due to amplification (estimated in Appendix C), $I_{MQW,H}$ is the MQW driver high level current given by Equation 3.119, and the photodetector currents are the same as in the previous section. The minimum clock period of the interconnection can be estimated as:

$$T = t_{sb} + t_{dr,p} + t_{f\,opt} + t_{p,ph} + t_{p,det} + RC_{\min}, \tag{3.90}$$

where $t_{p,MQW}$ is the MQW modulator internal propagation delay, $t_{dr,p}$ is the modulator driver propagation delay which can be approximated as half the signal rise time t_r given by Equation 3.122.

As in the PLZT case, there is an extra energy requirement for an external system laser to optically drive the modulators. The electrical power requirement in the system laser due to a single MQW modulator is calculated by plugging

$$P_{opt,in} = (\eta_{MQW,H} - \eta_{MQW,L})^{-1} \cdot DR_{opt} \tag{3.91}$$

in Equation 3.85. The efficiency terms in Equation 3.91 are given by Equations 3.117 and 3.118, and DR_{opt} was given by Equation 3.68.

The technology parameters used for numerical illustrations in the next sections are presented in Table 5.

1-to-1 connection

Figure 19(a) illustrates the speed comparison between MQW-based optical and off-chip electrical 1-to-1 interconnections. The behavior of the curves is similar to the PLZT case, except for a slightly better optical speed performance due to the smaller buffer requirement of the MQW modulator as a result of the smaller device capacitance. In fact, up to a certain fan-out, even minimum-sized logic can drive the MQW modulator driver. On-chip energy dissipation is comparable to the PLZT case, whereas the overall system energy is less due to the higher transmission efficiency of MQW modulators at 5 V driving voltage. The break-even line length for equal energy requirement is around 3 cm compared to parallel or series terminated electrical interconnects. From both overall system energy and only processing plane energy requirement point of views, MQW-based optical interconnects offer a simultaneous speed/energy advantage for one-to-one connection. For long interconnects, the MQW optical interconnects require almost 6 times less processing plane energy and yet operate faster than the fastest electrical interconnect. From an overall system energy point of view, the energy efficiency of the optical interconnects drop to less than twice better than their electrical counterparts.

Symbol	Description	Value
VTRMQW	MQW modulator supply voltage	10 V
r_{MQW}	MQW modulator responsivity	0.53 A/W
V_L	MQW modulator low level logic voltage	0.5V
$I_s(V_m)$	MQW modulator saturation intensity at the modulation voltage	800 W/cm^2
$I_s(0)$	MQW modulator saturation intensity at zero voltage	244 W/cm^2
K_m	MQW modulator absorption slope ratio	4
$k(0)$	MQW modulator absorption slope at zero voltage	0.2
C_{MQW}	MQW modulator device capacitance per unit area	0.12 fF/μm^2
$t_{p,MQW}$	MQW modulator internal propagation delay	30 ps

Table 5 MQW Modulator technology Constants

Note that by operating the optical light detectors 10 times slower than the speed of minimum inverter, optical interconnects could be more energy efficient than even series terminated electrical interconnect for long interconnections, and yet operate faster.

Comparison To Off-Chip Electrical Interconnects

As in the PLZT case, the biggest contribution to the system energy requirement comes from the system light source. The biggest contribution to the processing plane energy requirement comes from the steady-state modulator current due to absorption.

Fan-out connection

For the fan-out connection (Figure 20), the speed difference between the two interconnect technologies increases slightly. The energy advantage of the optical interconnect, however, reduces as in the PLZT case. Up to 20 cm length interconnections, the optical interconnect processing plane energy requirement is less than that of the parallel terminated electrical interconnection, and yet it is considerably faster. Again, optical fan-out connections can offer simultaneously better speed and energy efficiency by operating the detectors 10 times slower than maximum.

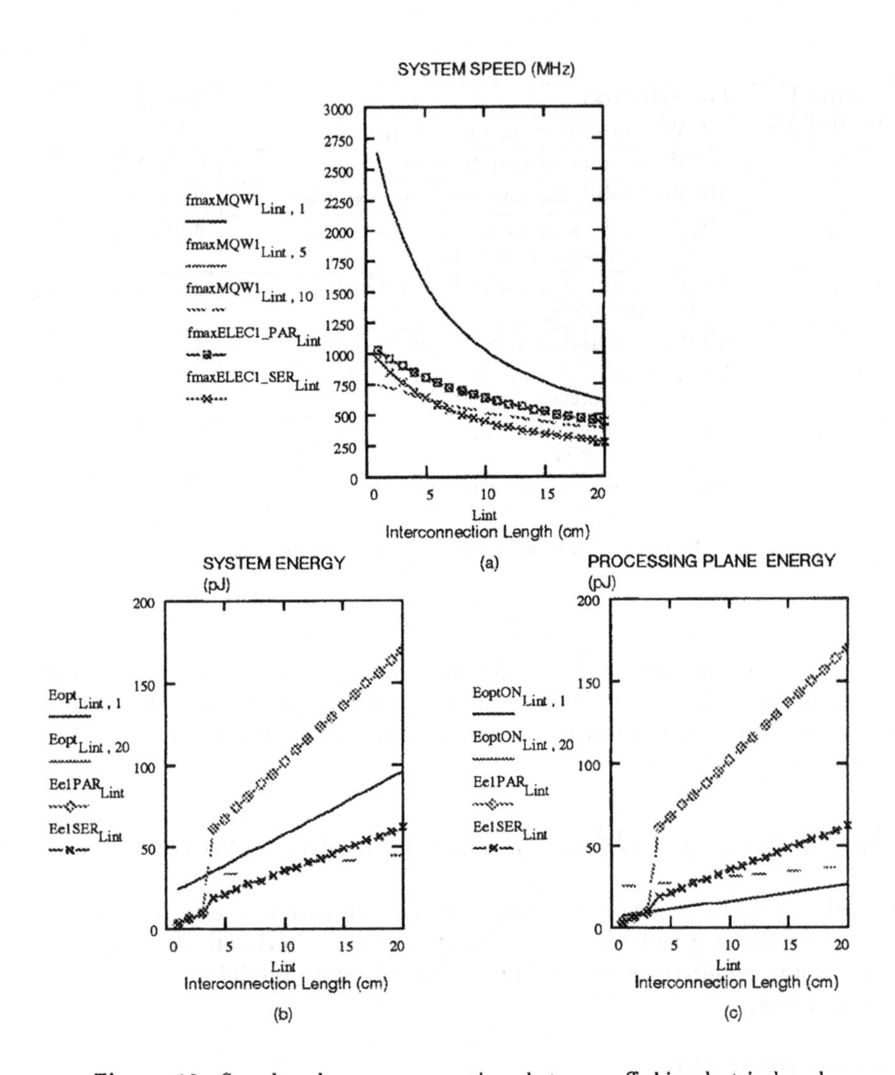

Figure 19 Speed and energy comparison between off-chip electrical and MQW-based optical interconnects for fan-out connection, a) speed, b) total system energy requirement, c) processing plane energy requirement. Curves related to electrical interconnect are illustrated with symbols. The higher energy and speed curves are those representing parallel terminated lines. Performance of optical interconnects is illustrated at different detection speeds to allow comparison to electrical interconnects running at the same speed.

In the graphs, the sudden drop (or increase) in speed (or energy) at around 10 cm is due to the addition of a superbuffer stage when the fan-out of a

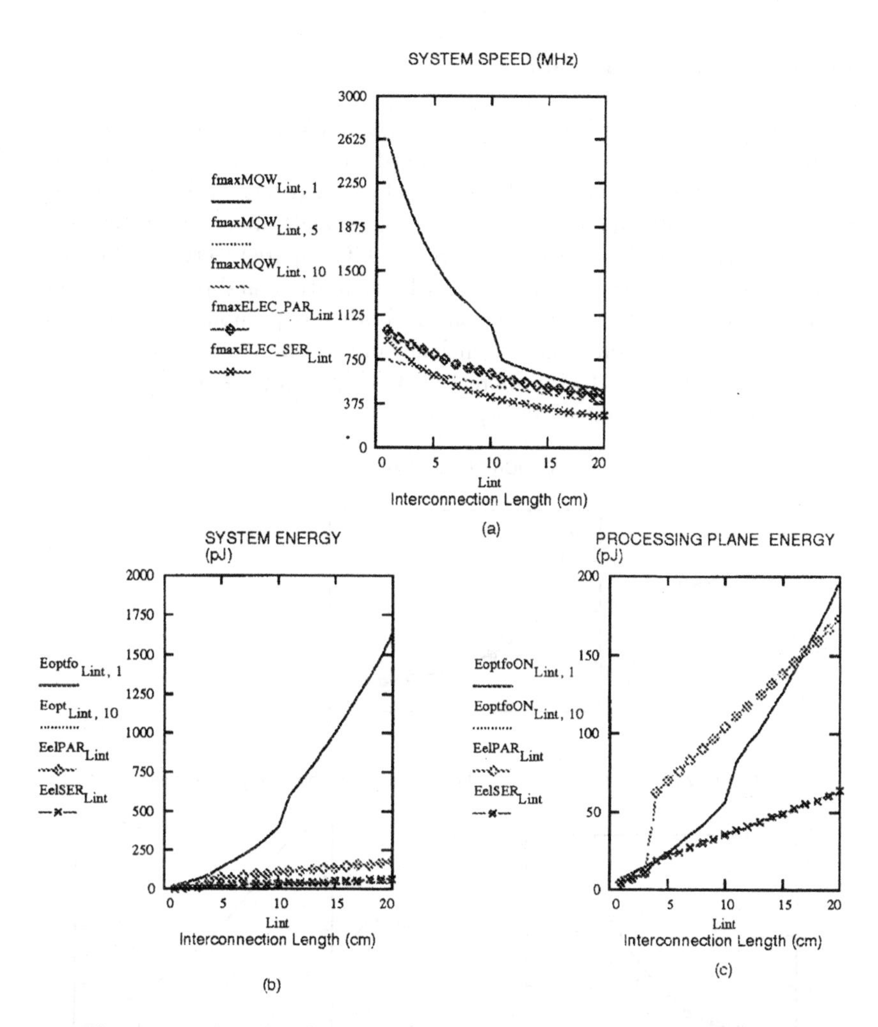

Figure 20 Speed and energy comparison between off-chip electrical and MQW-based optical interconnects for fan-out connection, a) speed, b) total system energy requirement, c) processing plane energy requirement. Curves related to electrical interconnect are illustrated with symbols. The higher energy and speed curves are those representing parallel terminated lines. The performance of optical interconnects is illustrated at three different detection speeds; as fast as minimum logic, 5 times slower, and 10 times slower.

single modulator reaches 10 (note that an additional receiver is added to the interconnection at every centimeter), thereby requiring a considerably larger size modulator.

Clock distribution connection

Figure 21 shows the comparison between off-chip and MQW-based optical clock distribution connections. Although the speed performance is close between the two technologies, the electrical off-chip clock distribution is more energy efficient in terms of overall system energy requirement but much less efficient in terms of processing plane energy requirement. Compared to the PLZT case, the system energy advantage of the electrical distribution, and processing plane energy advantage of optical distribution are both less pronounced. The first one is due to the higher transmission efficiency of the PLZT modulator at 10 V, whereas the second is due to the increased on-chip energy requirement of the MQW modulator as a result of absorption.

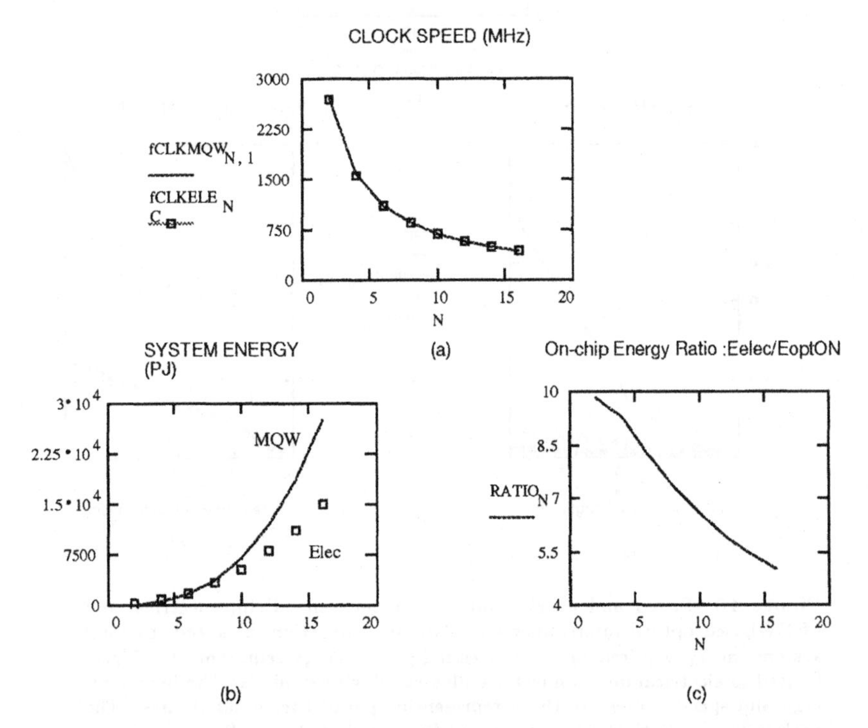

Figure 21 Speed and energy comparison between off-chip electrical and MQW-based optical interconnects for clock distribution, a) speed, b) system energy c) electrical-to-optical ratio of processing plane energy requirements.

Comparison To On-Chip Electrical Interconnects

1-to-1 connection

Figure 22(a) illustrates the speed comparison between MQW-based optical and on-chip electrical 1-to-1 interconnections. The characteristic is similar to the PLZT case, except for a slightly better optical speed performance. There is a break-even line length of about 4 cm for equal system energy. It is interesting to note that, faster optical interconnects require less on-chip energy per bit. This is due to the smaller contribution of the steady-state components. In this case, the break-even line length is only about 2 cm. For long wafer-scale interconnects, the simultaneous speed/energy advantage of the MQW-based optical interconnect exceeds an order of magnitude.

Fan-out connection

For the fan-out connection (Figure 23), the speed difference between the two interconnect technologies increases slightly. While the faster operation of optical interconnects requires much higher system energy, even the slowest case of MQW interconnect operation is still more than twice faster than long wafer-scale connections and yet requires less system, and much less processing plane energy. The fastest optical interconnect still offer less processing plane energy, but a high system energy cost must be paid.

Clock distribution connection

Figure 24 shows the comparison between on-chip and MQW interconnects for optical clock distribution connections. Figure 24 shows that for large networks, optical distribution is much faster, but a high price in terms of processing plane and system energy must be paid.

6.3 VCSEL as transmitter

Compared to light modulators, the use of Surface-Emitting Lasers (SELs) as the optical transmitters in a free-space optical interconnection system significantly simplifies the optical system design by eliminating the requirement for an external laser source and associated optics. Thus, SELs have the potential for improving the optical link efficiency and the system stability and robustness. In particular, vertical-cavity surface-emitting lasers (VCSELs) are very promising for 2-D array applications [4, 12]. A great amount of work has been

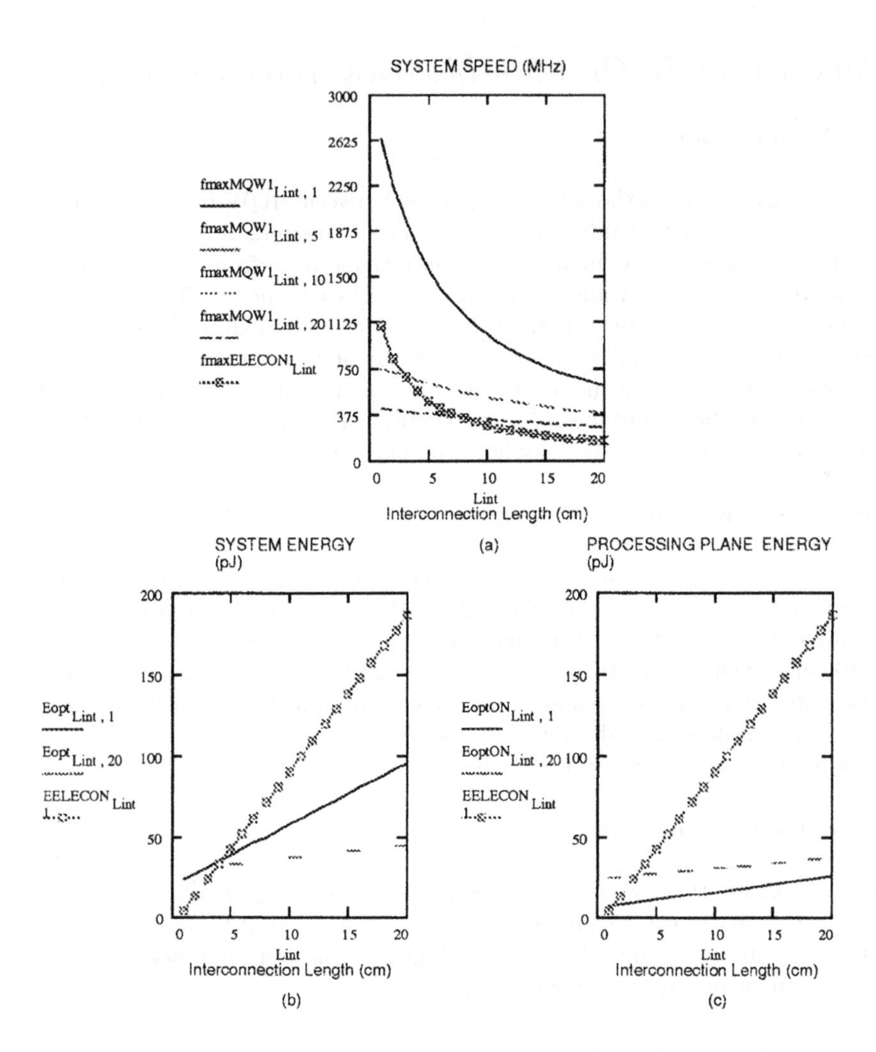

Figure 22 Speed and energy comparison between wafer-scale electrical and MQW-based optical interconnects for 1-to-1 connection, a) speed, b) total system energy, c) processing plane energy. Curves related to electrical interconnect are illustrated with symbols.

performed lately to improve the uniformity of laser arrays, reduce the threshold currents, and increase the maximum power output [4, 8, 28]. The laser threshold voltage is a major concern in a VCSEL due to the large series resistance from the layered mirrors. The threshold voltage thus affects the electrical to optical power conversion efficiency as well as the threshold power penalty. When the

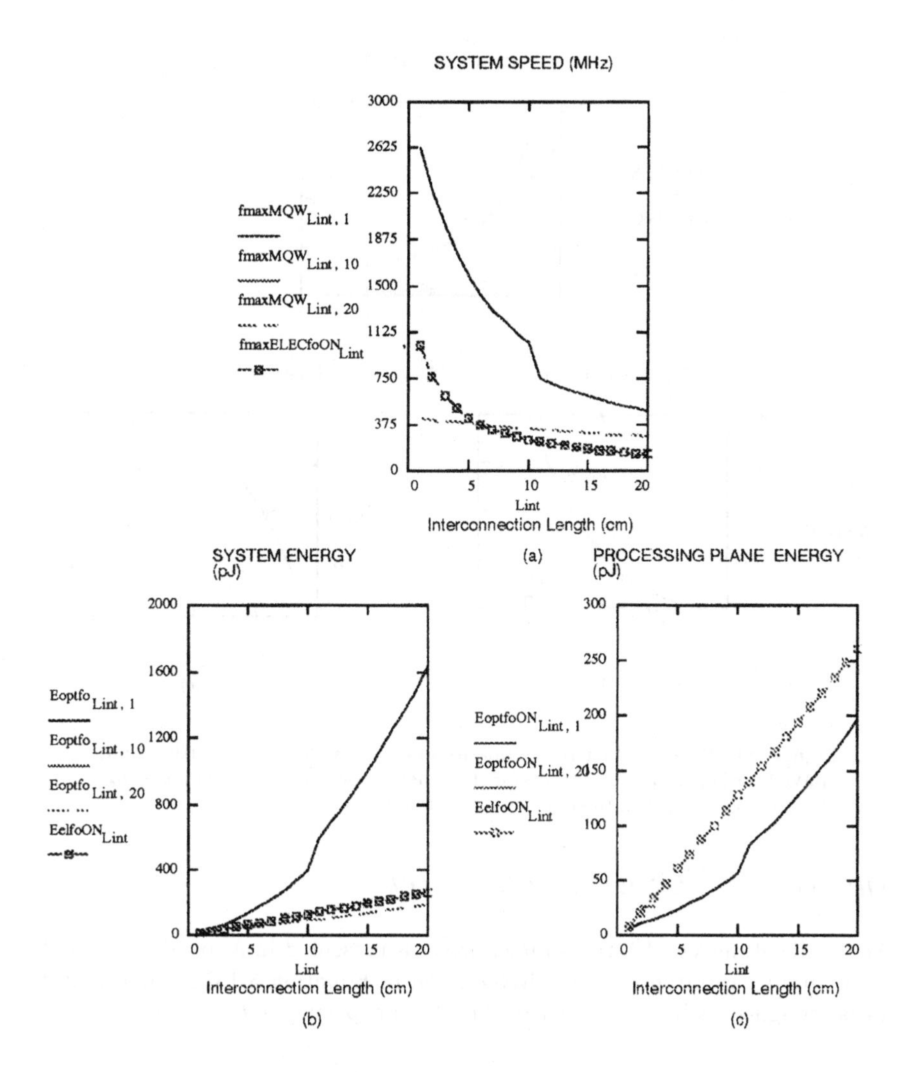

Figure 23 Speed and energy comparison between wafer-scale electrical and MQW-based optical interconnects for fan-out connection, a) speed, b) total system energy, c) processing plane energy.

threshold voltage is higher, the electrical to optical power conversion efficiency is lower for a given laser slope efficiency, and the threshold power penalty is higher for a given threshold current [5].

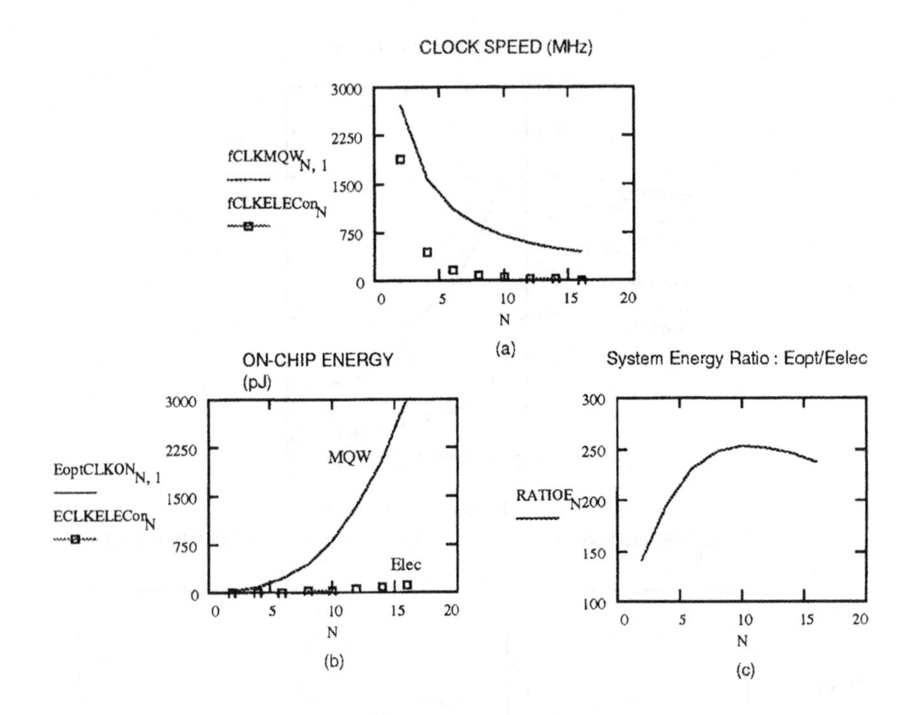

Figure 24 Speed and energy comparison between wafer-scale electrical and MQW-based optical interconnects for clock distribution, a) speed, b) processing plane energy, c) optical-to-electrical ratio of system energy.

Analysis of VCSEL interconnect

The details of the VCSEL transmitter driver is presented in AppendixF. Based on the input capacitance of the driver given in Equation 3.137, a superbuffer can be designed as before if needed. The total capacitance C_{tot} is then:

$$C_{tot,VCSEL} = (C_{sb} + C_{TR,i} + C_{TR})A_{TR} + NC_{rc}, \tag{3.92}$$

where C_{sb} is the superbuffer total parasitic capacitance, $C_{TR,i}$ and C_{TR} are the transmitter driver input and transmitter capacitances given by Equations 3.137 and 3.135 respectively, and C_{rc} is the receiver capacitance given by Equation 3.79. The effective interconnection transconductance is:

$$k_{eff} = k_{eff}^{sb} + k_n, \tag{3.93}$$

where k_n is the transmitter driver transconductance given by Equation 3.133. The high and low level steady-state currents are calculated as:

$$I_H = I_{TR} + I_{VCSEL,H} + I_{Load} + I_{RC} ; \quad (3.94)$$

$$I_L = I_{TR} + I_{th} + I_{ph,L} + I_{RC} , \quad (3.95)$$

where the VCSEL transmitter high level current is given by Equation 3.132, VCSEL threshold current I_{th} is given by Equation 3.131, and the receiver currents are the same as in the previous section. The minimum clock period of the interconnection can be estimated as:

$$T = t_{sb} + t_{dr,p} + t_{p,VCSEL} + t_{fopt} + t_{p,ph} + t_{p,det} + RC_{\min} , \quad (3.96)$$

where $t_{p,VCSEL}$ is the VCSEL device propagation delay, t_{dr} is the laser driver propagation delay which can be approximated as half the signal rise time t_r given by Equation 3.134.

Unlike in the modulator cases, there is no need for an external system light source since each VCSEL is a light source itself. VCSEL parameters are shown in Table 6.

Symbol	Description	Value
VTRlas	VCSEL supply voltage	10 V
φ	VCSEL, slope of threshold current/laser diameter characteristic	0.7 mA/μm
γ	VCSEL, slope of output power/laser diameter characteristic	0.5 mW/μm
η_{LI}	VCSEL, slope of output power/laser current characteristic	0.3 W/A
V_{th}	VCSEL threshold voltage	2 V
$t_{p,VCSEL}$	VCSEL internal propagation delay	30 ps
C_{las}	VCSEL device capacitance per unit area	0.2 fF/μm^2

Table 6 VCSEL Constants

Comparison To Off-Chip Electrical Interconnects

1-to-1 connection

Figure 25(a) illustrates the speed comparison between VCSEL-based optical and off-chip electrical 1-to-1 interconnections. The speed of the VCSEL-based interconnection is slightly lower than in the PLZT or MQW case due to the larger driver requirement of the VCSEL as a result of high laser current. The break-even line lengths for equal energy is around 1 cm. For 20 cm long interconnects, VCSEL-based interconnect requires an order of magnitude less system energy and operates two to four times faster. Note that with improvements in the VCSEL technology, the numbers reported in this chapter are expected to improve further [24]. Also not that compared to the MQW and PLZT modulator cases, the fastest VCSEL-based optical interconnect requires less system energy.

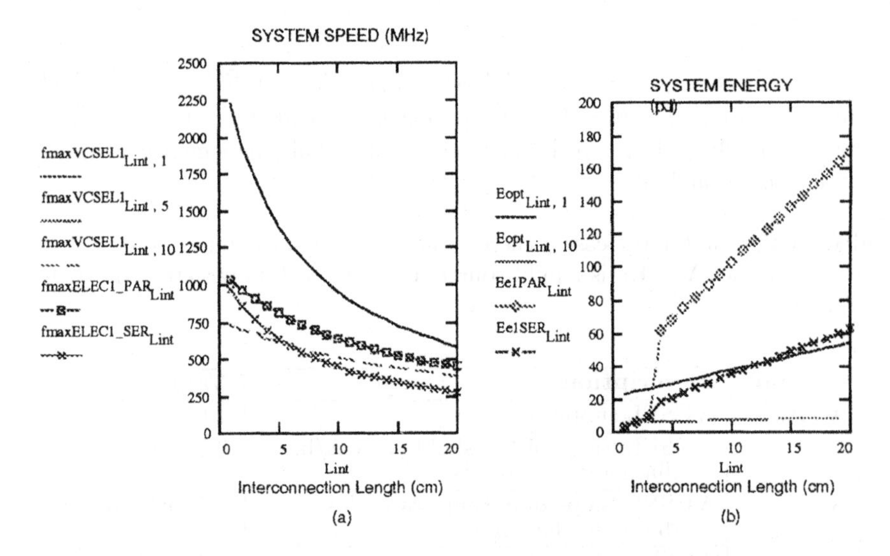

Figure 25 Speed and energy comparison between off-chip electrical and VCSEL-based optical interconnects for 1-to-1 connection, a) speed, b) system (processing plane) energy. Curves related to electrical interconnect are illustrated with symbols. The higher energy and speed curves are those representing parallel terminated lines. Performance of optical interconnects is illustrated at different detection speeds to allow comparison to electrical interconnects running at the same speed.

Compared to the MQW PLZT modulator cases, the fastest VCSEL-based optical interconnect requires less system energy.

Fan-out connection

When the interconnect is loaded (fan-out case), the speed advantage of optical interconnects degrades (Figure 26(a)) rapidly. This is due to the larger buffer requirement of the VCSEL driver as a result of drastically increased optical laser power and thus driver current. Still, at low detector speeds, comparable speed of operation can be achieved at a smaller energy cost than electrical interconnects.

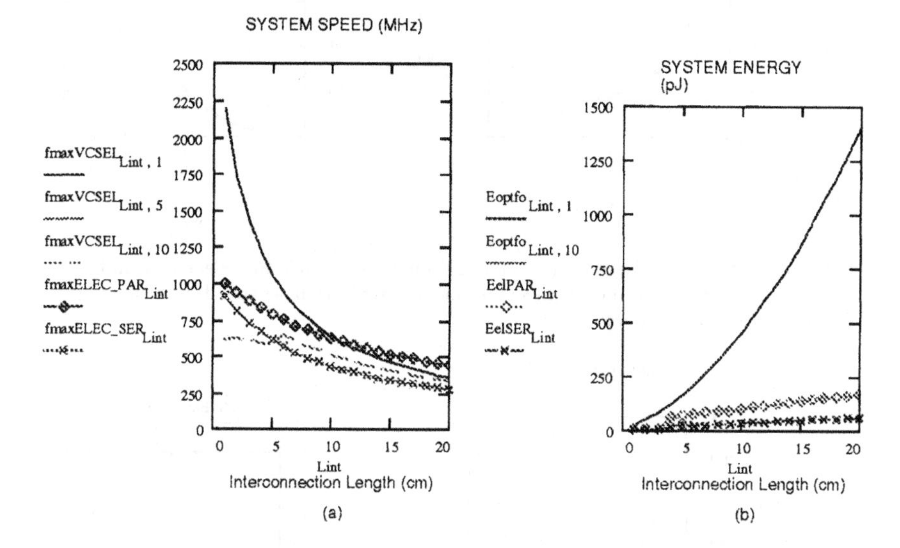

Figure 26 Speed and energy comparison between off-chip electrical and VCSEL-based optical interconnects for fan-out connection, a) speed, b) system (processing plane) energy. Curves related to electrical interconnect are illustrated with symbols. The higher energy and speed curves are those representing parallel terminated lines.

Clock distribution connection

As observed from Figure 27, VCSEL-based optical interconnect provides same clock distribution speed as electrical interconnect, but at a much smaller energy cost.

Comparison To On-Chip Electrical Interconnects

1-to-1 connection

The break-even line length is less than a centimeter. Compared to the wafer-scale electrical connections, VCSEL-based optical interconnects provide a drastic speed and energy advantage: in the case of long interconnects, optical interconnect is 4 times faster and yet 10 times more energy efficient. Note that even for very short interconnects, optics provide faster operation.

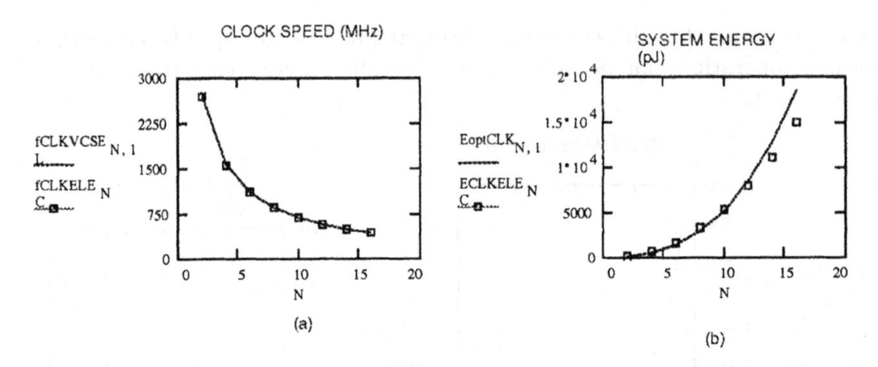

Figure 27 Speed and energy comparison between off-chip electrical and VCSEL-based optical interconnects for clock distribution, a) speed, b) system (processing plane) energy.

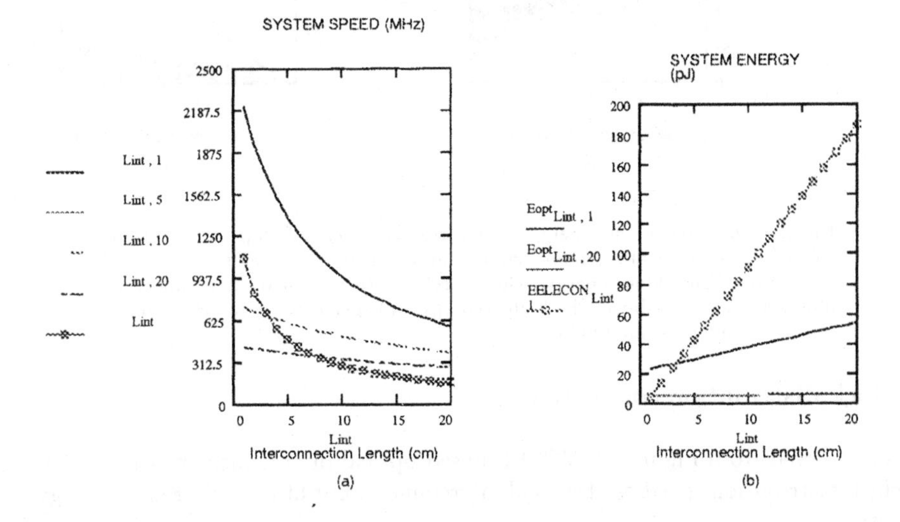

Figure 28 Speed and energy comparison between wafer-scale electrical and VCSEL-based optical interconnects for 1-to-1 connection, a) speed, b) system (processing plane) energy.

Fan-out connection

In the fan-out case, the optical interconnects are still faster than their electrical counterparts but require a comparable cost in energy. It is interesting to note that, running the detectors too fast may result in an overall lower system speed due to the increased requirements from the transmitter side (Figure 29). For interconnects shorter than 5 cm, running the detectors faster pays off in terms of overall system speed while for longer interconnects, it does not. In the case of 10 times slower detector operation, optical interconnects are still faster for long lengths, but require only about 10% of the electrical interconnect energy.

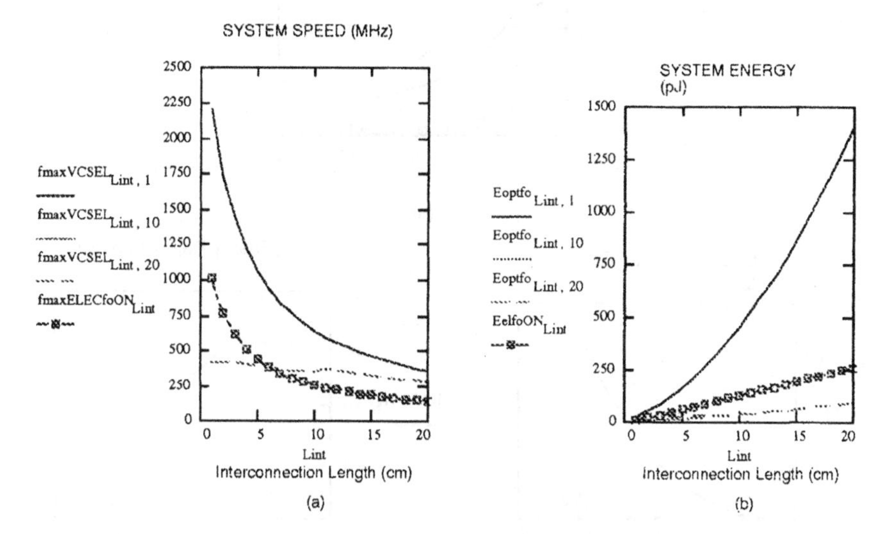

Figure 29 Speed and energy comparison between wafer-scale electrical and VCSEL-based optical interconnects for fan-out connection, a) speed, b) system (processing plane) energy.

Clock distribution connection

At a high energy cost, VCSEL-based optical interconnect distributes the clock much faster than its electrical counterpart (Figure 30).

7 EFFECTS OF TECHNOLOGY SCALING

This section covers, mostly from a qualitative point of view, the potential effects of VLSI scaling (reduced minimum feature size) on different cases of interconnect technologies considered in this chapter. From a very general point of view, scaling down the VLSI technology usually yields faster circuits that consume less power. It also yields reduced parasitics as the transistor areas get smaller. However, it also yields increased resistance of the metal wires on a VLSI chip.

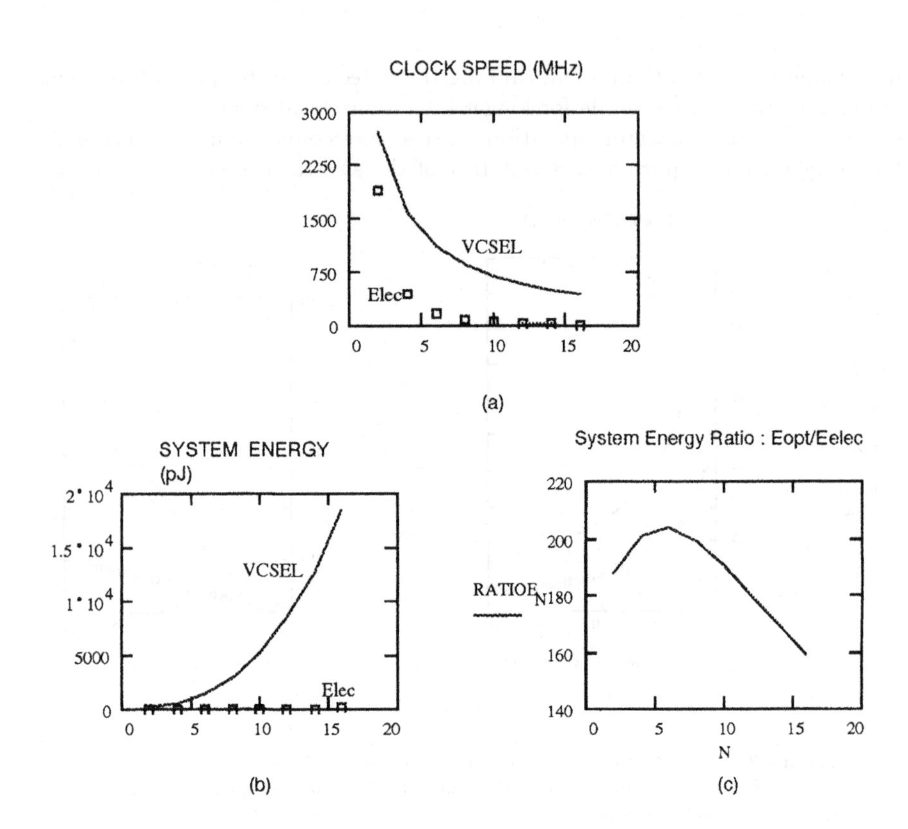

Figure 30 Speed and energy comparison between wafer-scale electrical and VCSEL-based optical interconnects for clock distribution, a) speed, b) system energy, c) optical-to-electrical system (processing plane) energy ratio.

7.1 Electrical Interconnects

Assume two VLSI chips interconnected via an off-chip interconnection line of impedance Z_0. As the technology scales down (assuming ideal scaling) the transistor resistance remains constant to a first order because although the transconductance increases linearly with the scaling parameter s, the supply voltage V_{DD} decreases linearly with s. The minimum gate delay decreases as s, because the resistance remains constant but the parasitic capacitances decrease as s. The interconnect technology scaling allows the off-chip interconnection line width to be decreased to a first order by s. If this is done along with VLSI scaling, then the line impedance Z_0 increases by s. An increased line impedance makes a line easier to drive, but because the number of stages in a superbuffer is a logarithmic function of the output load, the effect of increased line impedance on the superbuffer driver performance is minimal. However, because the minimum gate delay scales down as s, the overall superbuffer propagation delay decreases as s. Scaling Z_0 up, however, may increase the propagation delay through the line if the line connects to large lumped capacitances, since the charging capability of the line is proportional to its impedance. If the line is practically unloaded, then the propagation delay is equal to the inherent time of flight, which is independent of the line impedance. Therefore, to a first order approximation, we can say that technology scaling (s) decreases the interconnect propagation delay by s if the line is unloaded, whereas the propagation delay remains roughly constant (or decreases less than s) if the line is heavily loaded.

On the other hand, the energy of data transmission decreases significantly with scaling. The superbuffer as well as the interconnect capacitance decrease as s. Because the supply voltage also decreases by s and because the dominant capacitive component of the energy is proportional to the square of the supply voltage, the energy requirement scales down as s^3 (Note that the short-circuit component scales as s^3, and the parallel termination component scales between s and s^3).

In the case of on-chip interconnects, the scaling of a repeater-based interconnect needs to be considered. When the interconnect is scaled down, the line capacitance decreases but the line resistance increases. Thus the RC time constant of the line remains constant (unless other measures are taken to keep the resistance low). The repeater propagation delay however, scales down as s due to the decreased gate capacitances. The overall interconnection propagation delay, which is a function of both line RC delay and the repeater propagation delay, decreases only as $s^{0.5}$. If, while the line width is scaled to reduce the capacitance, the line thickness is increased to keep line resistance constant, then, the propagation delay decreases as s. This approach, however, is in contradiction with the VLSI trends to increase the number of interconnect layers. Increased line thickness also increases line to line as well as fringe capacitances and results in a less than s decrease in line capacitance with line width.

As in the off-chip interconnect case, the energy requirement of on-chip interconnects scales down as s^3, due to the decrease in both supply voltage and capacitances. Therefore, the energy-delay product of electrical data transmission scales down somewhere between s^3 and s^4.

7.2 Optical Interconnects

The energy requirement of optical interconnect is mostly dominated by static currents. As the VLSI technology scales down, the parasitic capacitance driven by a photodiode decreases by s. The gain of the amplifier following the photodiode increases as s as a result of increased transconductance and reduced current. Because the supply voltage also reduces by s, the voltage swing requirement at the output of the photodiode reduces as s^2. Therefore, for the same speed, the required photocurrent reduces as s^3 as a result of reduced capacitances along with a reduced voltage swing. Following the photocurrent, all the transmitter optical power and currents scale as s^3. Because the voltage supply reduces by s, the energy requirement of optical interconnect scales down by s^4 (better than in the electrical interconnect case) without assuming any improvement in the efficiency of transmitters and interconnect optics. Because in the above argument we assumed constant speed, the energy-delay product scales as s^4, as good as the best electrical case. The comparison of the scaling effect is summarized in Table 7.

Interconnect	Energy/Delay Improvement
Off-Chip	s^3
On-Chip	$< s^4$
Optical	s^4

Table 7 Effect of scaling the VLSI on the various interconnect technologies.

8 CONCLUSIONS

In this chapter, we compared the electrical and free-space electro-optical interconnects in terms of speed performance and energy cost for digital transmission in large scale systems. Free-space optical interconnects, using PLZT modulators, MQW modulators, or VCSELs as transmitters, offer a significant speed advantage over both off-chip and wafer-scale on-chip electrical interconnects. Compared to the fastest off-chip electrical interconnects (parallel terminated lines) and for lengths up to a few centimeters, optical interconnects provide as much as twice better speed performance. For medium length off-chip interconnections (5 to 15 cm), the speed advantage of the optical interconnect reduces to about 50%. Finally, compared to long off-chip interconnections, the speed advantage reduces to around 20-40%. Compared to series terminated off-chip electrical interconnects which require much less power than their parallel terminated counterparts, optical interconnects provide at least twice better transmission speed even for very long interconnects (up to 20 cm). MQW-based optical interconnect provides the best one-to-one speed performance while VCSEL and PLZT-based interconnects follow with small differences. This is due to the smaller driver requirement of the MQW modulators owing to their smaller device capacitance compared to PLZT modulators, and their much smaller transmitter current compared to the VCSEL.

In comparison to wafer-scale VLSI connections, optical interconnects provide increasingly better speed performance as the line length gets longer, reaching 4 times faster transmission speed. In case of long on-chip connections of length up to 2 cm, optical interconnect provides twice faster interconnect speed. The speed advantage of optical interconnects is slightly higher in the fan-out connection case than in the one-to-one connection case.

The processing plane (on-chip) energy requirement of one-to-one optical interconnects is generally less than 50pJ per bit transmitted. For electrical interconnects, it is on the order of several hundred picojoules. The break-even line length between optical and electrical interconnects for equal energy is on the order of a few centimeters. For applications where a large processing plane is needed, optical interconnects provide simultaneous speed and energy advantage over electrical interconnects reaching a combined factor of 20.

Modulator-based optical interconnects require more overall system energy than VCSEL-based optical interconnect due to the requirement for an external system light power source. PLZT modulators offers the best on-chip energy requirement due to their non-absorptive nature, but they also requires the highest overall system power as a result of their lower optical modulation coefficient at 10 V modulation voltage. VCSEL-based optical interconnects offer the best system energy requirement, but also require the highest on-chip energy.

For fan-out connection, PLZT-based optical interconnects provide the best speed performance due to their non-absorptive nature: larger fan-out does not require larger transmitter driver with longer delays. MQW modulator provides the second best speed performance, while VCSEL is the slowest due to a large buffer requirement of the transmitter resulting from the large laser current requirements.

Fan-out considerably increases the energy requirement of optical interconnects. This is due to the fact that the energy dissipation of a free-space optical interconnection of fan-out N is essentially equal to the sum of the energy dissipations of N 1-to-1 connections. In other words, each fan-out receiver requires its own energy resource in the transmitter. In the electrical interconnections, however, distributed fan-out loads require only a square root growth of the line driver on fan-out. The fastest optical interconnect, in this case, also requires much more overall system energy. Still, at comparable speeds, optical interconnects offer comparable or slightly better system energy efficiency. From processing plane energy requirement point of view, modulator-based optical interconnect still provides considerably better energy efficiency, and in these cases, optical interconnect maintains its simultaneous speed/energy advantage. Compared to off-chip electrical connections, VCSEL-based optical interconnect looses its simultaneous speed/energy advantage, while it still holds this advantage compared to wafer-scale VLSI connections.

For fan-out application, therefore, modulator-based optical interconnects seem to offer the best overall performance.

In off-chip clock distribution application, optical interconnects do not provide a speed advantage, and yet require much more energy than electrical interconnects. In wafer-scale clock distribution application, optical distribution is much faster, but also requires much more energy. PLZT-based optical distribution is the most efficient one in terms of on-chip energy, whereas MQW-based clock distribution network offers the lowest overall system energy.

Our analysis in this chapter did not include the area requirement of the devices in the interconnection. Our main emphasis in this chapter was to design the interconnects to operate at the fastest speed they can, and then compare the energy requirements. This approach naturally results in a larger area requirement by the electrical interconnects compared to the optical ones. In the off-chip electrical interconnect case, line drivers as well as line terminators, and in the wafer-scale connection case, the repeaters occupy a substantial amount of area on the processing plane. For comparable interconnect area, there is therefore enough room on the optical interconnect side, to, for example, improve the light detector circuitry in order to reduce the input optical power requirement and considerably reduce the energy requirement of the optical interconnect. While more complicated detector circuitry may increase the detector propagation delay, the overall link delay may benefit since this may decrease the transmitter site propagation delay as a result of reduced transmitter power requirement [24]. Also remember that, in some systems, slower than maximum speed optical interconnects can be designed to provide much less energy requirement than their electrical counterparts, while still operating faster then them. Especially in such cases, there is room for increased detector delay, to obtain even better energy efficiency. Improved light detection, possibly with more complicated circuitry is therefore an important issue that should be addressed in the future [23].

REFERENCES

[1] H. B. Bakoglu. *Circuits, Interconnections and Packaging for VLSI.* Addison-Wesley Publishing Company, 1990.

[2] L. A. Bergman et al. Holographic optical interconnects in VLSI. *Optical Engineering*, 25:1109, 1986.

[3] D. S. Chemla, D. A. B. Miller, P. W. Smith, A. C. Gossard, and W. Wiegmann. Room temperature excitonic nonlinear absorption and refraction in GaAs/AlGaAs multiple quantum well structures. *IEEE Journal on Quantum Electronics*, QE-20:265–275, 1984.

[4] L. Coldren, S. Corzine, R. Feels, A. C. Fonard, K. K. Law, J. Merz, J. Scott, R. Simes, and R. H. Yan. High efficiency vertical cavity lasers and modulators. *Proc. Soc. Photo-Opt. Instrum. Eng.*, 1362:79–92, 1990.

[5] C. Fan, B. Mansoorian, D. A. Van Blerkom, M. W. Hansen, V. H. Ozguz, S. C. Esener, and G. C. Marsden. A comparison of transmitter technologies for digital free-space optical interconnections. *Applied Optics*, 34(17):3103–3115, June 1995.

[6] C. Fan, D. W. Shih, M. W. Hansen, S. C. Esener, and H. H. Wieder. Quantum-confined stark effect modulators at 1.06 mm on GaAs. *IEEE Photonics Technology Letters*, 5:1383–1385, 1993.

[7] M. R. Feldman, S. C. Esener, C. C. Guest, and S. H. Lee. Comparison between optical and electrical interconnects based on power and speed considerations. *Applied Optics*, 27(9), May 1988.

[8] R. Geels and L. Coldren. Submilliamp threshold vertical cavity laser diodes. *Applied Phsics Letters*, 57:1605–1607, 1990.

[9] R. Geiger, P. Allen, and N. Stroder. *VLSI design techniques for analog and digital circuits.* McGraw Hill, New York, 1990. pp 590–593.

[10] J. W. Goodman, F. I. Leonberger, S. Y. Kung, and R. A. Athale. Optical interconnections for VLSI system. *Proceedings of the IEEE*, 72:850, 1984.

[11] G. Haertling and C. Land. Hot pressed PLZT ferroelectric ceramics for electro-optics applications. *Journal of the American Ceramic Society*, 56:1–11, 1971.

[12] J. Jewell and G. Olbright. Vertical cavity surface emitting lasers. *IEEE Journal of Quantum Electronics*, 27:1332–1346, 1991.

[13] F. Kiamilev, P. Marchand, A. Krishnamoorthy, S. Esener, and S. H. Lee. Performance comparison between optoelectronic and VLSI multistage interconnection networks. *Journal of Lightwave Technology*, 9:1674–1692, Dec. 1991.

[14] R. K. Kostuk, J. W. Goodman, and L. Hesselink. Optical imaging applied to microelectric chip-to-chip interconnections. *Applied Optics*, 24:2851, 1985.

[15] A. Krishnamoorthy, P. Marchand, F. Kiamilev, K. S. Urquhart, S. Esener, and S. H. Lee. Grain-size study for a 2-D shuffle-exchange optoelectronic multistage interconnection network. *Applied Optics*, 31:5480–5507, Sept. 1992.

[16] A. V. Krishnamoorthy et al. Operation of a single-ended 550 Mbits/sec, 41 fJ, hybrid CMOS/MQW receiver-transmitter. *Electronics Letters*, 32(8):764–765, Apr. 1996.

[17] N. C. Li, G. L. Haviland, and A. A. Tuszynski. CMOS tapered buffer. *IEEE Journal on Solid-State Circuits*, 25(4):1005–1008, Aug. 1990.

[18] B. Mansoorian, G. Marsden, V. Ozguz, C. Fan, and S. Esener. Characterization of a free-space optoelectronic interconnect system based on Si/PLZT smart pixels. In *Spatial Light Modulators and Applications*, pages 128–131, Washington, DC, 1993. Optical Society of America.

[19] F. B. McCormick. *Photonics in Switching*, volume II, chapter Free-space interconnection techniques, pages 169–250. Academic Press, New York, 1993. John E. Midwinter, Ed.

[20] B. Pezeshki, D. Thomas, and J. S. Harris, Jr. Optimization of modulation ratio and insertion loss in reflective electroabsorption modulators. *Applied Physics Letters*, 57(15):1491–1492, Oct. 1990.

[21] P. J. Stevens and G. Parry. Limits to normal incidence electroabsorption modulation in GaAs/(GaAl) as multiple quantum well diodes. *Journal of Lightwave Technology*, 7:1101–1108, 1989.

[22] K. Urquhart, P. Marchand, Y. Fainman, and S. H. Lee. Diffractive optics applied to free-space optical interconnects. *Applied Optics*, 33:3670–3682, 1994.

[23] D. Van Blerkom, C. Fan, M. Blume, and S. C. Esener. Optimization of smart pixel receivers. In *IEEE LEOS Summer Topical Meeting on Smart Pixels*, Keystone, Aug. 1996. to be published in Journal of Lightwave Technology.

[24] D. A. Van Blerkom, O. Kibar, C. Fan, P. J. Marchand, and S. C. Esener. Power optimization of digital free-space optoelectronic interconnections. In *OSA Topical Meeting on Spatial Light Modulators*, Lake Tahoe, Mar. 1997. to be published in Journal of Lightwave Technology.

[25] H. J. Veendrick. Short-circuit dissipation of static CMOS circuitry and its impact on the design of buffer circuits. *IEEE Journal on Solid-State Circuits*, SC-19(4):468–474, Aug. 1984.

[26] T. H. Wood, J. Z. Pastalan, C. A. Burrus, Jr., B. C. Johnson, B. I. Miller, J. L. de Miguel, U. Koren, and M. G. Young. Electric field screening by photogenerated holes in multiple quantum wells: A new mechanism for absorption saturation. *Applied Physics Letters*, 57:1081–1083, 1990.

[27] W. H. Wu et al. Implementation of optical interconnections for VLSI. *IEEE Transactions on Electron Devices*, ED-34:706, 1987.

[28] D. B. Young, J. W. Scott, F. H. Peters, M. G. Peters, M. L. Majewski, B. J. Thibeault, S. W. Corzine, and L. A. Coldren. Enhanced performance of offset-gain high-barrier vertical-cavity surface-emitting lasers. *IEEE Journal of Quantum Electronics*, 29(6):2013–2022, June 1993.

A SUPERBUFFER DESIGN

A superbuffer, that is, a chain of inverters with increasing sizes is widely used to reduce propagation delays when driving large capacitive loads. In this appendix, we use some of the results of a superbuffer design reported in [17]. A superbuffer circuit with a tapering factor β is illustrated in Figure 31.

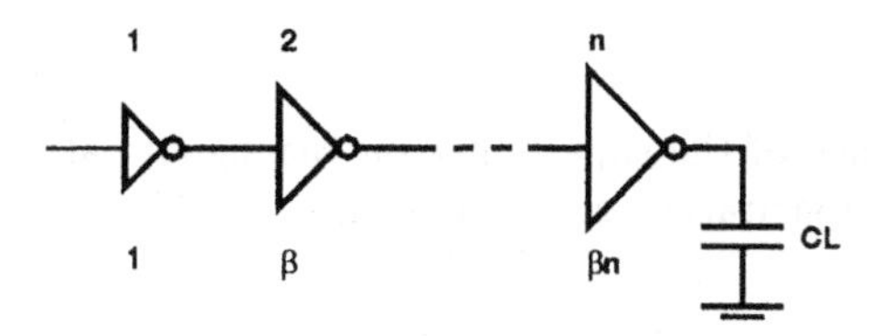

Figure 31 An n-stage superbuffer used to drive large capacitive loads.

The minimized propagation delay through the superbuffer is given as [17]:

$$t_{sb,p} = n \cdot RC_{min} \cdot \alpha, \tag{3.97}$$

where n is the number of inverter stages, RC_{min} minimum logic RC time constant, and α is defined as:

$$\alpha \equiv 1 + p(\beta - 1), \tag{3.98}$$

where β is the tapering factor, $p = \frac{C_{\min,o}}{C_{\min,i}+C_{\min,o}}$, $C_{\min,i}$ is the minimum inverter input capacitance, and $C_{\min,o}$ is the minimum inverter output capacitance.

The number of stages n (excluding the minimum first stage) can be calculated as citeli90:

$$n = \frac{1}{\ln \beta} \ln \left\{ \frac{C_L}{C_{\min,i}} \right\} - 1 \tag{3.99}$$

where C_L is the output load capacitance of the last superbuffer stage.

Because the signals within the superbuffer can be treated using the lumped capacitor approximation, the rise time t_r of the signals in the superbuffer is approximately twice the propagation delay of a single stage:

$$t_r = 2 \cdot RC_{\min} \cdot \alpha. \tag{3.100}$$

The output capacitance of the last superbuffer stage, which is β^{n-1} times larger than minimum, is calculated as:

$$C_{sb,o} = \beta^{n-1} \cdot C_{\min,o}. \tag{3.101}$$

The total capacitance of the superbuffer on the signal path is the sum of the input and output capacitances of all the stages:

$$C_{sb} = (C_{\min,i} + C_{\min,o}) \sum_{i=1}^{n-1} \beta^i = (C_{\min,i} + C_{\min,o})\beta \frac{\beta^{n-1} - 1}{\beta - 1}. \tag{3.102}$$

The effective transconductance of the superbuffer, which we define as the sum of the transconductances of all the stages, is calculated as:

$$k_{eff} = k_{min} \sum_{i=1}^{n-1} \beta^i = k_{min} \frac{\beta^n - \beta}{\beta - 1}, \tag{3.103}$$

where k_{min} is the transconductance parameter of minimum geometry inverter.

B DETECTOR REQUIREMENTS

In this appendix we will express the detector photocurrent dynamic range requirement as a function of detector parasitics and speed of operation. Figure 12 illustrates the detector circuit considered. Figure 32 shows the presumed detector transfer characteristic, where the optical input transition between a low and a high intensity level is linear, thus resulting in a linear photocurrent signal transition through the photodiode.

Assuming that the optical interconnection system does not alter the timing characteristics and the photodiode device limits are not pushed, the photocurrent rise time is equal to the rise time of the optical signal at the transmitter output. This is illustrated in Figure 32, where $t_{r,TR}$ denotes the transmitter (or photocurrent) 0% to 100% rise time. Figure 32(a) illustrates the detector operation where the photodiode output voltage rise time $t_{r,det}$ is larger than

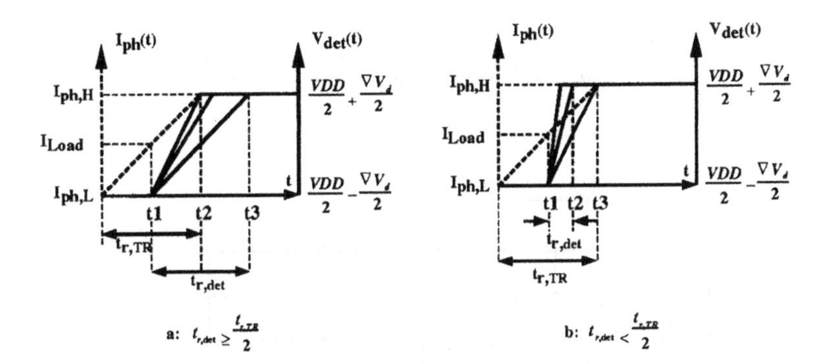

Figure 32 Photodetector input/output waveforms used in the calculations. a) transmitter rise time is less than twice the detector rise time, b) otherwise.

or equal to the half transmitter rise time. Applying the capacitance charging equation to the photodiode output node yields:

$$\int_{\frac{V_{DD}}{2}-\frac{\nabla V_d}{2}}^{\frac{V_{DD}}{2}+\frac{\nabla V_d}{2}} dV = \frac{1}{C_{det}}\left[\int_{t_1}^{t_2}(i_1(t)-I_L)dt + \int_{t_2}^{t_3}(i_2(t)-I_L)dt\right], \quad (3.104)$$

where $i_1(t) = I_L + \frac{t-t_1}{t_2-t_1}(I_{ph,H} - I_L)$, $i_2(t) = I_{ph,H} - I_L$ and ∇V_d is the photodiode output voltage swing, C_{det} is the total photodiode output capacitance and I_{Load} is the photodiode load current. integrating Equation 3.104 and using the definitions of rise times (in Figure 32(a)) gives:

$$I_{ph,H} - I_{ph,L} = \frac{2\nabla V_d \cdot C_{det}}{t_{r,det} - \frac{t_{r,TR}}{4}} \, ; t_{r,det} \geq \frac{t_{r,TR}}{2}. \quad (3.105)$$

Similarly, for $t_{r,det} < t_{r,TR}/2$:

$$\int_{\frac{V_{DD}}{2}-\frac{\nabla V_d}{2}}^{\frac{V_{DD}}{2}+\frac{\nabla V_d}{2}} dV = \frac{1}{C_{det}}\left[\int_{t_1}^{t_2}(i_3(t)-I_L)dt\right], \quad (3.106)$$

where $i_3(t) = I_L + \frac{t-t_1}{t_3-t_1}(I_{ph,H} - I_L)$.

Integrating Equation 3.106 with the help of Figure 32(b) results in:

$$I_{ph,H} - I_{ph,L} = 2\nabla V_d \cdot C_{det}\frac{t_{r,TR}}{t_{r,det}^2} \; ; \; t_{r,det} < \frac{t_{r,TR}}{2}. \tag{3.107}$$

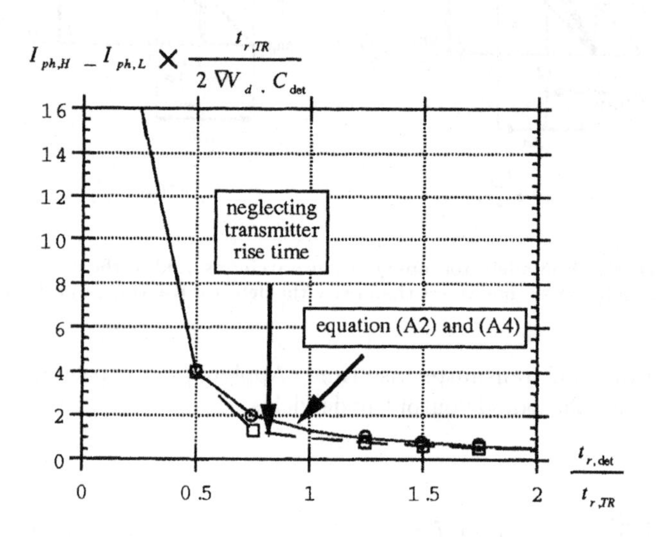

Figure 33 Plot of Equations 3.105 and 3.107.

In Figure 33, we plotted Equations 3.105 and 3.107. As observed from Figure 33, operating the detector at a shorter rise time than that of the transmitter requires increasingly large optical power dynamic range, whereas operating the detector slower than the transmitter results only in a small power saving. Therefore, operating at equal transmitter and detector rise times is nearly optimal in terms of speed and energy. This is the region of operation defined by Equation 3.105.

Also plotted in Figure 33 is Equation 3.105 when neglecting $t_{r,TR}$. We observe that in the interested region of operation, neglecting the rise time of the input signal to the detector causes only a small error while simplifying the calculations. Therefore, for practical purposes, we can assume that the photocurrent dynamic range DR_I can be represented simply by:

$$DR_I = I_{ph,H} - I_{ph,L} = \frac{2\nabla V_d \cdot c_{det}}{t_{r,det}}, \tag{3.108}$$

and the photodiode propagation delay can be approximated as

$$t_{p,det} \approx \frac{t_{r,det}}{2}. \tag{3.109}$$

C TRANSMITTER AND RECEIVER STEADY-STATE CURRENTS DUE TO AMPLIFICATION

As illustrated in Figure 12, the first inverter of the superbuffer at the transmitter site has to amplify its input signal (driven by a minimum logic gate) from VLSI supply level to the transmitter supply level, which is generally around 10 V to achieve an acceptable optical modulation depth. Since this value is only a few times larger than the VLSI supply levels (3-5 V), a single inverter is sufficient to perform the amplification. Similarly, at the receiver site, the thresholding inverter consumes steady-state currents due to the amplification of the photodiode output voltage whose swing is limited to around 330 mV by the clamping diodes.

In any case, a steady-state current results from the fact that the input voltage swing of an inverter is limited around half supply level. If the input voltage swings by an amount around around mid-supply level, then it is possible to show that the average steady state current through the receiver site inverter is:

$$I_{RC} = \frac{k_{min}}{2}(V_{DD}/2 - \Delta V - V_T)^2. \tag{3.110}$$

For the receiver site, where we assumed 330 mV voltage swing, $\Delta V = \Delta V_d/2 = 165\ mV$. For the transmitter site, we replace V_{DD} in Equation 3.110 with V_{TR}:

$$I_{TR} = \frac{k_{min}}{2}(V_{TR}/2 - \Delta V - V_T)^2, \tag{3.111}$$

where $\Delta V = V_{DD}/2$..

D ESTIMATION OF OPTICAL TIME-OF-FLIGHT DELAY

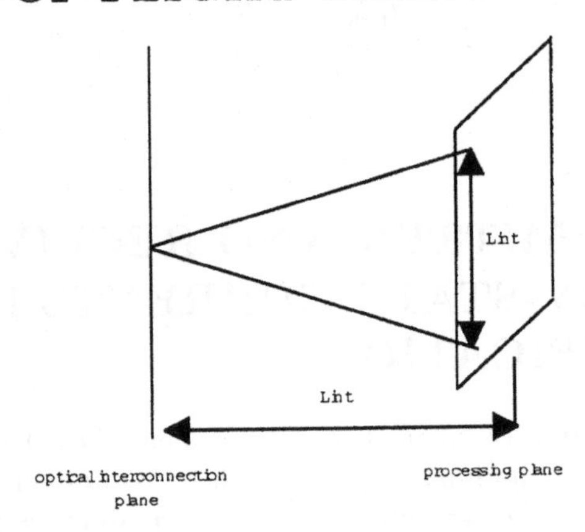

Figure 34 Geometrical assumption of the optical interconnect scheme.

If we assume that, two points on a plane with a spacing of L_{int} can be interconnected optically via an optical plane at a distance L_{int} from the plane in the third dimension (see Figure 34), then, the length of the optical path between the two points is calculated as:

$$L_{opt} = 2\sqrt{L_{int}^2 + \left(\frac{L_{int}}{2}\right)^2} = 2.2L_{int}, \tag{3.112}$$

which results in an optical time-of-flight delay of:

$$t_{f\,opt} = \frac{L_{opt}}{c} \tag{3.113}$$

where c is the speed of light propagation in the medium.

E MQW MODULATOR DRIVER DESIGN

In this appendix, we design an inverter stage to drive the MQW light modulator, as illustrated in Figure 35.

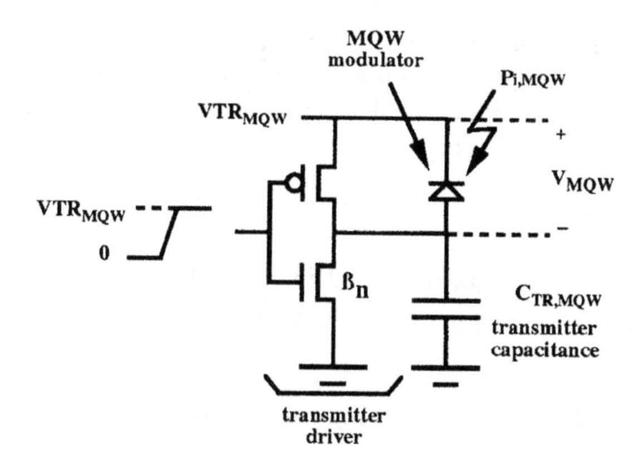

Figure 35 CMOS MQW modulator driver circuit.

The size of the inverter depends on MQW current due to optical absorption, device and integration parasitics and speed of operation. The MQW modulator input optical-to-output optical power efficiency (for high and low voltage states) is given as [5]:

$$\eta_{MQW,H} = \frac{k(0) \cdot K_m}{1+\frac{P_{i,MQW}}{A_{MQW} \times I_S(V_H)}} \;; \quad V_{MQW} = V_H \;; \tag{3.114}$$

$$\eta_{MQW,L} = \frac{k(0)}{1+\frac{P_{i,MQW}}{A_{MQW} \times I_S(V_H)}} \;; \quad V_{MQW} = V_L = 0. \tag{3.115}$$

where V_H and V_L are the high and low level voltages across the modulator, $k(0)$ is the absorption slope at zero voltage, K_m is the absorption slope ratio, $P_{i,MQW}$ is the input optical power to the modulator, A_{MQW} is the modulator area, and $I_S(V)$ is the saturation intensity as a function of the modulator voltage. To render the efficiency independent of the input optical power or modulator area, we will make the following practical assumption:

$$\frac{P_{i,MQW}}{A_{MQW} \times I_S(V_H)} = 0.2. \tag{3.116}$$

Note that reducing the above ratio further results in a larger modulator area for a given power with only a small gain in efficiency. Therefore, when the output power requirement from the modulator increases (due to fan-out or speed increase), the area of the modulator is also increased according to Equation 3.116 in order to keep the modulator efficiency constant. Using Equation 3.116 in Equations 3.114 and 3.115 yields:

$$\eta_{MQW,H} = 0.83 \cdot k(0) \cdot K_m \; ; \tag{3.117}$$

$$\eta_{MQW,L} = 0.9k(0). \tag{3.118}$$

On the other hand, the modulator high level absorption current can be estimated as:

$$I_{MQW} = r_{MQW} \cdot P_{i,MQW} \cdot \eta_{MQW,H}, \tag{3.119}$$

where r_{MQW} is the modulator responsivity. The NMOS transistor of the driver inverter should be large enough to sink this current in order to satisfy the DC condition $V_{MQW} \geq V_H$. Equating Equation 3.119 to the NMOS current in linear region, and solving for NMOS transconductance yields:

$$k_n^{DC} = \frac{I_{MQW,H}}{(V_{TR} - V_T)V_{dslow} - 0.5 \cdot V_{dslow}^2}, \tag{3.120}$$

where V_{TR} is the transmitter power supply voltage, V_T the transistor threshold voltage, and $V_{dslow} = V_{TR} - V_H$.

In addition to the DC requirement, the driver should also satisfy the switching speed, or AC requirement. Assuming that the NMOS transistor is in the linear region of operation during switching, its resistance can be estimated as:

$$R_{NMOS} = \frac{2}{K_n^{AC} \cdot (V_{TR} - V_T)} \tag{3.121}$$

where we included a factor of two to better approximate the transistor resistance, since during switching, the average value of the input voltage is less than V_{TR}, and the PMOS current confronts the NMOS current. The rise time of the signal at the driver output can then be estimated as:

$$t_{r,dr} 2.3 R_{NMOS} C_{TR}, \tag{3.122}$$

where C_{TR} is the total transmitter capacitance composed of:

$$C_{TR} = C_{bond} + A_{MQW}C_{MQW} + \frac{k_n^{AC}}{k_{\min}}C_{\min,o}, \tag{3.123}$$

where C_{bond} is the flip-chip bond capacitance, C_{MQW} is the modulator capacitance per unit area, $C_{\min,o}$ is the minimum inverter output capacitance and $k_{\min}$ is the minimum transistor transconductance parameter. It is not meaningful to operate the driver faster than a minimum inverter stage. Thus, equating Equation 3.122 to the the rise time of the signals at the output of a minimum inverter (driving another minimum inverter), using Equations 3.121 and 3.123 in Equation 3.122, and solving for the transistor transconductance gives:

$$k_n^{AC} = \frac{\cdot\ C_{bond} + A_{MQW}C_{MQW}}{0.5RC_{\min}\cdot(V_{TR} - V_T) - \frac{C_{\min,o}}{k_{\min}}}. \tag{3.124}$$

The NMOS transistor is therefore sized based on the stricter of the DC and AC conditions given by Equations 3.120 and 3.124. To complete the driver design, the PMOS transistor is sized twice as big as the NMOS transistor to compensate for its lower mobility. The design of the driver gives the transmitter driver input capacitance as:

$$C_{TR,i} = \frac{k_{dr}}{k_{\min}}C_{\min,i}, \tag{3.125}$$

where k_{dr} is the larger of DC and AC transconductances, and $C_{\min,i}$ is the minimum-size inverter input capacitance.

F VCSEL DRIVER DESIGN

Laser driver circuitry considered is shown in Figure 36, where a single NMOS transistor is used to switch the laser diode.

The driver should satisfy the DC and the minimum switching speed requirements: the NMOS transistor should be large enough to conduct the high-level laser current, and switch the laser in no more than one minimum inverter propagation delay. In this analysis, we will assume that the laser is operated at the maximum output power level for a certain laser diameter. Therefore, the laser diameter is increased to achieve higher optical output power from the

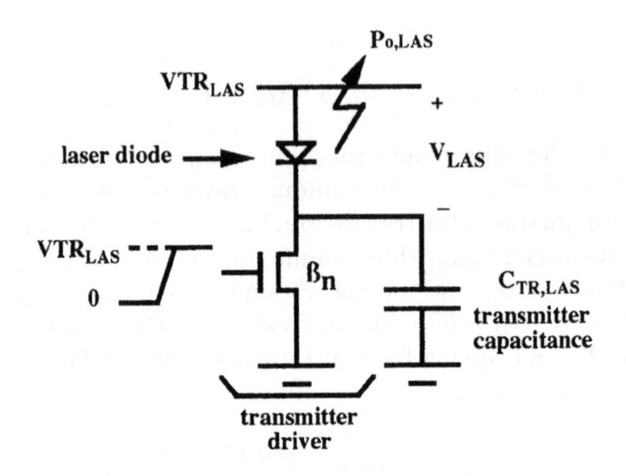

Figure 36 CMOS VCSEL driver circuit.

laser, as needed for larger fan-out or faster operation. We assume linear output power-laser diameter and threshold current-laser diameter characteristics:

$$P_{OH} = \gamma D \; ; \tag{3.126}$$

$$I_{th} = \varphi D, \tag{3.127}$$

where P_{OH} is the optical output high power level, I_{th} is the laser threshold current, D is the laser diameter, and γ and φ are the slopes of the characteristics. Laser Current to Power characteristic is assumed as:

$$P_{OH} = (I_{TR} - I_{th})\eta_{LI}, \; I_{TR} \geq I_{th} \tag{3.128}$$

$$P_{OH} = 0, \; I_{TR} < I_{th}, \tag{3.129}$$

where η_{LI} is the slope of the laser L-I curve.

For a given laser output power requirement, laser diameter, threshold current, and operating DC current are found using Equations 3.126–3.129 as follows:

$$D = \frac{P_{OH}}{\gamma} \; ; \tag{3.130}$$

$$I_{th} = \frac{\varphi}{\gamma} P_{OH} \; ; \tag{3.131}$$

$$I_{TR} = P_{OH}\left(\frac{1}{\eta_{LI}} + \frac{\varphi}{\gamma}\right) \tag{3.132}$$

The NMOS transistor should satisfy the DC condition $I_{DS} = I_{TR}$. Equating Equation 3.132 to the linear region transistor current, neglecting the small quadratic term, and solving for the transistor transconductance yields:

$$k_n = \frac{I_{TR}}{(V_{TRLAS} - V_T)V_L}, \tag{3.133}$$

where V_L is the NMOS drain voltage when it's "ON". Due to the low efficiency of the optical link, the resulting laser current is high. For this reason, we will assume that the design of the switch transistor based on Equation 3.133 also satisfies the minimum switching speed requirement even for a fanout of 1. For large fanout, however, the laser driver may in fact operate faster than a minimum inverter, as a result of the stricter DC requirement given by Equation 3.133. Assuming that transient capacitive current is approximately one forth of the high-level laser current I_{TR}, the rise time of the signal at the driver output can be estimated as:

$$t_r = 4C_{TRLAS}\frac{V_H - V_L}{I_{TR}}, \tag{3.134}$$

where V_H is the high-level driver output voltage; $V_H = V_{TR} - V_{th}$, and C_{TR} is the laser driver output capacitance composed of flip-chip bond capacitance, NMOS transistor junction capacitance and the laser device capacitance:

$$C_{TR} = C_{bond} + \frac{k_n}{k_{\min}}C_{\min N,i} + A_{las} \cdot C_{las}, \tag{3.135}$$

where $C_{\min N,i}$ and $k_{\min}$ are the minimum NMOS transistor junction capacitance and transconductance, C_{las} is the laser device capacitance per area and A_{las} is the laser area:

$$A_{las} = \pi \cdot D. \tag{3.136}$$

Finally, the laser driver input capacitance can be calculated as:

$$C_{TR,i} = \frac{k_n}{k_{\min}} C_{\min N,i}, \tag{3.137}$$

where $C_{\min N,i}$ is the minimum NMOS transistor input capacitance.

4

LOW LATENCY ASYNCHRONOUS OPTICAL BUS FOR DISTRIBUTED MULTIPROCESSOR SYSTEMS

L. Fesquet, J.H. Collet and R. Buhleier

Laboratoire d'Analyse et d'Architecture des Systèmes du CNRS
TOULOUSE, France

ABSTRACT

We present an investigation on how to reduce the internode communication latency in distributed multiprocessor systems. The strategy consists of replacing the "store and forward" mechanism that is usually carried out on each electronic node by a free diffusion of optical packets along an optical bus. No routing, no extraction of optical packets and no access arbitration into the bus are needed. Intermediate nodes between two processors engaged in a communication are optically transparent. These specifications considerably simplify the optoelectronic interface of the node and reduce the latency of communication. In the asynchronous operation mode that we consider, each node multiplexer inserts on the fly optical packets into the bus. The optical demultiplexer identifies the packet destination address before transmitting the data to the node electronics. The key point is to make the intranode latency shorter than the internode propagation time to attain a network latency close to ultimate limit imposed by the velocity of light. We discuss the utilization of all-optical logical gates and the specifications for the requested photonic devices.

Keywords: Optical bus, latency of communication, Asynchronous time division multiplexing

1 INTRODUCTION

The expression *"distributed multiprocessor system"* will be used in the following for describing both tightly coupled multiprocessor machines and clusters of close workstations. The critical parameter to classify these different systems

P. Berthomé and A. Ferreira (eds.), Optical Interconnections and Parallel Processing: Trends at the Interface, 129-147.

is the average internode distance L which ranges from a few centimeters for a parallel machine to a few tens of meters for a cluster of workstations. How to improve the performance of such a multiprocessor machine is a complex challenge because the global performance results from the optimization and the interaction of several software and hardware layers. In this work, we focus on the communication aspects. The role of communications is essential for all the distributed memory multiprocessor systems because whatever application we consider, the communication latency represents a bottleneck which limits the global performance. This bottleneck is due to the fact that the access time to remote and local data are extremely different. In the fastest multiprocessor systems [Cray T3D, Intel Paragon, IBM SP2], the local memory access time is typically 100 to 1000 shorter than that to a RAM on a remote node. Remote accesses are still slower in a cluster of workstations. With such a high communication penalty, it is impossible to strongly couple the processors to construct efficiently the single addressing space needed for the development of portable applications.

Several indirect software solutions are available to partly overcome the communication latency. For instance, it is possible to increase the size of the exchanged messages (i.e., increase of the communication granularity) or to anticipate as much as possible the calls to remote data. In this work, we consider the direct hardware solution. The goal is to reduce the access latency to remote memories T_R to a value comparable to the local RAM access time T_L with typically $T_R/T_L = 3 - 5$. This implies two critical changes in the architecture:

- It is necessary to replace the electrical networks generally based on the "store and forward" mechanism by a free diffusion of optical packets along an optical bus. The key point is to make the intranode latency shorter than the internode propagation time to attain a network latency close to the ultimate limit imposed by the velocity of light.

- It is necessary to directly connect the network interface to the each node **memory bus**. Naming and protection mechanisms must be provided by the processors with a small additional overhead. Providing these mechanisms by the operating system greatly increases the overhead and occupancy costs of communications which in turn reduces bandwidth and increases latency. Connecting the optical bus to the I/O bus (for instance the PCI bus) is surely simpler but dramatically degrades the communication performances, both the throughput and the latency.

We focus in this work on the definition and the analysis of the optical bus parameters (access times and latency) and describe each component layer in detail. Preliminary work has been published in [10].

2 ORGANIZATION OF THE OPTICAL BUS

2.1 Principle of Operation

The utilization of optical networks for connecting multiprocessor systems has been analyzed for many years by computer scientists [2, 9, 11, 14, 16, 21]. The principle of reconfiguration of interconnections in multistage networks was considered in [12, 19]. Optical synchronous busses were also considered in [4, 12]. Concerning the demonstration of optical interconnections, several point to point optical links operating around 0.5-1 Gb/s/channel have been constructed [15, 17, 23]. The communication width in these investigations ranges from 10 to 32-bits. These realizations are of great interest but do not address the general problem of interconnecting a network of processors with a short communication latency. Several low latency bus structures are possible. Figure 1 shows the connection of nodes via two unidirectional busses. The bus width is not drawn for simplicity.

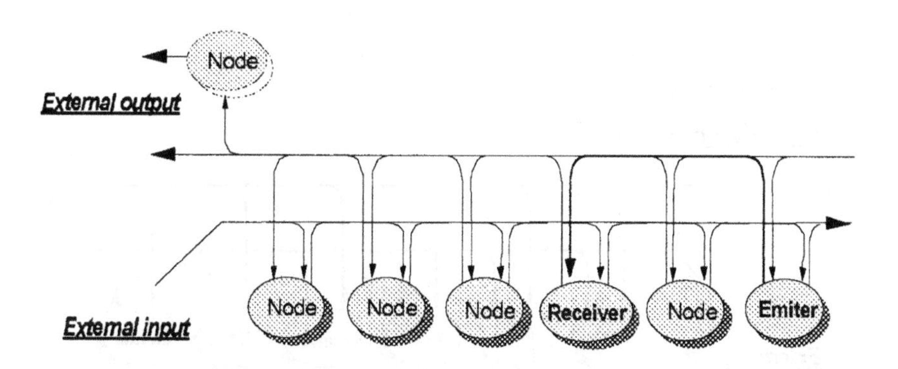

Figure 1 Double unidirectional bus

A single bidirectional bus may also be considered. Its structure is simpler than that of the double unidirectional bus shown in Figure 1. The bidirectional bus multiplexer is however more complicated because it is necessary to identify a hole in the packet flux propagating in the direction leading to the destination node. Processing the collision of packets intended to the same node is basically the same problem for the two types of busses. The final choice between a bidirectional and two unidirectional busses is open. Another possible 1D topology is the unidirectional ring network. However, it has several disadvantages:

- The dialog between 2 nodes implies the propagation of messages along the whole ring. This is obviously a drawback to get fast communications between adjacent nodes.
- By comparison with the linear bus shown in Figure 1, packets are no more automatically "killed" at the end of the line. Therefore, each destination node must be able to remove the optical packets it receives with a very large operation contrast. Otherwise, noise will emerge in the ring due to the accumulation of packet residues. Removing subnanosecond packets by all-optical techniques (to operate at the bus repetition rate) with a high contrast is a real challenge.

Another solution is the U-bus shown in Figure 2. Its structure is simpler than the two unidirectional bus. However, it is not space invariant and does not favor the locality of communications.

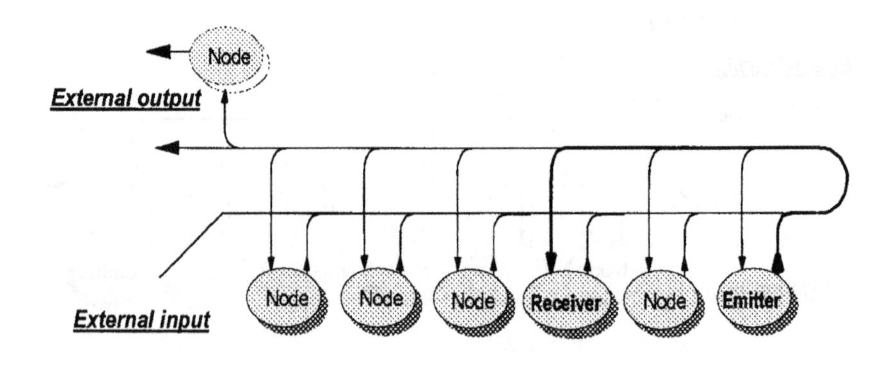

Figure 2 U Bus (i.e., folded bus)

Whatever topology we consider, the bus operates in the mode of Asynchronous Time Division Multiplexing (ATDM). ATDM is an efficient method to share the resources of a network when transmission needs of customers (here, node computers) are heterogeneous. Moreover, the asynchronous transmission mode may easily satisfy the need for scalability of multiprocessor machines or of distributed systems. The bus conveys optical packets with headers containing the addresses of communicating nodes (emission and destination). The optical bus is connected with the electronic memory bus of each node via a specific electronic interface which controls an optical multiplexer and an optical demultiplexer (Figure 3).

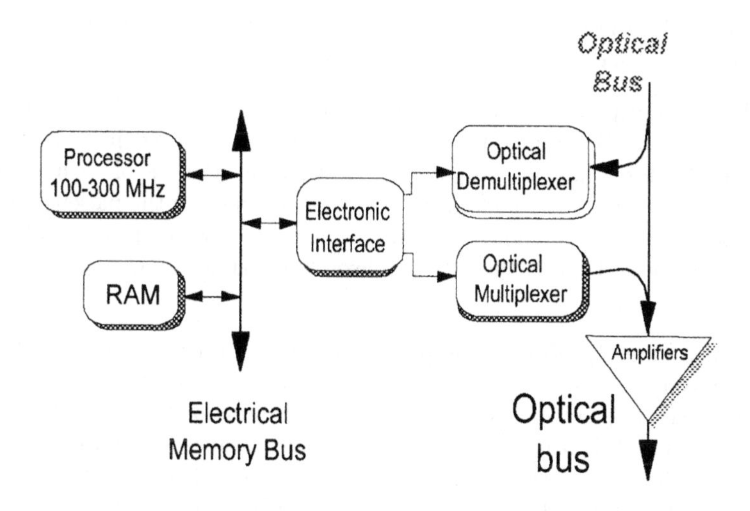

Figure 3 Bus-node interface

Each node contains two optical beam splitters. One forwards T% of the incoming packet amplitude to the local demultiplexer (DMUX). The second inserts the packet generated by the multiplexer into the bus. Optical amplification is needed to balance the losses. Whatever bus topology we consider, the following points are critical for an ideal performance of the asynchronous optical bus:

- The multiplexer inserts **on the fly** one optical packets into the bus as soon as an open time slot is detected in the flow of incoming packets. The demultiplexer (Figure 3) compares the destination address of each incoming packet to that of the node. The result of the address recognition is a logical bit that authorizes the transmission of the identified packet to photodiodes which interface the optical bus with the node electronics.

- Optical packets freely propagate along the bus with no risk of collisions or contentions as compared to 2D or 3D networks. This is very important because solving contention of packets in 2D or 3D networks or defining packet routing paths even by Time Division Multiplexing (TDM) [19, 20] is time consuming. Solving packet collisions generally implies to temporary store data. That would be very difficult for optical packets without opto-electronic conversion because, to our knowledge, the operation at several hundreds of MHz of fast access optical RAMs has not yet been demonstrated. The latency of the optical data processing of this bus is just limited by the light propagation time through the bus and the electronic time necessary to emit and insert a packet.

- The demultiplexer operates on the fraction of the packet that is forwarded to the node electronics by the beam splitter mentioned just above. The demultiplexer has to compare the destination address of each incoming packet to that of the node in order to decide whether the packet has to be transmitted to the node electronics. The packets which progress along the bus waste no time in the demultiplexer stage (Figure 7).

The transit time T_T of incoming packets trough each node equals the latency time necessary to generate and insert a packet into the bus. If T_T is shorter than the internode propagation time T_{NN}, the communication latency is determined by the velocity of light in fibers! Considering for instance an internode distance L of 1-2 m (case of a workstation in a single room), then $T_{NN} \approx$ 5-10 ns. A packet insertion delay of 1ns is sufficient to attain a communication latency very close to the limit imposed by light velocity. Considering now $L = 10$ cm (case of a multiprocessor system in a cabinet), then $T_{NN} \approx 0.5$ ns. A trigger delay of the order of 100-200 ps is necessary to approach the ultimate communication latency.

2.2 Comparison of Remote and Local Access Latencies

We now compare the access latencies to data in the local RAM and in a remote RAM. As the optical bus interface and the RAM are connected to the memory bus (Figure 3), we analyze the elapsed time after the processor has transmitted the access request to the bus. The access latency to the local RAM access is

T_{RAM}. The access latency to a remote RAM is $T_R = T + T_{RAM} = (T_I + T_A + T_{NN} + T_D + T_I) + T_{RAM}$ where T_I, T_A, T_{NN}, T_D are the latency time of the electronic interface, the access time to the unsaturated network (i.e., multiplexer latency), the internode propagation time along the bus and the demultiplexer latency respectively. Thus, the time penalty to access remote data is $T = 2 * T_I + T_A + T_{NN} + T_D$ that must be compared to the local access time T_{RAM} which yields typically 50-100 ns. The propagation time T_{NN} depends critically on the internode distance. In a multiprocessor system, it is typically less than 1-2 m, so that $T_{NN} < 10$ ns. T_A may be of the order of a few nanoseconds (say $T_A < 5$ ns) with a fast trigger electronics for the laser diode driver. T_D is also of the order of T_A (see Section 4.3). Therefore $T_{NN}+T_A+T_D$ is on the order of 10 ns. The electronic interface is basically a sequential circuit that reads (or writes) the memory bus and stores the data intended to remote nodes in SRAMs. We must consider here the utilization of GaAs digital ASICs which can operate at frequency as high as 2.5 GHz. Available arrays of gates with embedded megacells such as RAMs, register files, phase locked loops are available. Even if the processing time T1 needed by these circuits is of the order of a few hundreds of ASIC clock cycles (i.e., $T_1 \leq$SS100 ns), the remote access penalty T_R will be typically lower than 200 ns. This time must be compared to the remote access time in current multiprocessor machines, which is on the order of several tens of microseconds using portable implicit communication libraries. This value, as given by the manufacturers is very optimistic. In fact, the reality is close to hundreds of microseconds for a high performance multiprocessor operating under normal conditions. Therefore an optical bus directly interfaced to the memory bus is able to reduce the latency times and to increase dramatically the number of processors connected to the bus. Short latency also authorizes fine grain communications, makes synchronization easier between processors and improves parallelism for the system.

3 30-100 GB/S OPTICAL BUS WITH FULL ELECTRONIC PACKET PROCESSING

Several implementation strategies are possible. In the first approach, **the throughput per channel T_C is deliberately maintained under 3 Gb/s to ensure that all the boolean operations involved in the demultiplexing of packets can be carried out by fast digital electronic circuits** (see Section 2.2). In this strategy, the role of Optics is reduced to the emission of the packet, the propagation along the bus, the detection of the packets by photodiodes as shown in Figure 4. The thick upper lines show the optical circuits. The advantages of this approach are important:

- This is a cost effective strategy for short term developments.
- The internal organization of the packet is not essential. It will be crucial in section 4 when considering higher throughput which require optical boolean operations.
- A fast electronic interface that operates at the bus throughput is able to spy on all the messages conveyed by the bus. It is therefore compatible with several hardware protocols of memory coherence for the global management of shared data.

The main drawback of this approach is the relatively low throughput of the bus. The global throughput is $T_B \approx N_B T_C$ where N_B is the bus width. With $T_C < 3$ Gb/s.and $N_B \approx$ 10-32 (as in the current parallel optical communication programs [24]) we get a global bus throughput T_B of the order of 30-100 Gb/s. Given the **top** throughput per node T_N, the global throughput limits the number of nodes that can be connected to the multiplexed bus to $N = T_B/T_N$. With $T_N \approx 1$ Gb/s, we get $N \approx$ 30-100. In practice, the number of nodes that can be connected may larger than this estimation if the average throughput is smaller than the peak throughput because the Asynchronous Multiplexing enables to share efficiently the communication resources. It is extremely important to stress that the bus transmission latency is almost independent of the bus throughput so long as the optical pulse duration T_P is short compared to the internode propagation time T_{NN}. For instance, if the internode distance is $L = 1$ m, the internode propagation time is 5 ns so that using ultrashort pulses as short as 1 ps or 500 ps pulses do not change the communication latency along the bus.

Generating and multiplexing the optical packets using VCSEL arrays is especially attractive. Current VCSELs operate with a threshold current of about 1 mA and can be directly driven by fast electronic circuits to generate 100-200 ps optical pulses. The demultiplexer interface is simply an array of subnanosecond photodiodes connected to fast electronic input buffers.

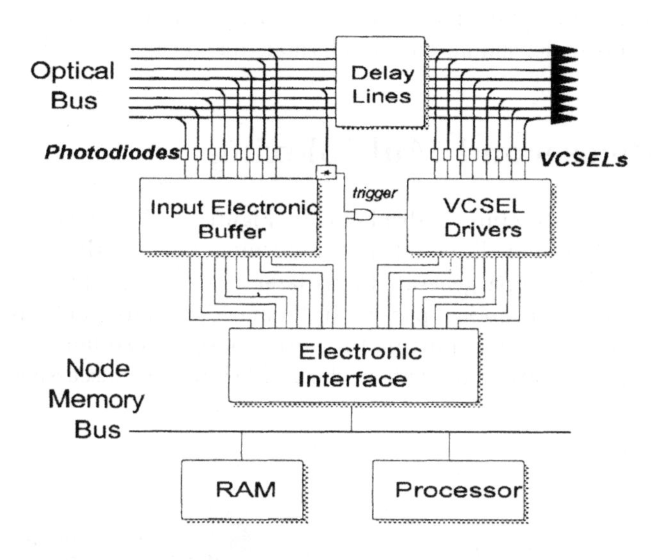

Figure 4 Node interface with VCSELs and Photodiodes. The optical bus width is 8 bits. The width of the memory bus is not shown

4 0.5-1 TB/S OPTICAL BUS WITH PHOTONIC GATES

Connecting a large number of nodes, say for instance 500, with a top node throughput T_N of the order of 1 Gb/s, requires a global bus throughput larger than 1000 Gb/s to avoid bus congestion. The electronic version of Section 3 cannot easily satisfy these requirements. It becomes necessary to increase the bus width or (and) to include Wavelength Division Multiplexing (WDM) or (and) to use shorter pulses which enable to increase the channel throughput. Thus, a second strategy considers the utilization of picosecond pulses in the range 10-20 ps to attain a channel throughput T_C around 50-100 Gb/s. The main problem in this approach is that all the demultiplexing operations cannot be carried out by fast electronic circuits. A possible solution consists of connecting the electronic layer to the optical bus via simple but ultrafast photonic gates that could be integrated in monolithic optical circuits as discussed below. For definiteness we consider a channel throughput of 40 Gb/s. This means a bit to bit period of 25 ps. Consequently, the transmission typically has to operate with 10-12 ps long optical pulses (Full Width at Half Maximum) possibly with monomode fibers depending on the internode distance.

4.1 Picosecond Multiplexing

The multiplexer structure is displayed in Figure 5. We consider the development of optical monolithic integrated circuits. One **single** optical pulse of duration 10-20 ps is first generated by a laser diode, then split to generate the different pulses of the optical packet. In Figure 5, the initial pulse is divided in 16 pulses. The reason for splitting the initial pulses in two families of 8 pulses with orthogonal polarizations will be clarified below when discussing the packet structure.

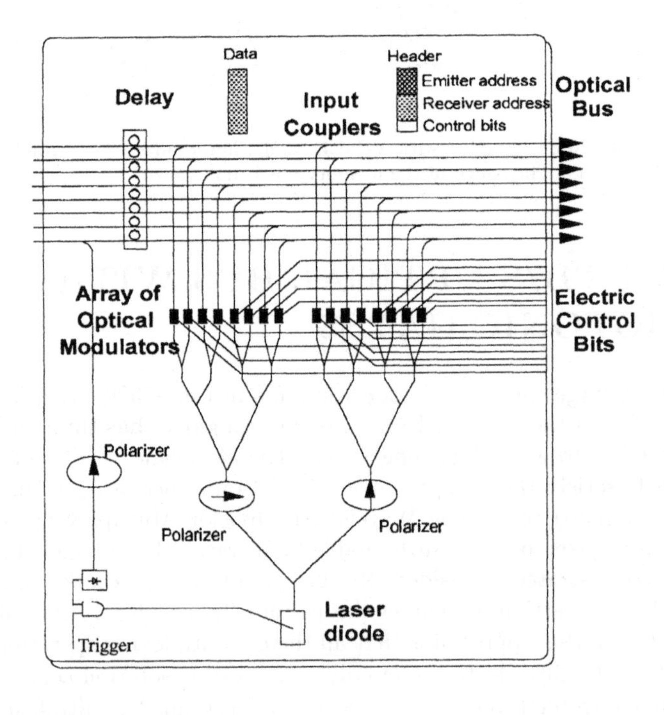

Figure 5 Multiplexer architecture with a 8-bit wide optical bus

Generating the packet from a single pulse eliminates any jitter and any wavelength dispersion among the pulses of the same packet and also simplifies the operation condition of the optical gates in the demultiplexer. This point is extremely important. The utilization of VCSELs (as in Section 3) is not satisfactory for generating such short pulses because of the wavelength dispersion of the VCSELs emission and because of the jitter due to the separate triggering of the different diodes. The generation of pulses in the range 1-30 ps is very common today, both with triggered or mode-locked diodes laser and with mode-locked solid state lasers. The laser diode is triggered by a fast ASIC circuit which measures the time difference ΔT between two successive incoming packets. One header bit may be specialized for the detection of the packet by an ultrafast photodiode and an ASIC circuit. Trigger occurs if $\Delta T > T_{PD} + T_J + T_S$, where T_{PD} is the packet duration, T_J the insertion jitter of the diode driver and T_S an optional security delay. Trigger occurs because the multiplexer can find, on the fly, a free slot to insert one packet in the packet flow. For correct insertion, incoming packets must be delayed (see the "delay block" in Figure 5) to balance the latency time necessary to insert the packet into the bus. In other words, the packet transit time in each node is limited by the insertion latency of the optical packet.

Each pulse is encoded by an integrated modulator (Array of Optical Modulators, Figure 5). Notice that the diode laser emission must be compatible with the operation wavelength of modulators. The control of the electro-optical modulators is parallel, not serial. In other words, each modulator of the AOM is separately controlled by one driver directly connected to memory output. The state of each optical modulator (open or closed) must be set up before triggering the diode laser. The parallel packet is thereafter injected in the optical bus via 2-input couplers (Figure 5). The diode laser and the modulators must be able to run at the electric bus frequency, i.e., typically 100 MHz. This condition is easily satisfied. The time difference between two packets emitted by the same node is typically larger than 10 ns. Of course, the bus conveys during this period many packets emitted by other nodes. We may consider absorptive modulators (e.g. Stark effect modulators [1, 18, 22]) but Traveling Wave Semiconductor Laser Amplifiers (TWSLA) [3, 8] are likely preferable because they provide signal gain and balance the losses.

4.2 Packet Structure

So far we have not discussed the internal structure of the optical packet. This point was not essential for the 30-100 Gb/s bus considered in Section 3 because

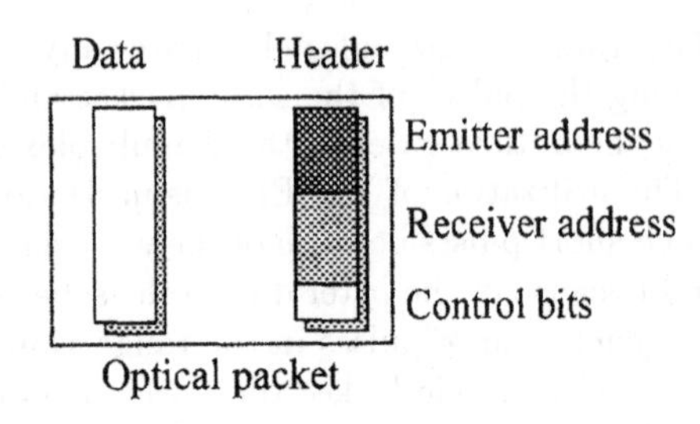

Figure 6 Packet structure

the channel throughput below $T_C = 3$ Gb/s is compatible with the electronic processing. The packet structure becomes crucial for the Terabit bus because there is no simple solution to separate serial (optical) pulses at a throughput T_C of 40 Gb/s (or higher). This is the general challenge of the serial to parallel conversion for optical pulses beyond the top operation frequency of electronic circuits. The packet structure shown in Figure 6 has been designed to simplify the packet demultiplexing. The packet is composed of 2 bits/line, namely one header bit and one payload bit. The first bit is the header bit. The great advantage of this simple structure is that header and payload bits can be very easily separated with passive optics using for instance orthogonal polarizations for each one[1]. For this reason, the multiplexer architecture in Figure 5 splits the pulse generated by the laser diodes in two channels with orthogonal polarizations. One channel generates the header and the other one the payload. The header contains the addresses of the emission and destination nodes. The parallel encoding of headers enables a simple and extremely fast recognition of the destination address by optical means (see Section 4.3). Considering a 32-bit wide bus, the header may contain for instance 12 bits for the emission address, 12 bits for the destination address and 8 free bits for transmission control that will be discussed later. Depending on the encoding method, 12 bits enable to address up to 4096 nodes. That seems sufficient for most multiprocessor systems.

The packet duration T_{PD} is 2 bits/line, i.e., 50 ps. The payload throughput is only 20 Gb/s/channel. A 32-bit wide bus provides a top bus throughput of 20x32 = 640 Gb/s. Of course, this top throughput cannot be used in reality because it implies a complete bus congestion and great difficulties to access the

[1] Maintaining the polarization through a fiber on few tens of meters for a cluster of workstations of a multiprocessor system is not a technical problem.

transmission resources. Such a bus can likely operate around T_B = 300-400 Gb/s with no significant increase of the access time.

4.3 All-optical Address Recognition

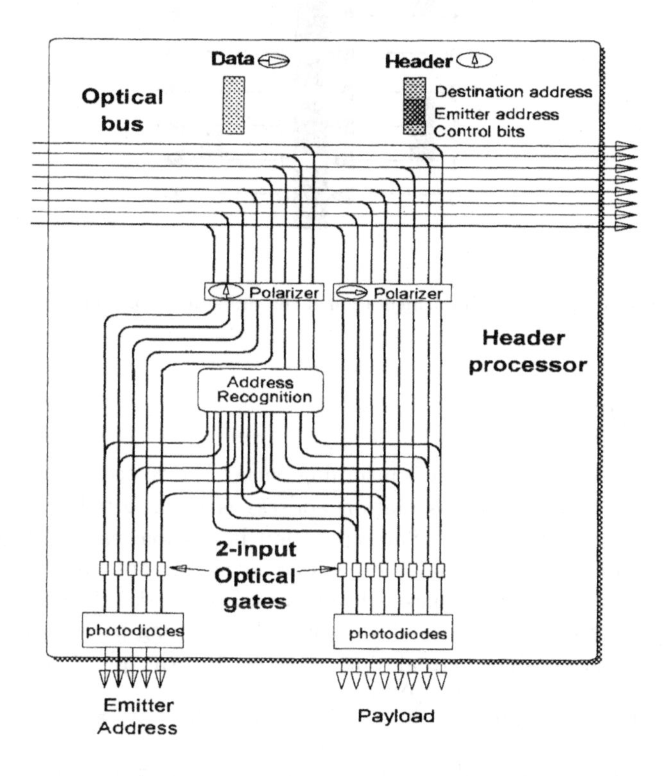

Figure 7 Photonic part of the demultiplexer

The demultiplexer is composed of three main subsystems, namely, the header processor, the logical gate and the optoelectronic interface. (Figure 7) shows the photonic part of the demultiplexer including the header processor and the photonic gates.

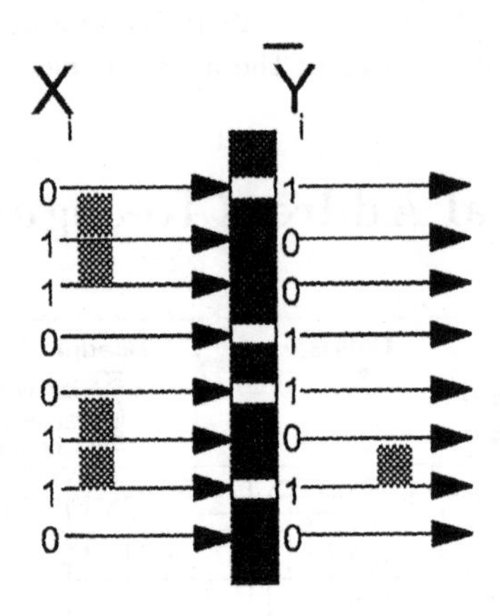

Figure 8 Simple implementation of the recognition for 8-bit addresses

Header processor.

The destination address of each incoming packet is directed to the header processor. In Figure 7, the destination address is transported by the 3 upper channels of the optical bus. The separation is possible using orthogonal polarizations for the header and the payload. Another possible solution would be to use wavelength multiplexed headers and payloads. The header processor compares the destination address of each incoming packet to that of the node. We must consider here all-optical gates. In principle, It is necessary to compute the logical bit $S = \sum_{i=1}^{N}(\bar{X_i}Y_i + X_i\bar{Y_i})$ where X_i and Y_i are the bits of the destination and node addresses respectively. $S = 1$ as soon at two bits X_i and Y_i are different. Computing S implies to invert the incoming bits X_i of the destination address. This operation is not impossible but optically complicated. We therefore follow the treatment proposed by D. Cotter et al. who showed that the truncated operation $S' = \sum_{i=1}^{N}(X_i\bar{Y_i})$ enables to recognize the addresses which are encoded with a constant number of high bits, say $N/2$ for a N-bit address [5, 6]. This coding enables to identify up to 924 nodes with 12-bit addresses. Additionally the Boolean function S' can be realized with passive optics because the node address $\{Y_i | 1 \leq SSi \leq SSN\}$ is static. This result is remarkable because it means that the comparison of the destination address to that of the node, that seemed at first sight a very complex challenge for optics, is simple. Figure 8 shows a simple implementation of the recognition for 8-bit addresses. In this example, the input address $X = \{0,1,1,0,0,1,1,0\}$ is compared to that of the node $Y = \{0,1,1,0,0,1,0,1\}$. Notice that X and Y contain the same number of high bits (here 4) in agreement with the encoding principle of [5]. Each operation $X_i\bar{Y}_i$ is achieved by transmitting or blocking the optical bit X_i. As the bits Y_i are static, this operation reduces to blocking or transmitting the different beams through an array of modulators or even simpler with holes in a screen! The header processor generates one (or more) optical bits if the addresses X and Y are different (Figure 8).

Logical gates.

The logical gates (Figure 7) must be able to operate at several tens of GHz. If an incoming packet is intended to the node, the payload and the emitter address must be transmitted, via the optical gates, to the node electronics that operates in the range 100-300 MHz. The logical gates realize the operation $B_i\bar{S}'$ where S' is the optical bit delivered by the header processor and B_i the data payload. If the destination address does not match the node address, then $S' = 1$, so that no data are transmitted. Concerning the physical realization of fast subnanosecond all-optical logical gates, several tracks are possible. The most promising solution is likely the saturation of gain in Semiconductor Laser Amplifiers [7,8] that can operate up to 40 Gb/s. We may consider Four Wave Mixing in fibers or microcavities [18]. The subnanosecond reflectivity of microcavities is also interesting [13]. Note that the optical gates must be able to operate even in the case of a slight dispersion of the packets wavelength (say 1-2 nm) because the packets are emitted by the diodes of different multiplexers.

Optoelectronic Interface.

Fast photodiodes convert the emitter address and the payload in the electronic domain for processing by the electronic interface (bottom of Figure 7). The bandwidth of the electronic circuits is lower than that of the optical bus, even if we consider an intermediate stage operating around F = 1-2 GHz between the optical bus and the memory bus. Therefore, the electronic interface cannot distinguish two packets that are transmitted by the optical logical gates within a time interval shorter than $1/F$. The electronic interface will therefore integrate the bits of two (or more) " colliding " packets and in particular the emission addresses. Note that these packets are colliding in the electronic interface, not in the bus. Now, colliding packets cannot originate from the same node because the emission bandwidth of each node is that of electronic circuits operating around 100-200 MHz. Following Cotter et al., the emitter addresses are also encoded with a constant number of high bits (say N_H). The integration of several emission addresses originating form different nodes therefore produces a new address which contains (at least) $N_H + 1$ high bits. This address is illegal. In other words, it is possible to identify the collision of packets in the demultiplexer electronic circuits by counting the number of high bits of the emission address. This processing can be quickly carried out by ASIC circuits, providing a fast validation bit. The subsequent control of communications can be hard or soft (or mixed) depending on the complexity of the involved communication protocol.

5 CONCLUSION

We investigated the architecture of fast multiplexed optical busses for reducing communication latency and increasing the bandwidth in distributed multiprocessor systems. The basic characteristics of optical bus architecture are the following ones:

- Direct connection of the optical bus interface to each node **memory bus**. Naming and protection mechanisms must be provided by the processors with no additional overhead. Providing these mechanisms by the operating system greatly increases the overhead and occupancy costs of communications which in turn reduces bandwidth and increases latency.
- Free propagation of the optical packets along the bus. No storage, no routing, no processing.
- Usage of asynchronous time division multiplexing because it is an efficient method to share the communication resources of a network when transmission needs of customers are heterogeneous.
- Fast access time to the bus, on the fly, by each node multiplexer. The key point is to make the intranode latency shorter than the internode propagation time to attain a network latency close to ultimate limit imposed by the velocity of light.

Considering that the access time to remote memories is currently in the range of some hundreds of μs with standard compilers that generate a portable code, lowering the access time to several hundreds of ns (or even less) would dramatically improve the overall performance of multiprocessor systems. Because the internode propagation time ΔT cannot be compressed ($c = 5$ ns/m), our conclusions are restricted to compact systems or local area network. This also implies to favor the locality of communications in local networks. The final performance of the asynchronous bus will depend on the operation of the multiplexer, of the demultiplexer and of the electronic interface.

Two implementations are considered.

- The 30-100 Gb/s optical bus is compatible with a full processing of the optical packets by fast electronic circuits operating around 2-3 Gb/s. The role of Optics is reduced to the internode transmission of data. This bus may potentially connect typically 30-100 nodes with an access latency of a few hundreds of ns.
- The 0.5-1 Tb/s optical requires the development of fast optical AND gates. The architecture is much more complicated than that of the 30-100 Gb/s bus. It may connect several hundreds of processors.

Acknowledgements

We are pleased to thank Dr. A. Munoz-Yague (LAAS-CNRS), Pr. D. Litaize (IRIT, Toulouse), Drs. J.P. Bouzinac, P. Churoux, D. Comte and N. Hifdi (CERT-ONERA, Toulouse) for many valuable discussions.

REFERENCES

[1] G. D. Boys and G. Livescu. Electro-absorption and refraction in Fabry-Pérot quantum well modulators: A general discussion. *Optical and Quantum Electronics*, 24:147–165, 1992.
[2] K.-H. Brenner, A. Huang, and N. Streibl. Digital optical computing with systolic substitution. *Applied Optics*, 25:3054, 1986.
[3] J. Bristow et al. Packaged AlGaAs waveguide modulator array at 830 nm wavelength. *Journal of Lightwave Technology*, 13:1041, 1995.
[4] D. M. Chiarulli, S. P. Levitan, R. G. Melhem, M. Bidnurkar, R. Ditmore, G. Gravenstreter, Z. Guo, C. Qiao, M. F. Sakr, , and J. P. Teza. Optoelectronic buses for high-performance computing. *Proceedings of IEEE*, 82:1701–1710, 1994.

[5] D. Cotter and S. C. Cotter. Algorithm for binary word recognition suited to ultrafast nonlinear optics. *Electronic Letters*, 29:945, 1993.

[6] D. Cotter, J. K. Lucek, M. Shabeer, K. Smith, D. C. Rogers, D. Nesset, and P. Gunning. Self routing of 100 Gb/s packets using 6-bit "keyword" address recognition. *Electronic Letters*, 31:1475, 1995.

[7] A. D. Ellis and D. M. Spirit. Compact 40 Gb/s optical demultiplexer using a GaInAsP optical amplifier. *Electronic Letters*, 29:2115–2116, 1994.

[8] D. Ellis, D. M. Patrick, D. Flannery, R. J. Manning, and R. J. Davies. Ultra-high speed OTDM networks using semiconductor amplifer-based processing nodes. *Journal of Lightwave Technology*, 13:761–770, 1995.

[9] M. Feldman, S. Esener, C. Guest, and S. Lee. Comparison between optical and electrical interconnects based on power and speed considerations. *Applied Optics*, 27:742–751, 1988.

[10] L. Fesquet and J. Collet. Low latency optical bus for multiprocessor architecture. In *ICAPT*, 1996.

[11] J. Goodmann, F. Leonberger, S. Kung, and R. Athale. Optical interconnections for VLSI systems. *Proceeding IEEE*, 72:850–866, 1984.

[12] Z. Guo, R. G. Melhem, R. W. Hall, D. M. Chiarulli, and S. P. Levitan. Pipelined communications in optically interconnected arrays. *Journal of Parallel and Distributed Computing*, 12(3):269–282, July 1991.

[13] J. L. Iehl, R. Grac, J. H. Collet, M. Pugnet, R. Buhleier, V. Bardinal, C. Fontaine, and R. Legros. Subnanosecond density dynamics of the electron-hole plasma in GaAs/AlAs microcavities. *Physica B*, 222:76–79, 1996.

[14] J. Jahns and Murdocca. Crossover networks and their optical implementation. *Applied Optics*, 27:3155–3160, 1988.

[15] H. Karstenen, C. Hanke, M. Honsberg, J. R. Kropp, J. Wieland, M. Blaser, P. Weger, and J. Popp. Parallel optical interconnection for uncoded data transmission with 1 Gb/s/channel capacity, high dynamic range and low power consummation. *Journal of Lightwave Technology*, 13:1017, 1995.

[16] S. Levitan, D. Chiarulli, and R. Melhem. Coincident pulse techniques of multiprocessor interconnection structures. *Applied Optics*, 29:2024–2039, 1990.

[17] J. Nishikido, S. Fujita, Y. Arai, Y. Akahori, S. Hino, and K. Yamasaki. Multigigabit multichannel optical interconnection module for broadband switching system. *Journal of Lightwave Technology*, 13:1104–1110, 1995.

[18] G. Parry, M. Whitehead, P. Stevens, A. Rivers, P. Barnes, D. Atkinson, J. S. Roberts, and C. Button. The design and application of III-V multiquantum well optical modulators. *Physica Scripta*, T35:210–314, 1991.

[19] C. Qiao and R. Melhem. Reconfiguration with time division multiplexed MIN's for multiprocessor communications. *IEEE Transactions On Parallel and Distributed Systems*, 5(4):337–352, Apr. 1994.

[20] C. Qiao, R. G. Melhem, D. M. Chiarulli, and S. P. Levitan. Dynamic reconfiguration of optically interconnection networks with time division multiplexing. *Journal of Parallel and Distributed Computing*, 22:268–278, 1994.

[21] J. Sauer. A multi-Gb/s optical interconnect. In SPIE, editor, *Proc. Digital Opt. Comp. II*, volume 1215, pages 198–207, 1990.

[22] K. Wakita, K. Sato, I. Kotaka, M. Yamamoto, and T. Kataoka. 20 Gbit/s 1.55 μm strained InGaAsP MGW modulator integrated DFB laser module. *Electronic Letters*, 30:302, 1994.

[23] Yiu-Man et al. Technology development of a high density 32-channel 16 Gb/s optical data link for optical interconnection applications for the optoelectronic technology consortium. *Journal of Lightwave Technology*, 13:995, 1995.

[24] S. X. Zhu. Breaking the data bottleneck with speedy optical interconnects. *Photonic Spectra*, page 95, July 1996.

5

OPTICAL FREE-SPACE INTERCONNECTIONS INSIDE PARALLEL ARCHITECTURES: ONERA-CERT ACTIVITIES

P. Churoux, J.-P. Bouzinac, S. Kocon, M. Fracès, D. Comte*, N. Hifdi*, T. Collette and P. Scheer****

ONERA-CERT, Optical Department, Toulouse, France

** ONERA-CERT, Computing Department, Toulouse, France*

*** LETI (CEA-Technologies Avancées)*
DEIN - DEA Saclay, France

ABSTRACT

Researchers are increasingly considering optics as a means of relaying electronic interconnections applied to parallel computer architectures. This paper presents the ONERA-CERT point of view by way of the MILORD, OEDIPE architectures studied in our laboratories.

1 INTRODUCTION

Parallel architectures are studied to get rid of bottleneck effects brought about by classical computers. The purpose is to distribute the task to be executed over several processors working at the same time.

Electronics has produced impressive upgrades during the last decade, with an extra 10 to 100 Gigaflop systems launched on the market. This represents about a 3 order of magnitude over the last ten years [13]. Mircroprocessor circuits of the 1990's have the power of the Cray-1 system of the 1980's. Supercomputer designers are talking about introducing teraflop levels in a few years' time.

P. Berthomé and A. Ferreira (eds.), Optical Interconnections and Parallel Processing: Trends at the Interface, 149-171.

But progress in electronics will not be infinite, especially due to interconnection problems. Figure 1 derived from [11] presents the conventional electronic packaging hierarchy. The length ℓ of an interconnection is well related to the packaging level. The lower scale is the maximum bit rate R in the line before a transmission line structure is required, where $R \leq \frac{c}{50\sqrt{\varepsilon_r}}$.

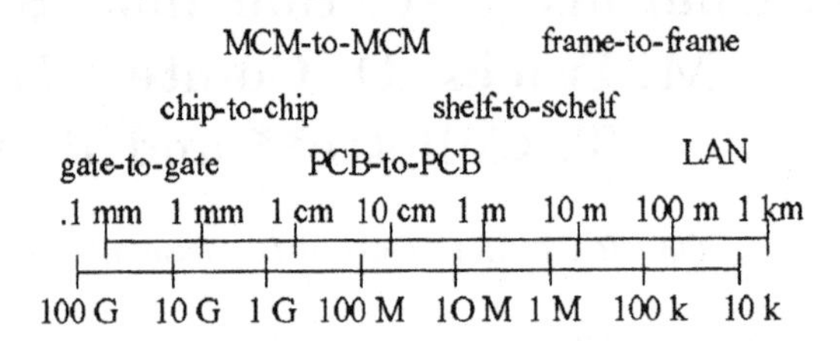

Figure 1 Interconnection hierarchy and their typical length from [11]

Optics has been studied for the last fifteen years as an alternative to electronics technology in massively parallel architectures. The properties of optics are well known, especially in the telecommunication domain. Moreover, optics could be a good candidate to replace the sequential Von Neuman computing unit of a computer.

Many computing models have emerged in the last decade, such as the symbolic substitution [1], optical array logic [12] and PLA logic models [10]. Their objective is to build an all optical computer instead of electronics. The main merit of these approaches is to take advantage of the two dimensional aspects. Optical materials are studied to achieve operations such as Boolean functions on optical beams. The advantage of this technology is that the operating structure is directly compatible with free space optics. But, up to now, they suffer too many limitations to be used in an efficient way. Electronic technology is becoming more and more integrated and offers wider versatility for making processors and memories.

On the other hand, data exchanges in a supercomputer are the barrier to obtaining effective performance. Future computing systems will have a very efficient communication system where optics will play a specific role. Optical interconnections have low loss in line and losses are independent of the temporal bandwidth. Moreover, they are insensitive to electromagnetic interferences. For such short distance interconnections, typically less than one meter, the temporal bandwidth is only limited by the optoelectronic and front-end electronic devices.

Time latency is low and is due to the propagation delay. It is equal to $\frac{\ell n}{c}$ where n is the refractive index. For example, with $ell = 0.2m$ and $n = 1.5$, time latency is 1 ns.

From the integration point of view, three-dimensional interconnections using free space optics offer high interconnection density. Data are arranged on a plane and light propagates perpendicular to the plane direction. Passive optical technology is compatible with collective planar technology derived from micro-electronics circuit designs. Optoelectronics such as VCSEL [8] and SEED [9] is also becoming increasingly available in linear or matrix array form and is compatible with electrical and optical characteristics. Moreover, integration capabilities, illustrated by the CMOS-SEED devices [7], are nearly compatible with system specifications.

Taking into account these basic arguments for using optics, what are the best parallel architectures which gain the most benefit from this technology?

Interconnections are present everywhere in parallel architectures. In particular, interprocessor communications and processor to memory communications are two critical sites where they have to be parallel. This chapter will concentrate on interprocessor communications.

2 TIME AND SPACE MULTIPLEXING

We define N as the number of processors. Processor data of B bits wide. For reconfiguration operations, each input channel can be connected to M output channels. For example, if M equals N the network is a CROSSBAR.

The interconnections network need at least four dimensions:

- one dimension for the N channels (possibly two dimensions for large N^2 parallelism),
- one dimension for the B word width (possibly two dimensions for binary page data),
- at least one dimension in case of reconfigurations,
- one for the optical link itself.

Moreover, the whole network cannot be spatially implemented on the three space dimensions.

Wavelength multiplexing [3] is an advantage in theory because it maintains parallelism. In practice, this technology is suited to telecommunication applications with only few multiplexed channels and relatively large physical dimensions. Thus, for the moment, we consider that this technology is not usable for short distance communications with large parallelism.

Time multiplexing is necessary for obtaining large values either of N, M or B. Table 1 shows different ways of mixing space and time multiplexing for various network topologies.

input channels	reconfiguration	word width	network topologies
N	N	time mux	CROSSBAR
N^2	2	time mux	OMEGA
N	1	B	RING
time mux	N	B	BUS

Table 1 Space and time multiplexing according to network topologies

Each of the first three configurations correspond to a project studied in our laboratory. The last one is exposed in a specific chapter of this book (See Chapter 6).

Each of the first three configurations corresponds to a project studied in our laboratory. The last one is exposed in a specific chapter of this book.

3 THE MILORD ARCHITECTURE

"architecture Multiprocesseurs Interconnectés par Liens Optiques Reconfigurables Dynamiquement"

3.1 Concept

The architecture is built around the CROSSBAR network which interconnects the processors. This powerful network allows all communication topologies between processors without any conflict. Several systems are possible [3].

In free space, using the Spatial Light Modulator SLM, we choose the so called matrix-vector inner product configuration illustrated in Figure2.

Each emitter in the input plane broadcasts data across all the detectors in the output plane. The signal of i-th emitter crosses the N switches of the ith column of the two dimensional SLM. For a connection between the input i and the output j, the transmission of the c_{ij} switch is equal to one. All the other switches on the same row are in the off state with a transmission of zero.

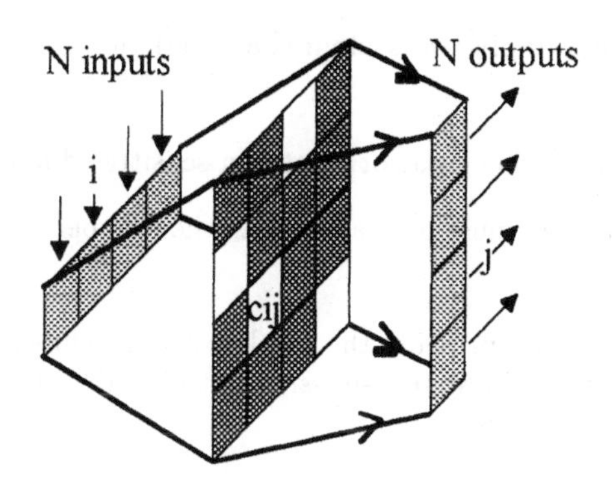

Figure 2 N^2 parallel matrix-vector inner product

3.2 Power budget

We note P the power of each input channel. The useful power P_u corresponds to the input power P divided by N and multiplied by the real transmission T_{on} of the switch in the "on" state.

$$P_u = \frac{P}{N} \cdot T_{on} \tag{5.1}$$

In the same time, the $N-1$ other input channels contribute to optical crosstalk. This maximum total noise signal P_c is, at the worst, case,

$$P_c = (N-1) \cdot \frac{P}{N} \cdot T_{off} \approx P \cdot T_{off} \text{ for large value of } N. \tag{5.2}$$

The effective amplitude modulation is:

$$P_u - P_c = P \cdot T_{on} \left(\frac{1}{N} - \frac{1}{C} \right), \tag{5.3}$$

with $C = T_{on}/T_{off}$.

Equations 5.1 and 5.3 imply two system characteristics:

- Input power have to be great or reception sensitive due to the $1/N$ factor.
- The contrast ratio C must be greater than the number N of input channels.

The factor $T_{on} \cdot (\frac{1}{N} - \frac{1}{C})$ represents the total optical attenuation assuming that T_{on} involves not only the SLM transmission but also all optical absorption and reflection effects in the links.

3.3 Reconfiguration time

The reconfiguration time T of the network is determined by the time of controlling the SLM and the response time of the SLM. The latter is widely dependent on the technology. With nematic liquid crystal devices, this time is relatively long, in the order of a few ms. Faster modulators such as the Multiple Quantum Well modulator are available, but for this kind of application they suffer a poor contrast ratio. Moreover, even with a short response time for one switch, when controlling a N^2 array sequentially, the total time is too long. The ferroelectric liquid crystal SLM offers a good compromise, with a contrast ratio of 200 and a reconfiguration time which can be as low as 64 ms with a parallel control.

3.4 Experimental results

We performed a first experiment [5] illustrated in Figure 3.

Characteristics are as follow:

The 16 INMOS T414 transputers each have 2 of their 4 links on the optical network; the SLM is a Hughes 4060 IR liquid crystal light valve with a contrast ratio of 160; the optical fibers are 100-140 μm core with a 180 μm pitch; there are Sharp LT 015 MD laser diodes (λ = 830 nm) and 34 UDT elements in photodiode linéar array. Data rate is limited to 10 Mbit/s by the processor. The 160 ms reconfiguration time is limited by the SLM.

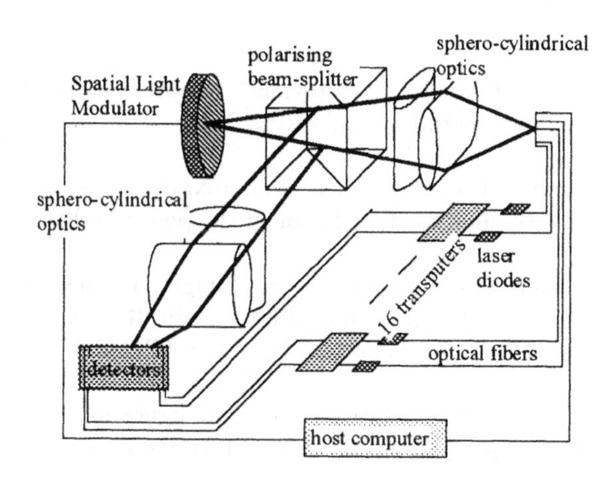

Figure 3 Schema of the experimental setup

The aim of the second experiment was to obtain the best characteristics we could expect from such a system. We replaced:

- the laser diodes by CQL 61A Philips multimode laser diodes, at a wavelength of 810 nm, to minimize modal noise effects,
- the SLM with a STC ferroelectric SLM to lower reconfiguration time,
- the detector by a -31 dBm sensibility PIN receiver to raise sensibility and bandwidth.

Under these conditions, we achieved a data rate of 150 Mbit/s for a Bit Error Rate of 10^{-9}.

3.5 Synthesis on optical Crossbar networks

The three main parameters of the network are:

- N: the number of input channels,
- T: the reconfiguration time,
- R: the data rate.

The lowest reconfiguration time, as previously mentioned, is 64 ms and can reach a few ms for a sequential control of the SLM.

The projected number of input channels is approximately 100. It is limited by both the contrast ratio and the $1/N$ factor in the power budget.

The data rate is limited by the optical power budget. For instance, with $N = 100$; $T_{on} = 25\%$; $C = 200$ the attenuation is -29 dB. This attenuation is compatible with data rates in the order of 100–200 Mbit/s.

Data flow is independent of the relatively long reconfiguration time. This means that when the channel is open one can send a long message.

The system based on the inner product configuration offers broadcast capability.

We have evaluated several algorithms on MILORD architecture. This evaluation revealed that for a wide class of algorithms such as data bases, vectorial products, image processing and numerical analysis, few network configurations are used compared to the $N!$ configurations offered by the CROSSBAR.

So rises the question: Can we increase the parallelism N up to $10^3 - -10^6$, the data rate R up to several Gbit/s and decrease the reconfiguration time T to a few nanoseconds by reducing the number of network configuration? These points are the driving force behind the OEDIPE architecture.

4 THE OEDIPE ARCHITECTURE

"architecture OptoElectronique Digitale à Processeurs Elémentaires"

4.1 Architecture

To map a large class of parallel algorithms, processors are arranged in a N^2 dimension array.

The control unit exchanges instructions and data with the elementary processors which operate in a SIMD (Single Instruction Multiple Data) or SPMD (Single Program Multiple Data) mode. Scalar processes are not suited to the matricial structure of the processors, and they are treated in the control unit with a specific scalar processor. Initially and for slow access data are stored in the mass memory.

For fast access, data are stored in a deep planar N^2 memory. Processors exchange data between themselves. The required communications are suited to a large variety of numerical applications as listed at the end of the previous section.

Optics can play a role at each interconnection level where parallelism is needed.

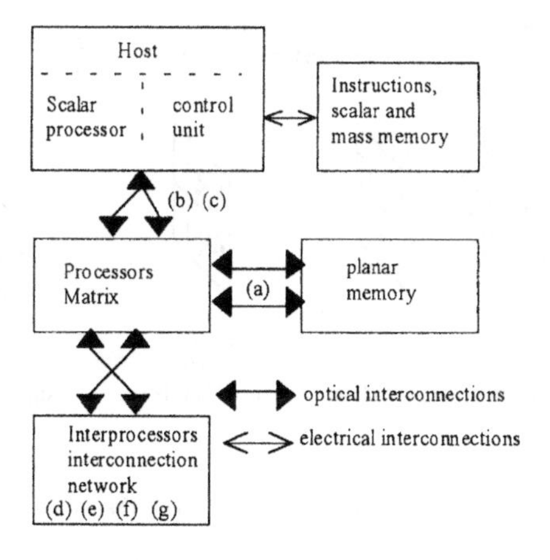

Figure 4 OEDIPE architecture

4.2 Parallel communications

Parallel communications are present at three levels in the architecture: between the control unit and the processors, between processors and the planar memory and between processors. They are illustrated in Figure 5.

The processing unit array communicates with the planar memory through N^2 point to point bi-directional interconnections (a).

Exchanges between the control unit and processors are of the broadcast function type (1 to N^2) when the Control Unit sends instructions and data to all the processors (b). For high level synchronization, mainly in the SPMD mode, processors send an acknowledgement message to the control unit via the N^2 to 1 communications (c).

Communications between elementary processors are efficient for basic data transfers as those needed in numerous algorithms. We selected four schemes: transposition (d), neighbourhood communications (e), column (or row) broadcast over the whole matrix (f) and translation (circular if possible) (g).

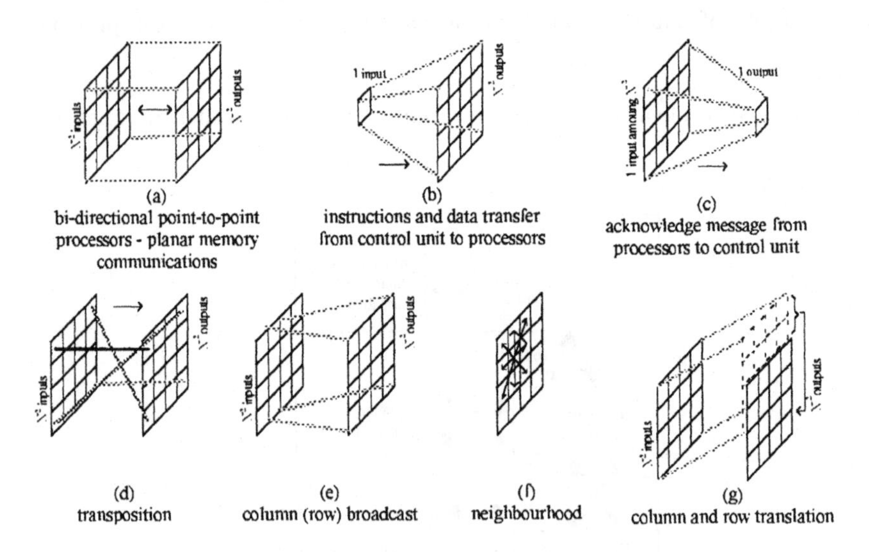

Figure 5 Parallel communications in the OEDIPE architecture

4.3 Interconnection parallelism

Parallelism degree comparison

We define the parallelism as the product of the number of inputs times the number of outputs which are addressable by each input. Table 2 gives the results, where each communication system is referred by the letter of Figure 5.

Communications	(a)	(b)	(c)	(d)	(e)	(f)	(g)
input number	N^2	1	N^2	N^2	N^2	N^2	N^2
output number	1	N^2	1	1	N^2	8	N^2
parallelism	N^2	N^2	N^2	N^2	N^4	$8N^2$	N^4

Table 2 Parallelism of communication

Free space optics offers three dimensions. One dimension is occupied by the light propagation direction. Under this condition, the (a), (b), (c), (d) and (f) communications, which have a N^2 parallelism, are straightly compatible with free space optics.

The N^4 parallelism of (e) and (g) is too high. For these two cases, the network cannot map physically the communications. Communications must be previously split up into subcommunications with a N^2 parallelism.

To reach the maximum N^2 parallelism, when data are words B bits wide, time multiplexing of the bits is necessary. Each input is physically a one bit input.

Reducing the parallelism of the column (or row) communication system

A solution for case (e) consists in performing the communications in two sequential steps (see Figure 6). The first one (h) consists in selecting a column (or a row) in the array and addressing an intermediate register. In the second step, data, now in the register, are than broadcasted toward the matrix. Parallelism is equal to $N^2 \cdot 1 = N^2$ in the first step and to $N \cdot N = N^2$ in the second one.

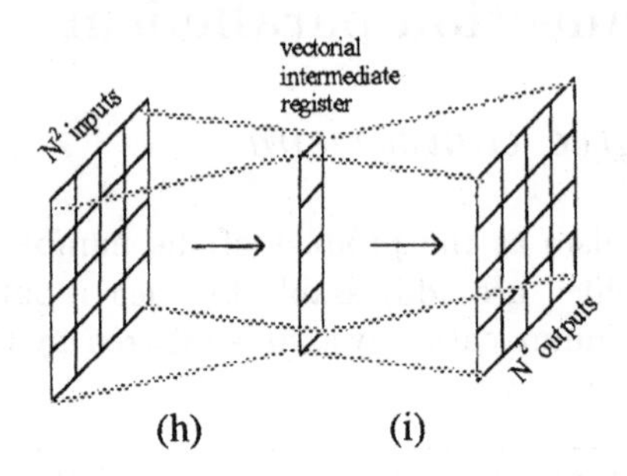

Figure 6 Two sequential N^2 communications instead of the N^4 communication system

Reducing the parallelism of the translation communication system

For the translation communication, we were able to identify the following incompatibility. With optics, image deflection is possible using several technical means such as shadows casting or acousto-optic deflectors. Theoretically, these systems have a N^4 parallelism.

From a practical point of view, when we retain a large input number N^2 with shadow casting the amplitude of the translation is weak and equivalent to neighbourhood communications. With acousto-optics, bearing in mind the constraint that the translation must be equal to the matrix size, the resulting number of inputs N^2 is low.

Of course, the translation can be achieved with any of the other communication systems. For instance, t iterations of the neighbourhood communications is equivalent to a translation of t columns. But for large value of t, the number of iterations is too high. A more efficient solution is to decompose any translation over a base of static translations. For instance, assuming that N is a power of 2 with $N = 2^n$, we define n static translations corresponding to a translation of 1 for the first one, two for the second, of 2^{i-1} for the i-th and of 2^{n-1} for the last one. Any translation can be decomposed sequentially on these specific translations. This is done in the two directions. The parallelism of each translation is equal to $N^2.1 = N^2$ and is compatible with free-space optics.

The maximum number of steps is equal to 2^{n-1} for any translation. The factor 2 is for the two directions, column and row.

4.4 Network topologies and system design

For processors to planar memory interconnections, the point-to-point N^2 parallel communication is simply achieved by a plane to plane imaging system. Optics is well suited for this function.

Reconfigurable network or sequential iterations

For interprocessor communications there seems to be a number of solutions. Using the communication requirements listed as above in Section 4.2, a large variety of network topologies are possible. In Section 4.3, all communications have been reduced to a number of basic communications with a N^2 parallelism. So, each of them is achievable with a specific static parallel optical system. This can lead to a global network made up of several elementary networks placed in parallel. Switches along the data bus select the desired network (see Figure 7). The reconfiguration is ensured by controlling the switches.

As another possible solution, we can choose only one network to minimize the system volume. Any communication is then sequentially decomposed over the unique network (see Figure 8).

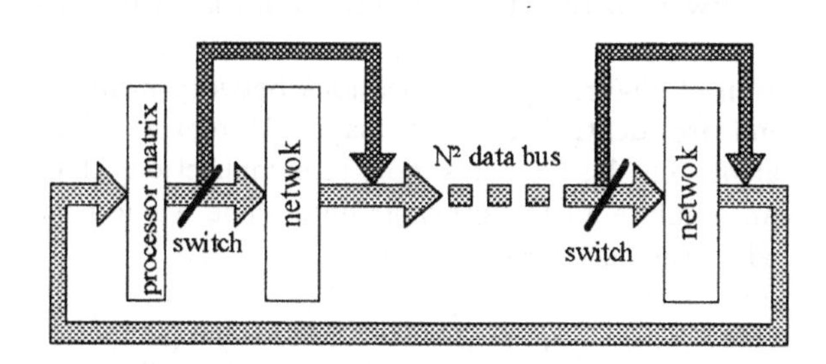

Figure 7 Choice of the network by switches

The solution in Figure 7 uses optical switches along the high parallelism data bus. After reconfiguration, the desired communication is quickly performed. But reconfiguration with optical devices is relatively slow and the global system is quite large.

The option shown by Figure 8 is attractive for compactness and in the case of a small number of iterations.

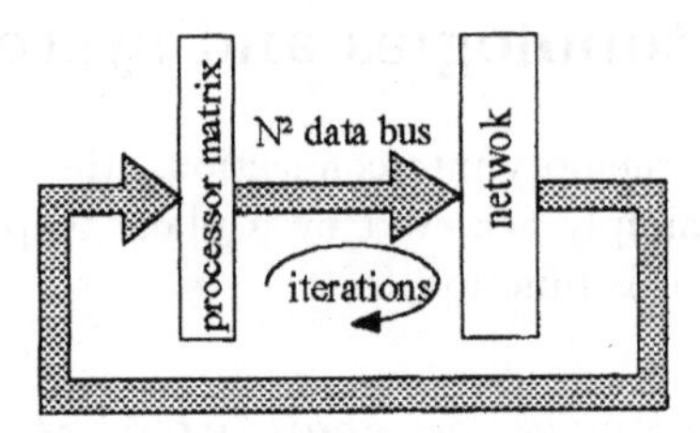

Figure 8 Iterations on an unique network

Let us investigate this last configuration.

Choice of networks

The translation communication system is a specific case. In Section 2, it has been split up into sequential steps. A one-dimensional illustration of the natural basic structure is illustrated in Figure 9(a).

Figure 9(b) is the circuit representation of the same network. It looks like a Shuffle-Exchange network with extra connections in the switch planes.

In the configuration of Figure 9(b), connections between switch planes are a Perfect Shuffle and are identical for each stage. Moreover, all switch planes are nearly equivalent. The last one has more interconnections than the others. As a result, it can emulate all the previous ones. The whole network can be physically reduced to the last stage.

Switch planes are no longer required, since the switching function is directly achieved by the processors interconnected with a neighbourhood network. $n(=\log_2 N)$ iterations are required for the translation in one direction.

It can be shown that all the other required communications (transposition, column or row broadcast) are carried out as n iterations by this network.

An other interesting characteristic of this network is that all the communications are of SIMD type. It means that a single command, broadcasted over all the processors, configures the emulated switch planes.

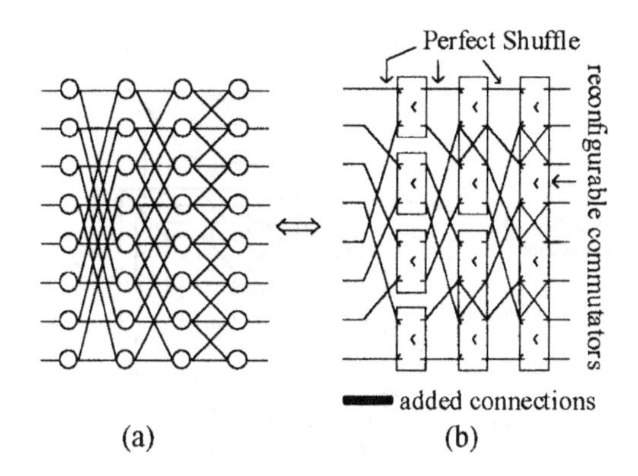

Figure 9 Multistage network for translations. (a) graph representation; (b) circuit representation. Black lines are added connections compared to classical Shuffle-Exchange

System design

Two physical interprocessor networks are needed: The neighbourhood and the Perfect Shuffle networks. The former is classically handled by electronics. The latter requires long-distance interconnections, making optics a suitable solution.

Figure 10 illustrates the whole system structure. In the processing plane, electronic processors are interconnected with an electrical neighbourhood network. All devices are N^2 planar technology. The optical Perfect Shuffle network above the Processing plane is bi-directional. Bi-directional interconnections are better for optical technology than the ring structure. Planar components are crossed by the bi-dimensional and bi-directional bus. Technology for this bus is open. Both optics [2] and electronics with flip-chip technology have the required assets for this function.

Figure 11 presents the optical bi-directional network using a microprism array to deflect the beams. Each elementary processor is linked to VCSEL (Vertical Cavity Surface Emitting Laser) and a detector. The VCSEL beam is collimated by a microlens. Light is deflected by a microprism. The angle of the microprism determined by the reciever. A mirror reflects the light back to the matrix. Light is then reoriented to the selected detector through a microprism and a microlens for focalisation.

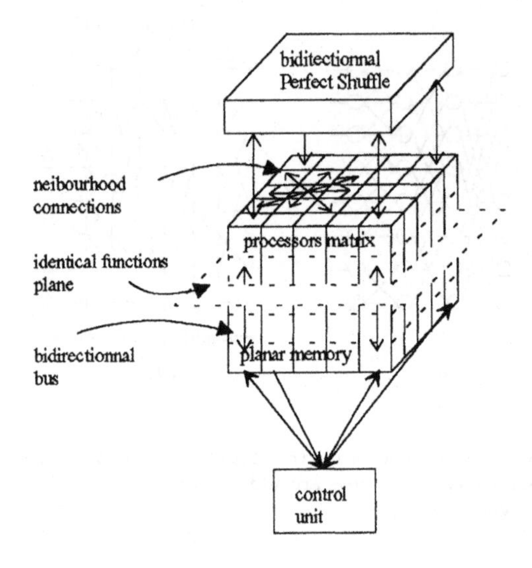

Figure 10 General system structure.

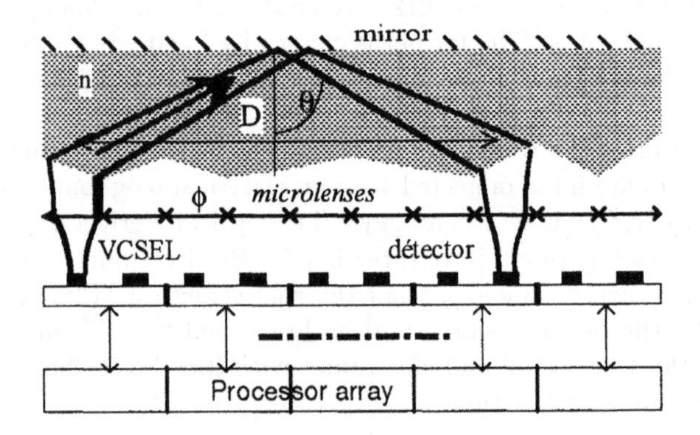

Figure 11 Optical bi-directional connections.

Physical characteristics

The evaluations of the system characteristics are based on the evaluation of the matrix dimension N^2. The largest value of N is limited by the longest connection that can be made with free-space optical technology in the Perfect Shuffle network.

Hypothesis The VCSELs have a waist diameter w, their illumination profile is single mode at the wavelength λ. The step between two consecutive processors is equal to $2 \cdot \phi$, where ϕ is the microlens diameter. D is the maximum distance which separates the two farther processors to be connected. n is the refractive index of the optical medium, f is the focal length of the microlenses and θ is the deflection angle.

Criteria

Image quality: The waist source of the VCSEL is imaged onto the back mirror. After reflection, the spot light size increases. To obtain a good optical link, the size of the spot light at reception must be equal to the diameter of the lens.

Optical budget: To limit optical crosstalk, losses are limited to -2.5 dB.

Theoretical results

Maximum angle θ: The total optical transmission T is defined by the relation

$$T = t_1^4 \cdot t_2 \cdot t_3 \cdot t_4^2, \tag{5.4}$$

where t_1 is the transmission of the four optical surfaces of the two microlenses, t_2 is the beam occultation by lens aperture, t_3 is the reflection on the back mirror and t_4 is the transmission of the two prisms.

Without antireflection coating, elementary transmissions are estimated at: $t_1 = 0.96$; $t_2 = 0.87$; $t_3 = 0.97$.

Maximum losses are limited to 2.5 dB, so the minimum transmission T is 0.56. The minimum value of t_4 is derived from Equation 5.4 which gives $t_4 = 0.88$.

This value of t_4 corresponds to a maximum angle θ of 27°.

Maximum parallelism N: Calculations are based on the scalar theory of Gaussian beam propagation across circular aperture in an aberration-free optical system.

Table 3 gives the N values for different configurations: 2 wavelengths and 2 refractive index.

We note that N increases when ϕ and n increase or when λ decreases.

The grey italic corresponds to the maximum N value. Assuming the desired value of N is a power of 2, we have $N = 128$, and the useful parallelism is $N^2 = 128 \times 128$.

material		BK7			
wavelength	λ μm	0.80		0.98	
refractive index	n	1.511		1.508	
lens diameter	ϕ μm	200	400	200	400
focal lens	f μm	769	1555	628	1269
distance	D mm	26.79	107.2	21.76	87.12
number	N	**92**	**188**	**74**	**152**
material		SF11			
wavelength	λ μm	0.80		0.98	
refractive index	n	1.764		1.757	
lens diameter	ϕ μm	200	*400*	200	400
focal lens	f μm	769	*1555*	628	1269
distance	D mm	31.78	*127.2*	25.73	103.0
number	N	**110**	***220***	**90**	**180**

Table 3 Configurations

Experimental and simulation results

Several factors (anamorphose effect and aberrations) are not taken into account in the theoretical calculation. However, they are evaluated by simulation and experiment.

Using refractive microprisms: Figure 12 illustrates the experiment. VCSEL is simulated by a single mode optical fiber. Microlenses are NPL components of focal length $f = 970$ μm and diameter 250 μm. Prism is a BK7 plate. The mirror is not included and does not influence the optical quality.

Figure 13 illustrates the output image in the detection plane and shows that the simulation is well representative of the experiment. The anamorphose effect is in the order of 30% between the two mean directions for an angle of 27°. This effect is easily integrated by the detector in the output plane. Theoretical results are well representative of optical transmission at $\pm 10\%$.

Unsing diffractive optics: Diffractive optics is a good candidate for such an application. To collimate and deflect beams, the corresponding diffractive component is an off-axis lens as illustrated by Figure 14. Two A and B type

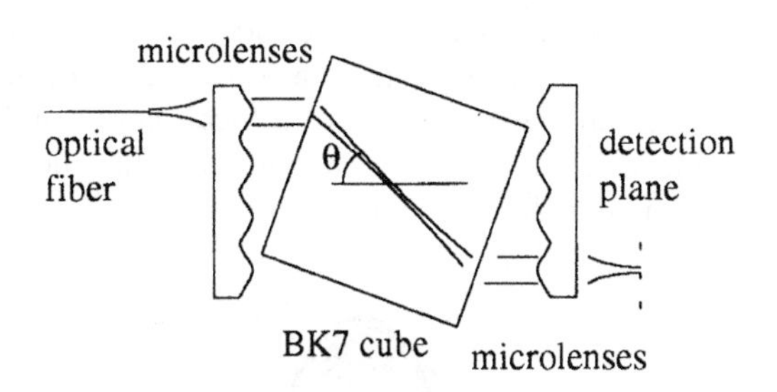

Figure 12 Experimental setup using refraction

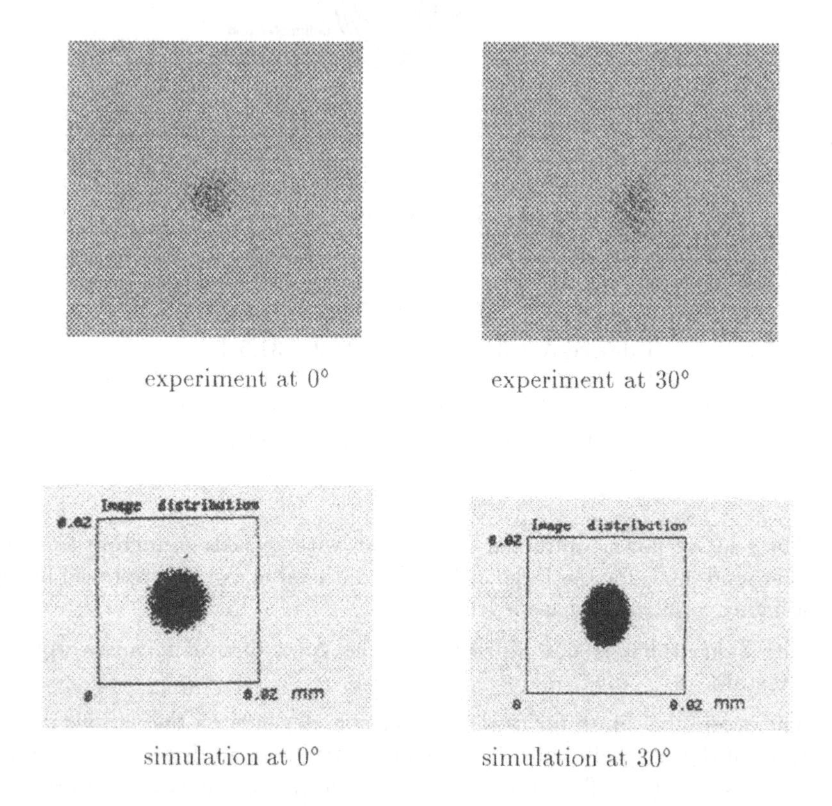

Figure 13 Spatial irradiance in the detection plane

DOC devices with a two-level optical phase coding have been characterized for angles up to 30°. The A types have a focal length of 920 μm and a diameter of 200 μm and the B type have a focal length of 1870 μm and a diameter of 400μm. Table 4 shows transmission efficiency according to the diameter and for four different angles.

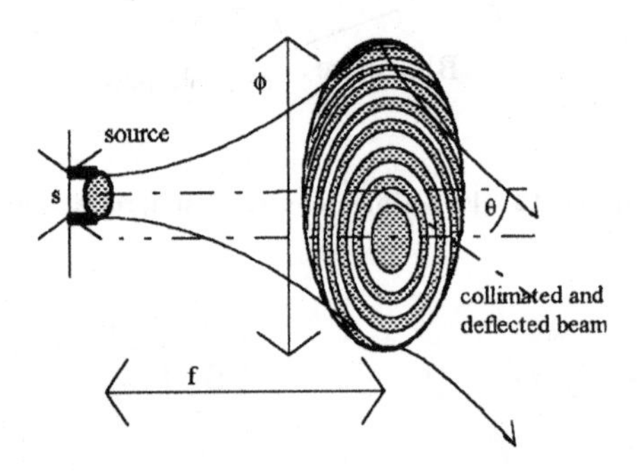

Figure 14 Off-axis diffractive optics.

Angles °	0	10	20	30
device A	31.5	27.9	31.2	31.5
device B	34.5	34.2	34.7	34.2

Table 4 Relative power diffracted in %.

The relative power diffracted is less than with refractive microprisms. It is around 30 % for the large 30° angle. This value can be improved using a higher number of phase levels.

The light which is not diffracted in the good direction causes optical crosstalk.

The acceptable limit, for precise evaluation, depends on the various directions of the crosstalk light. The wider it spreads in all directions in space, the less crosstalk effect there is. Figure 15 presents measured diffracted light power according to the angle. In this figure, the desired diffraction angle is 30°. Signals in other directions creates noise in the system. The crosstalk effect is relatively low and less than -25 dB in all directions. The impact on system performance must be investigated.

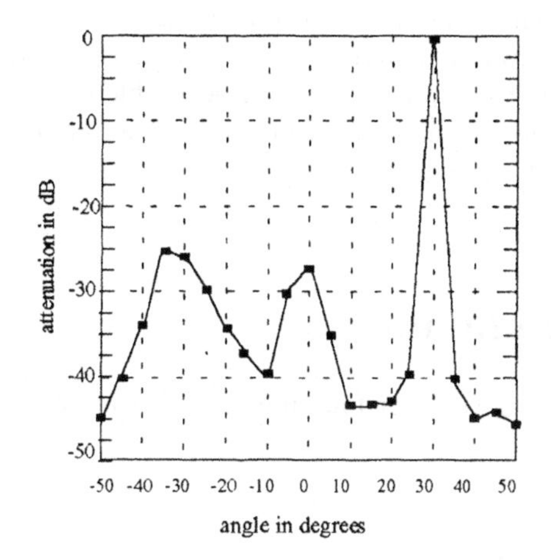

Figure 15 Relative diffracted power versus angle for the 30° deflection angle.

4.5 Synthesis on the Oedipe system

In a Perfect Shuffle network, optical interconnections are point to point links, which is an advantage for the optical power budget. The data rate is only limited by interface circuits. Several Gbits/s have been demonstrated [4].

The interconnection time is, as in OMEGA networks, $\log_2 N$ multiplied by the elementary cycle time. This elementary time is short and defined by integrated electronics.

The parallelism N^2 of the network is large and in the order of 128×128.

With an integration density of 2 Mtransistors/cm^2 [6] that can be foreseen in the near future, a matrix of 128×128 processors having a complexity equivalent to an Intel 8080 type processor (5000 transistors) should be feasible over an area of 70×70 mm^2.

For more efficient processors, such as floating point processors, the processor plane can be physically separated from the optical interconnection plane. Signals are transmitted from the processors to the optical network by optical fibers.

In OEDIPE architecture, the bit serial multiplexing is a limitation for the system for large data flows. Furthermore, it is attractive to keep the natural bit parallelism for compatibility with common available technology. Optical free-space interconnections can also be useful for such an application. This will be explained in the next chapter.

Acknowledgements

We thank the DRET agency for founding part of this work.

REFERENCES

[1] K. H. Brenner, M. Kufner, and S. Kufner. Highly parallel arithmetic algorithms for digital optical processor using symbolic substitution logic. *Applied Optics*, 25(18), Sept. 1986.

[2] R. F. Carson, M. L. Lovejoy, K. L. Lear, M. E. Warren, O. Blum, S. P. Kilcoyne, T. Du, P. K. Seigal, D. C. Craft, and B. H. Rose. Low-power, parallel photonic interconnections for multi-chip module applications. In *45th Electronic Components and Technology Conference*, May 1995.

[3] P. Churoux, M. Fracès, M. Laug, D. Comte, P. Siron, and X. Thibault. Optical crossbar network analysis. volume 862, page 42, Cannes, 1987.

[4] U. Fiedler, E. Zeeb, G. Reiner, and J. Ebelink. 10 Gbit/s data transmission using top emitting VCSELs with high sidemode suppression. *Electronic Letters*, 31(19):1664–1665, 1995.

[5] M. Fracès, J.-P. Bouzinac, P. Churoux, M. Laug, D. Comte, P. Siron, and X. Thibault. A multiprocessor based on an optical crossbar network : The MILORD project. volume 963, pages 223–231, Toulon, Aug.–Sept. 1988.

[6] R. H. Katz and J. L. Hennessy. High performance microprocessor architectures. *International Journal of High Speed Electronic*, 1(1):7403, 1990.

[7] A. V. Krishnamoorthy, A. L. Lentine, K. W. Goossen, J. A. Walker, T. K. Woodward, J. E. Ford, G. F. Aplin, L. A. D'Asaro, S. P. Hui, B. Tseng, R. Leibenguth, D. Kossives, D. Dahringer, L. M. F. Chirovsky, and D. A. B. Miller. 3-D integration of MQW modulators over active submicron CMOS circuits : 375 Mb/s transimpedance receiver-transmitter circuit. *IEEE Photonics Technology Letters*, 7(11), Nov. 1995.

[8] M. Lebby, C. A. Gaw, W. Jiang, P. A. Kiely, C. L. Shieh, P. R. Claisse, J. Ramdani, D. H. Hartman, D. B. Schartz, and J. Grula. Characteristics of VCSEL arrays for parallel optical interconnects. In IEEE, editor, *Electronic Components and Technology Conference*, 1996.

[9] D. A. B. Miller et Al. The quantum well self-electrooptic effect device: Optoelectronic bistability and oscillation, and self-linearized modulation. *IEEE Journal of Quantum Electronics*, 21(9), Sept. 1985.

[10] M. J. Murdocca, A. Huang, J. Jahns, and N. Streibl. Optical design of programmable logic arrays. *Applied Optics*, 27(9):1651, 1988.

[11] R. A. Nordin, W. R. Holland, and M. A. Shahid. Advanced optical interconnection technology in switching equipement. *Journal of Lightwave Technology*, 13(6), June 1995.

[12] J. Tanida and Y. Ichioca. A paradigm for digital optical computing based on coded pattern processing. *Int. J. of Optical Computing*, 1:3819, 1990.

[13] A. Trew and G. Wilson. *Past , Present, Parallel. A Survey of Available Parallel Computing System*. Springer-Verlag, 1991.

6

MASSIVELY PARALLEL COMPUTERS USING OPTICAL INTERCONNECTS – THE SYNOPTIQUE PROJECT –

P. Scheer, T. Collette and P. Churoux*

LETI (CEA-Technologies Avancées), DEIN - CEA Saclay, GIF-SUR-YVETTE, France.

** CERT-ONERA, TOULOUSE, France.*

1 INTRODUCTION

The use of computers in industry, research centres, telecommunications and at home is growing very quickly. When the applications are time consuming, such as image or signal processing, embedded processing or CAD software, the use of powerful parallel computers becomes a necessity. For this reason, we began to design, in 1987, a parallel computer, named SYMPATI2, dedicated to low level image processing. Six years later, we designed another parallel computer, called SYMPHONIE, which is well suited for iconic and symbolic data processing. We are now working on a new generation of parallel computers which will use fast electronic ASIC based on 0.25(m standard cell technology. However, although the clock frequency can be very high in those components, this is not the case for the speed of the inputs and outputs. In order to fully take advantage of powerful components within parallel architectures, we started two years ago a research project named SYNOPTIQUE in which we are studying the advantages of optical interconnects in our domain. This project is carried out in collaboration with the Optical Department of ONERA CERT.

In the first part of this chapter, SYMPATI2 and SYMPHONIE computers are briefly described. The second part deals with the limitations which appear for the next generation of parallel computers. Then, the SYNOPTIQUE project is detailed in term of motivations, challenges and planning, followed by a conclusion and presentation of future work.

P. Berthomé and A. Ferreira (eds.), Optical Interconnections and Parallel Processing: Trends at the Interface, 173-194.

2 DESCRIPTION OF SYMPATI2 AND SYMPHONIE

2.1 Sympati2

In 1987, a SIMD (Single Instruction stream, Multiple Data stream) parallel computer was developed at the CEA/LETI. This computer, called SYMPATI2 and commercially available since 1991, is suitable for executing low level image processing algorithms. The SYMPATI2 processors are organised linearly and the data are distributed among the processors using a helicoidal scheme [9]. SYMPATI2 can thus be seen as a 1.5D parallel structure. The number of processing elements (PE) involved in SYMPATI2 scales from 32 to 256. Each processor is based on an ALU, registers and an interconnection system allowing access, within one machine cycle, to any one of the three, left or right, neighbouring processors. The ALU performs binary, arithmetic and shifting operations on 16-bit words. Each processor can address its own memory in a direct or indexed mode, using its own built-in index register.

An ASIC (Application Specific Integrated Circuit) has been designed to integrate four SYMPATI2 processors. The complete system consists of a single board which supports the control unit of the SIMD computer, and a maximum of 8 PE-boards, each of them integrating 8 ASIC and the memory modules.

This computer is very efficient for implementing the low level image processing class of algorithms. It has already been evaluated in industrial applications such as measurement, inspection and quality control, and has been classified as the most powerful system in Preston's Abingdon Cross Benchmark [17].

2.2 Symphonie

We have developed another massively parallel SIMD computer which will be installed in the French military aircraft, RAFALE, in 1997. In this computer, called SYMPHONIE, the number of included super scalar processing elements ranges from 32 to 1024. In the 1024-PE configuration, the performance of the system can reach 50 GOPS and 2,5 GFLOPS [5]. The topology and the external interfacing of this computer are depicted in Figure 1.

To obtain those performances, different concepts have been implemented in the system:

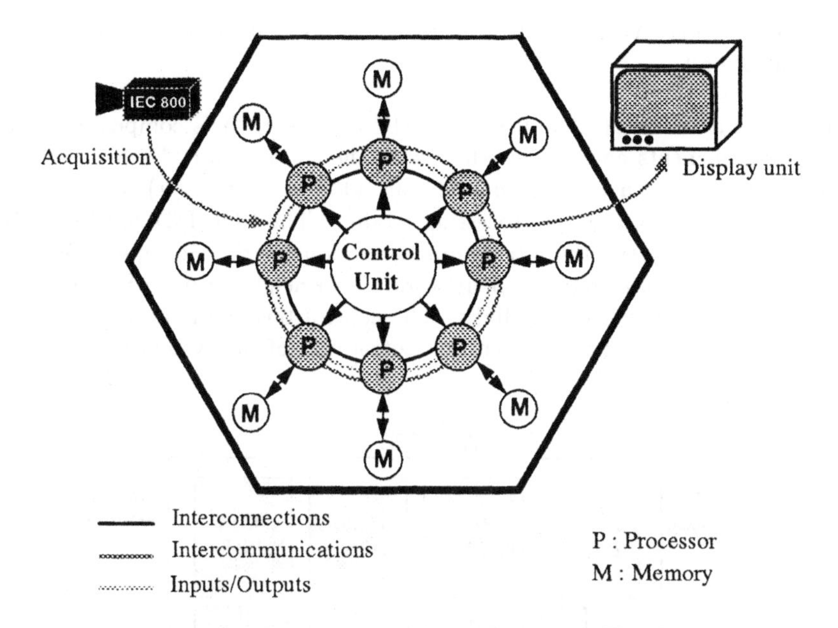

Figure 1 SYMPHONIE's topology. A control unit sends instructions to the processors. Each processor is directly connected to its local memory and to its two nearest neighbours with three communication links.

- a powerful addressing processor is integrated into each processing element. This unit, based on the addressing processor used in SYMPATI2, can manage many addressing possibilities in the memory which is connected to the PE;

- a message passing network is associated with all the PE, and allows the PE to exchange 32-bit wide data without slowing down the PE operation. With this network, SYMPHONIE is able to support the low and intermediate level classes of image processing algorithms [4].

To ensure a highly integrated system, the ASIC, which includes 4 PE, is constructed using 0.5μm Standard Cell ASIC technology. Since each PE is connected to a 128Kbits static RAM, the ASIC integrates about 200,000 standard cell gates and 64Kbytes memory. Furthermore, ASIC are integrated four by four in Multi-Chip Modules (MCM) and thus, 64 PE are available on a single board. The number of boards is limited to 16, which corresponds to the 1024-

PE configuration. As in the SYMPATI2 system, a specific board is dedicated to the control unit.

Figure 2 is a sketch of SYMPHONIE's ASIC processor. It is composed of four processing elements and one interface with the control unit. Each processing element consists of a memory module, an address module (@), an interconnection module (I/O) and a processing module. The ROI, I/O and VG (VD) buses are, respectively, the intercommunication network, the Inputs/Outputs network and the interconnection network. Two neighbouring ASIC are connected through point-to-point links. The control bus is one arm of a 1-to-N network, which is used by the central control unit of the system to distribute the micro-instructions.

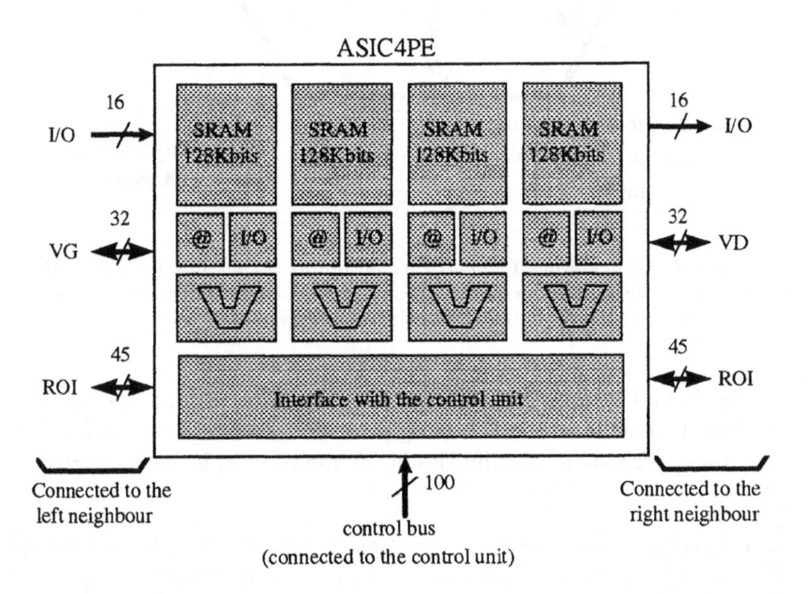

Figure 2 Sketch of SYMPHONIE's processor ASIC.

Table 1 sums up the main features of the two systems.

	Sympati2 System (1991)	**Symphonie System (1996)**
Performance	5 GOPS	50 GOPS
Architecture	SIMD Parallel Computer	SIMD Parallel computer with asynchronous data exchange network
Maximum number of PE	256	1024
ASIC	EPA4 • 4 SYMPATI2 PEs • 16-bit computing unit • 20 000 gates • Standard cell 1.2 μm AT&T • 225 I/O • 5 Volts • 16 MHz	ASIC4PE • 4 SYMPHONIE PEs • 32-bit computing unit + floating point registers • 200 000 gates + 64 Kbytes SRAM • Standard cell 0.5 μm Thomson TCS • 430 I/O • 3.3 Volts • 25 Mhz
Frequency I/O	8 Mhz	12.5 Mhz
Integration	8 ASIC per board 32 PE per board	4 ASIC per MCM 4 MCM per board 64 PE per board
Number of boards	2 to 9 boards (a board is dedicated to the control unit)	2 to 17 boards

Table 1 Main features of SYMPATI2 and SYMPHONIE calculators developed by CEA/LETI/DEIN

3 LIMITATIONS OF THE ELECTRICAL INTERCONNECTIONS

3.1 Introduction

Electronic systems are built around micro-processors or ASIC and are thus sensitive to the evolution of microelectronics. According to Moore's law, the integration density of micro-processors and ASIC is improved by a factor of 2 every 18 months. Simultaneously, the clock frequency of these devices increases, and we have gone up from a 68000 at 2 MHz in 1980 to the present Pentium-Pro at 200 MHz.

Signals at frequencies up to 100 MHz are rather easily managed within an electronic circuit - a die or a MCM -, but not outside it. As a result, data are quickly processed in the circuit but cannot be efficiently imported from, or exported to, another circuit. When the system is small enough to be fully integrated in a unique MCM, this drawback has no penalising effect on the performances, but in the case of large systems and especially massively parallel computers, low efficiency electrical interconnections lead to communication bottlenecks between the processing units [8]. As an example, Weems in [21] states that in SIMD computers, the clock period is limited to 20ns or 10ns when using only electronic interconnections.

In this section, we describe the limitations due to latency effects and Input/Output capabilities in parallel processing systems. The study is based on the two parallel computers, SYMPATI2 and SYMPHONIE, which have been developed in our laboratory. Finally, we estimate the necessary bandwidth of a future parallel computer by extrapolating the features of our previous systems.

3.2 Limitations due to the latency

In a complex system like a parallel computer, it is necessary to tightly interconnect the processing elements. When the system is physically distributed over several MCM and/or boards like massively parallel computers, an ASIC inside a MCM on the board number i must be directly connected to another ASIC inside a MCM located on the board number i+1. This leads to very long data paths with several buffer stages and thus significant signal delays. In the SYMPHONIE computer for instance, it takes about 40ns for signals to be stable on the bus of the receiving ASIC. This set-up time is known as the latency of the interconnection.

To eliminate the latency problem at the system level, designers either only slow down the Inputs/Outputs of the ASIC, or lower the global frequency of the ASIC. In the first solution, the high frequency and pipelined architecture of the ASIC are preserved, but in tightly coupled systems, several waiting operations are inserted in the program to force the processors to wait for the exchanged data. The second solution allows travel through long data paths between two clock edges, and external data exchange is possible in one ASIC cycle. To benefit from the technology, the processor must then be designed around a set of complex instructions which fully exploits the cycle time. In this case, the latency is the parameter that determines the operation frequency of the overall system.

This latter method has been used to set the clock frequency of both the SYMPATI2 and SYMPHONIE systems. In the first of these, the clock period is 120ns. In each cycle, 80 nanoseconds are spent passing through the logic and 40 nanoseconds are dedicated to the communications from one ASIC to another. In SYMPHONIE, the clock frequency is 12.5 MHz. All the 80 ns of the cycle are spent passing through the logic (standard cell and memory) for instructions using data stored in the local memory (the memory of each PE is included inside the ASIC). For the instructions requiring external data, the cycle is split into two 40 ns periods, one for passing through the logic, and the other for going from one ASIC to another.

Due to the integration of memory modules within the ASIC (which makes the ASIC very different from that of SYMPATI2), we have reduced the latency effects for local memory transfer cycles. Nevertheless, the latency problem remains for ASIC-to-ASIC data exchange cycles and still quenches the overall throughput of our system. We expect that with the next generation of ASIC dedicated to massively parallel computers, the time required for crossing the logic will decrease to less than 20ns for the next CMOS 0.35μm technology, and 10 ns for the 0.25μm technology in 3 years. However, the time required to exchange data between two MCM, especially when they are located on two different boards, does not follow the same scaling law and will remain very high (about 40 ns). Its impact on the cycle time will be unacceptable.

3.3 Limitations due to Input/Output capabilities

The throughput of an ASIC depends on both the data rate and the number of Inputs/Outputs (I/O). The latency is a limiting factor as far as the data rate is concerned, but its effect on the throughput can generally be lightened by an increase in the number of physical paths of a given link. However, the technology used up to now for implementing electronic systems limits the number of I/O at different integration levels:

- at the ASIC level, the number of links (pads) is a function of the size of the die (geometric constraint),
- at the MCM level, the type of packaging determines the number of pins,
- at the board level, standard backplane connectors restrict the number of I/O.

I/O counts are a major issue in the design of massively parallel computer: the efficiency of the system widely depends on the connectivity of the PE network, and implementing a large number of PE requires several levels of integration which all offer a limited number of I/O. The designer must always keep in mind that a new functionality added to the system requires increased silicon area but also implies new I/O. A mathematical expression of this last point is given by Rent's rules, where the number of I/O required by a chip can be estimated using the number of logic gates it contains [11]. Its form depends on the application which is under study and, for dies, the following formula is generally given [13, 14]:

$$N_{I/O} = KN_{gates}^{\alpha}, \tag{6.1}$$

where $N_{I/O}$ is the number of I/O and N_{gates} that of logic gates. K and α depend on the function of the die.

Once the number of I/O is estimated, the designer can check whether the packaging is able to provide enough interconnections. In some cases, difficulties can be expected [13, 14]. In the following paragraphs, we illustrate some of the I/O problems we have encountered during the design of our computers.

In SYMPATI2, the number of the ASIC I/O is restricted due to packaging limitations (the die is embedded in a 220-pin PGA). As a result, only 40 pins are dedicated to communications with the control unit, and a multiplexed bus was required. Data exchanges with the memory and the neighbouring ASIC are handled with 80 pins.

In SYMPHONIE, the number of I/O per ASIC is 420, with 70 pins used for grounds and power supply, 100 pins used for communications between the control unit and the processors, and 175 pins dedicated to interconnections with the neighbouring ASIC. The inflation in the number of pins dedicated to data exchanges between neighbouring ASIC is first caused by an increased data words format (16bits in SYMPATI2 versus 32bits in SYMPHONIE), and also by the multiplication of networks. In SYMPATI2, only one network interconnects the ASIC, whereas in SYMPHONIE, three different networks tightly couple the ASIC: one for direct interconnections (32-bit wide), one for intercommunications (45-bit wide) and one for the loading/unloading of data (16-bit wide). Once again, the limited number of I/O influenced the architecture of the computer: the memory has been integrated with the logic on the same die to avoid address and data memory buses outside the component. Moreover, I/O counts at the ASIC and backplane connector levels prevented us from duplicating certain links between PE.

Because the processors are increasingly powerful, they are able to support more complex image processing operations. But to feed those new processors, new means of communication must be implemented. As an example, in SYMPHONIE, the intercommunication network has been implemented to improve the processing capabilities of the system, allowing the system to efficiently support the low and intermediate levels of image processing [4]. In the future, taking into account the computing capabilities of the next generation of processors used in massively parallel computers will further increase the volume of communications. We can expect, for instance, that it will be necessary to have a 32-bit wide network dedicated to the data loading/unloading operations, possibly two 64-bit wide direct interconnection networks and two 80-bit wide intercommunication networks. In other respects, as we mentioned before, the cycle time of future processors will scale down to the 10ns range, and thus the processors will have an aggregate I/O throughput of 35GBit/s, considering that the 10ns period can be supported by the I/O. In Table 2, we compare this throughput and other figures with those of SYMPHONIE and SYMPATI2.

		SYMPATI2 processor	SYMPHOMIE processor	Future Parallel processor
Number of links with the neighborhood	for data laoading and unloading	8 I + 8 O	16 I + 16 O	32 I + 32 O
	for direct communications	16 I + 16 O	32 I + 32 O	(64 I + 64 O) × 2
	for intercommunications	-	45 I + 45 O	(80 I + 80 O) × 2
Period of the I/O		120 ns	80 ns	10 ns
I/O processing capabilities		**0.4 GBit/s**	**2.3 GBit/s**	**35 GBit/s**

Table 2 I/O throughputs achieved with SYMPATI2 and SYMPHONIE processors and desirable for the future processor of a new massively parallel calculator.

We have largely discussed the difficulties encountered with the finite number of I/O at the ASIC level. In fact, the trouble may be even worse at the board level. With SYMPHONIE for instance, the fact that the system is physically split into several boards implies that any signal coming out of a board passes through a backplane connector which has only 320 pins. This bottleneck source appeared in the early design of the system and solutions had to be found, at the expense of a few changes in the original architecture. Thus, we foresee that innovating interconnection technologies must be developed before new massively parallel computers with increasing demands in speed and I/O are built.

4 SYNOPTIQUE PROJECT

Designing a new generation of computers is always a challenge: the newcomer has to be more powerful, consume less energy and, if possible, be less expensive. From the above, the development of a new massively parallel computer supposes that solutions for implementing efficient interconnections are found. Therefore, we started, in the beginning of 1995, a research program named SYNOPTIQUE whose aim is to assess the ability of optical interconnections to improve our architectures [19]. This project is undertaken by the parallel software and architecture group of the CEA/LETI/DEIN in Saclay, in collaboration with the optical interconnections group of the ONERA-CERT/DERO in Toulouse [3]. The latter is in charge of the optical, optoelectronic and mechanical parts of the system, from the design to the realisation. Different project stages are identified and described in this section.

4.1 Introduction

The SYNOPTIQUE research program consists of three main parts. First, the potential role of optics is identified by studying the difficulties of implementation encountered in our current massively parallel computers. The investigation relies on bandwidth considerations (that required and that available at a given level of integration) and latency evaluations in the communication paths of the different networks. In a second stage, which started almost at the same time as the first, the various types of optical interconnections are reviewed and the best fit with present architectural and physical requirements (based on SYMPATI2 and SYMPHONIE features) is sought. Another and important stage of the project will be the study of the impact of optics at the system level. It concerns the enhancement of the communication features and also the potential improvement of the architecture due to implementation of new functions based on optics. Using optical and optoelectronic components is rather new for designers of parallel computers. Another interesting point is thus the integration of optical interconnections in design and simulation tools, as well as in assembling and test proceedings. When all these investigations are completed, we will concentrate on the design and the implementation of a massively parallel computer using optical interconnections (OI).

We are now dealing with the first two stages, which represent the main subject of this chapter. However, at the end of this paragraph, we will outline the future stages of the project we have just evoked.

4.2 Motivations for using optical interconnections

The aim of this section is not to deal with the advantages offered by optics over electronics in general since a great deal of work has already been published in this domain [6, 7, 16]. We more specifically concentrate our attention on the reasons why we believe that optics can improve our systems.

We have seen in Section 3 that a multi-level implementation of the networks leads to latency in the communications and limits the number of available I/O per level. Both contribute to a bandwidth reduction of the communication channels, which has a severe impact on the overall efficiency of the parallel computer. The main idea of our approach is to replace electrical wires wherever an optical path would improve the system performances.

The longest delays are generated by the slowest interconnection levels, namely the backplane and the printed circuit boards (PCB), and by crossing several buffer stages, as illustrated in Figure fig:synoptique.latency. In other respects, MCM provide efficient interconnections between ASIC up to hundreds of MHz. OI between MCM may thus be used to cancel and advantageously replace backplane and PCB interconnections (a comparison of optical and electrical interconnections at the board and backplane levels can be found in [1]). In other words, optics could handle communications outside the MCM with the same bandwidth and density characteristics as electrical interconnections within the ASIC and the MCM.

By cancelling the quenching effects due to backplane and PCB limitations, the computer efficiency will scale with electronics capabilities regardless of the

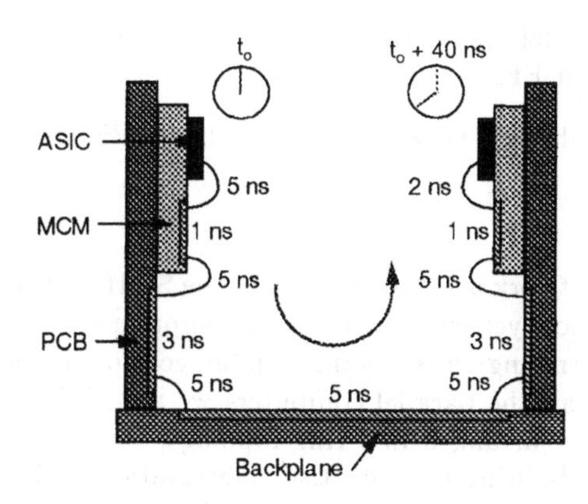

Figure 3 Worst case timing of inter-board electrical interconnections

physical distribution of the MCM. According to the estimated propagation delays in the present SYMPHONIE computer, reducing the number of levels, and accordingly the number of buffers, would almost double the clock frequency of the system without any other change. Moreover, we expect that OI will offer data rates high enough to handle communications between processing nodes designed with future CMOS 0.25μm technology (see end of Section 3.2. Lastly, if the number of I/O per MCM can be optically improved, new topologies may be implemented, opening the way to new architectures.

4.3 Optically interconnected multi-chip modules

Basically, the suppression of electrical interconnections at the board and backplane levels implies that MCM are interconnected by only optical links. Without other constraints, several optical interconnection approaches are practicable: optical fibres, integrated waveguides, planar optics and free-space optical systems. However, considerations such as scalability, compactness (for embedded systems), and friendly maintenance are important guidelines when dealing with massively parallel computers. Then, the nature of the interconnection networks may also influence the technological choice. To date, we are concerned by SIMD architecture in which there are at least two kinds of networks:

- a tightly, parallel, point-to-point interconnection network for data exchange between PE,
- a 1-to-N distribution network from UC to all PE for instructions broadcasting.

The distribution network is rather specific to the SIMD nature of the computer. Since we do not know yet whether the new computer will be entirely SIMD or not, we are concentrating on the point-to-point communication network, which will be present in all the parallel computers we are likely to develop. In this network, the PE are arranged in a ring topology, where each PE is physically bound to two neighbouring PE. Due to the integration of the PE within MCM, the same topology is reproduced at the MCM level. Consequently, optical interconnections have to handle parallel, point-to-point communications between MCM which form a ring network, and whose number can vary, for instance, from 4 to 32.

Free-space optical interconnections are well suited for dense communications within massively parallel computers [6, 20]. In our case, this optical technology is particularly attractive since it preserves modularity and offers a compact interconnection system. A simple way, at least in theory, of providing scalability is to use double-side optoelectronic planes which can be added behind each other as many times as required, as depicted in Figure 4.

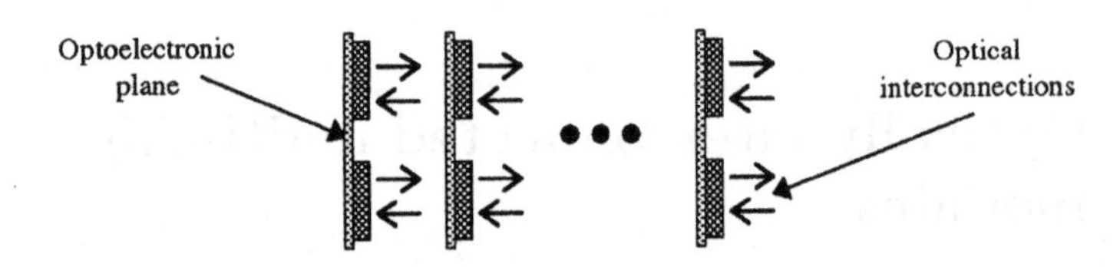

Figure 4 Concept of double-side optoelectronic planes

In 1995, NTT demonstrated the feasibility of this concept at the system level by interconnecting several electronic boards with LED-based bi-directional optical links [18]. In SYNOPTIQUE, optical interconnections must be located at the MCM level instead of the board level. The optoelectronic plane would then be a board supporting 4 MCM with emitters and detectors either on both sides or on the same side of each module, depending on the material and the packaging technology used. An appropriate disposition of the MCM on the different boards would equalise the data path between the nodes of the ring network, which naturally reduces the occurrence of skew (see 5).

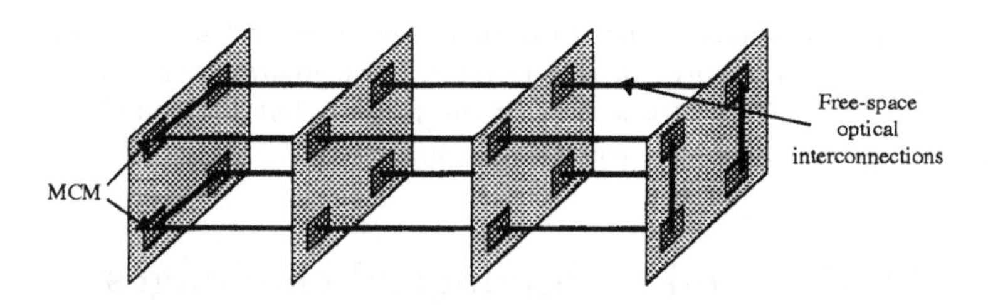

Figure 5 Optical implementation of the ring network in SYNOPTIQUE

A key issue is the way that the ring is completed. In Figure 5, MCM connections at the board level are assumed at both edges of the structure. The choice between free-space or guided wave interconnections is not made yet, and depends on both technical and architectural considerations. Figure 6 shows one possible implementation of the optical interconnects within the SYNOPTIQUE computer. Optical fibre ribbons emerging from the Unit of Command (UC) are used for instruction distribution and system control.

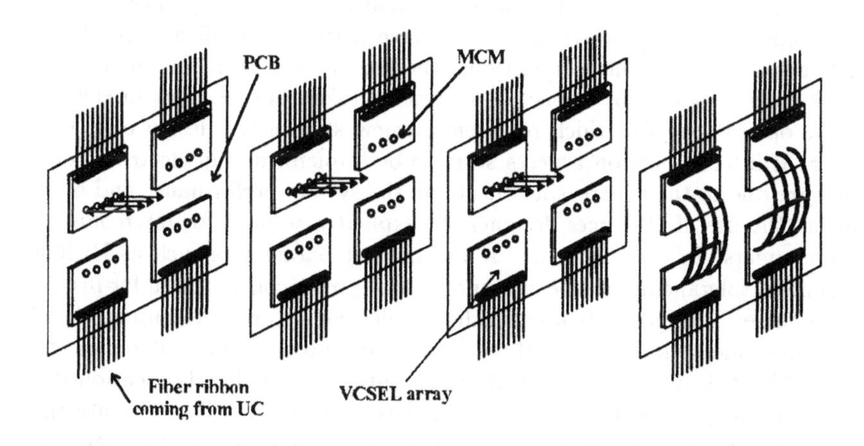

Figure 6 Outlines of a possible implementation of the SYNOPTIQUE computer

An advantage of the optical implementation depicted in Figure 5 is that we may eliminate the PCB if needed and directly stack the MCM. This would reduce the alignment constraints and even improve the compactness of the system. However, thermal aspects of this implementation have to be carefully studied. Optical interconnections between stacked MCM are already being investigated in several laboratories (see, e.g., [2, 22]).

Technical considerations in the realisation of such an optical interconnection system are not the subject of this chapter. The following section provides however some elements of this part of the project, but the complete design study will be presented elsewhere in the near future.

4.4 Design and technological challenges

Even if massively parallel computers can take benefit of OI, the designers of such systems may be hesitant to integrate a new, or more precisely an unknown, technology in their environment. In other words, using OI supposes that two challenges are taken up: design methodology of complex electronic systems is preserved, and technological solutions are found for minimising the impact of OI on the system assembling.

Whatever the medium of propagation is, optical signal transmission requires additive front end components, namely an transmitter and a receiver, which do not exist in an electrical transmission. These components differ from that used in digital electronics: they have a current and not voltage operation mode and some are non-linear (laser diodes have a threshold effect). In the case of free-space optical interconnections, the geometric features of these components have a significant impact and must be taken into account in the study of the optical system, which represents a new step in the design of the computer. If interconnection aspects start to be implemented in electronic design software due to their increasing effects on the system performance and optimisation, the use of OI forces designers to explicitly account for interconnection "components" in their systems. It means that parameters such as BER and alignment tolerancing will be as important as power consumption for instance. Since current design tools used in digital electronics are not compatible with the requirements imposed by the design of an optoelectronic interconnection, new CAD tools must be developed. Some authors have already mentioned this fact and have proposed methods for this development [12]. The ultimate aim is to insert optical interconnection models in simulation tools working at a high abstraction level, such as VHDL or Verilog language, and which are commonly used in electronic systems CAD approaches [10].

In other respects, a real estimate of the contribution of OI to the performance gain supposes an evaluation of their technical "cost". Implementing OI in an electronic system is a new challenge, requiring a new design and possibly a new assembling procedure. Some specific problems are raised, especially in the field of optoelectronics/electronics interface and, if free-space optical interconnections are used, in the mechanics and packaging of the system.

As far as the mechanical aspect is concerned, the main idea is that OI have to be compatible with electronics packaging. A well known problem with OI implementation is the alignment constraint. For free-space interconnections, the problem is worse and must be carefully studied. In particular, it can be readily imagined that a trade-off between high density interconnections and low mechanical constraints will be necessary. The two figures below give a simple description of the alignment constraint.

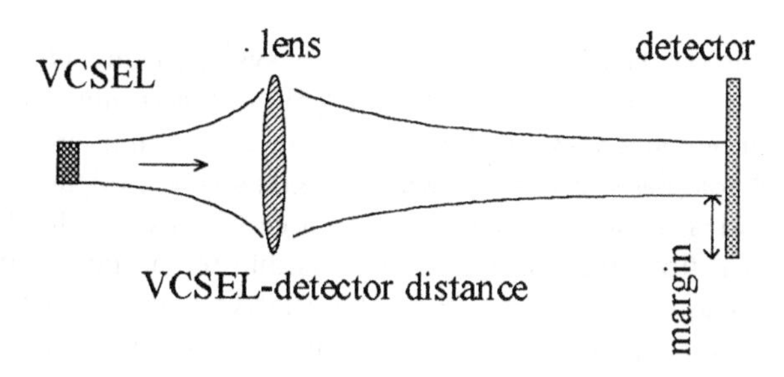

Figure 7 Point-to-point optical free space link

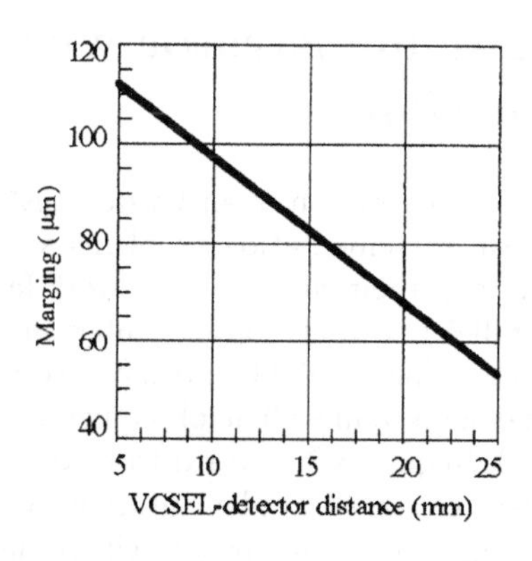

Figure 8 Alignment margin vs. VCSEL-detector distance

The parameters used for calculating the curve in Figure 8 are: a detector diameter of 250μm equal to the lens diameter and a VCSEL emitting at 850nm with a 5 μm-diameter waist. It is clear that the VCSEL-detector distance, i.e., the interconnection length, depends on the margin budget. This budget is defined by taking into account the alignment accuracy at each implementation level: hybridisation of micro-optical elements onto optoelectronic devices, placement of the hybrid devices onto the MCM, positioning of the MCM onto the PCB and finally PCBs relative placement.

As we mentioned above, connecting two ASIC with optical links requires additive components which are the optoelectronic devices, their interfaces with electronics and the optical system. For density, power consumption and latency considerations, the number of components per optical link must be minimised. Therefore, features such as zero-bias modulation of the emitter, low gain amplifier at the reception edge and fully parallel interconnections are sought [2, 15].

Predicting both the cost and reliability of a system using OI is not a simple affair. Today, these features in the domain of electronics can be assessed with good accuracy and very quickly, because processes in microelectronics and packaging technologies are well known and extensively modelled. The same is not true for optical technologies, except, to a certain extent, for those used in long haul optical fibre transmissions. This is due to the large variety of non-normalised optical components and to the relative newness of the associated technologies. As a result, this does not help designers in electronics to have confidence in the optical alternative.

4.5 Towards a demonstrator with optical interconnections

From the above section, the realisation of an OI demonstrator may be a good way to overcome potential implementation problems and to convince system designers that optics can help their computers to work faster and better. The aim is to implement parallel free space optical interconnections between two digital electronic modules, using available advanced components and custom mechanics for supporting the modules. The objective in terms of performance is to achieve a faster clock frequency for board-to-board communications than that offered by the current electrical implementation. Vertical-cavity surface-emitting laser (VCSEL) and microlens arrays will be used for the emission module, whereas the receiver module will include PIN photodiode arrays and transimpedance amplifiers.

A hybridisation of a microlens array onto a packaged 1x8 VCSEL array has been performed at ONERA CERT/DERO. Figure 9 shows the hybrid device after the alignment and gluing procedures. The microlens array is made of photoresist deposited on a 1mm thick silica substrate. The focal length in air is 1055μm, the lens diameter is 247μm and the pitch of both arrays is 250μm.

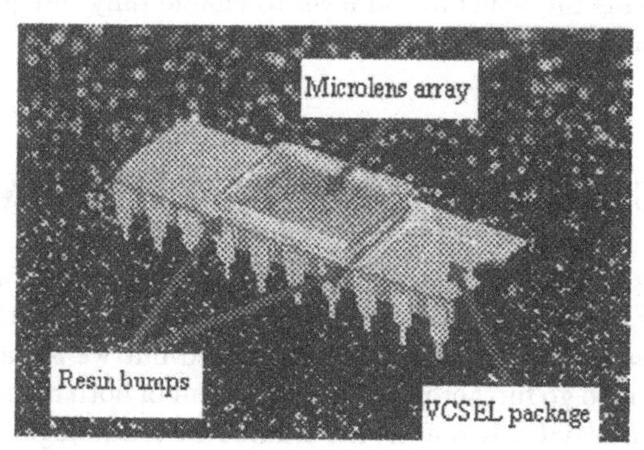

Figure 9 Hybridisation of refractive microlenses onto a 1x8 VCSEL array

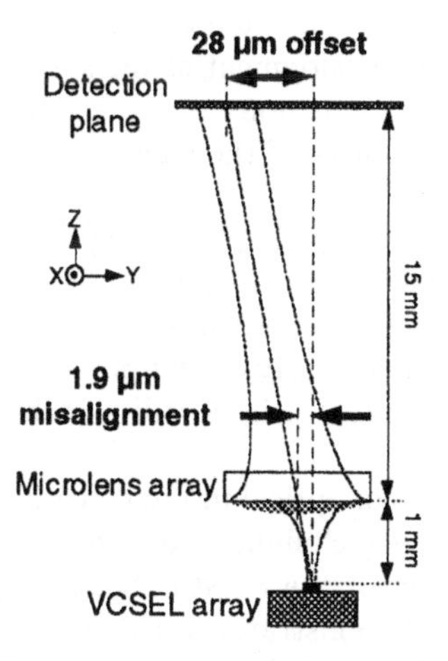

Figure 10 Alignment characteristics of the hybrid emitter

The experimental conditions are synthesised in Figure 10. The targeted objective was to align the optical axis of the microlens and the centre of the VCSEL active window within 2μm. A estimation of the misalignment has shown that the achieved accuracy was better than 1.9μm. According to these results, it seems that the total alignment requirements are achievable with classical integration techniques, except for the board-to-board positioning. Standard backplane mechanics offers roughly a 500μm placement accuracy, which is an order of magnitude higher than the required value. A specific technique has thus to be found for this implementation level to enable fully integrated free-space optical interconnects within electronic systems.

5 CONCLUSION AND FUTURE WORK

We have presented our motivations for using optical interconnections in our future massively parallel computers. The impact of optics on the performance of the system has still to be accurately quantified but we already foresee that optics may help us to go further in the improvement of both operation frequency and network connectivity. However, the studied OI technologies are not specific to the architecture and may be used in any system requiring parallel digital OI.

According to our experience in the development of parallel computers such as SYMPATI2 and SYMPHONIE, a better fit between the logical description of the computer and its physical structure is desirable for maintaining the original specifications during all the development of the system. This may be achieved by replacing electrical interconnections wherever they have too large adverse effects on the global performances of the computer.

We also emphasise that a methodology of integration of OI within electronic systems is required, especially at the simulation level. Thus, efforts have to be made in the domain of design tools such as CAD software to account for the behaviour of optical links in the early conception stages. Then, the design of a new massively parallel computer with OI will be achievable.

As far as the implementation of optical interconnections in a computer environment is concerned, innovative techniques must be developed and experimentally validated. The contribution of the SYNOPTIQUE project in this area mainly consists in building up a demonstrator with free-space optical interconnects between standard electronic boards located in a standard backplane structure.

REFERENCES

[1] P. J. Ayliffe, J. W. Parker, and A. Robinson. Comparison of optical and electrical data interconnections at the board and backplane levels. In P. SPIE, editor, *Optical Interconnections and Networks*, volume 1281, pages 2–15, 1990.

[2] R. F. Carson, M. L. Lovejoy, K. L. Lear, M. E. Warren, P. K. Seigal, G. A. Patrizi, S. P. Kilcoyne, and D. C. Craft. Low-power modular parallel photonic data links. In *46th ECTC Conference*, pages 321–326, Orlando, May 1996.

[3] P. Churoux, J.-P. Bouzinac, S. K. M. Fracès, D. Comte, N. Hifdi, T. Collette, and P. Scheer. *Optical Interconnections for Parallel Processing*, chapter Optical free-space interconnections inside parallel architectures: ONERA-CERT activities. Kluwer Academic Publishers, 1997. P. Berthomé and A. Ferreira, editors (This Issue).

[4] T. Collette, H. Essafi, D. Juvin, and J. Kaiser. Sympati X: a SIMD computer performing the low and intermediate levels of image processing. *Future Generation Computer System*, 10(1), Apr. 1994.

[5] T. Collette, C. Gamrat, D. Juvin, J.-F. Larue, L. Letellier, R. Schmit, and M. Viala. SYMPHONIE calculateur massivement parallèle : modélisation et réalisation. In *Journées Adéquation Algorithme Architecture en Traitement du signal et images*, Toulouse, Jan. 1996. in French.

[6] M. R. Feldman, S. C. Esener, C. C. Guest, and S. H. Lee. Comparison between optical and electrical interconnects based on power and speed considerations. *Applied Optics*, 27(9):1742–1751, 1988.

[7] J. W. Goodman, F. I. Leonberger, S.-Y. Kung, and R. A. Athale. Optical interconnections for VLSI systems. *Proceedings IEEE*, 72(7):850–865, 1984.

[8] A. Hussain, T. Lane, C. Sullivan, J. Bristow, and A. Guha. Optical backplanes for massively parallel processors and demonstration in the Connection Machine. In *GOMAC*, Las Vegas, Nov. 1990.

[9] D. Juvin, J.-L. Basille, H. Essafi, and J.-Y. Latil. SYMPATI2, a 1.5 D processor array for image application. In *EUSIPCO*, Grenoble, 1988.

[10] S. Koh, L. Ye, H. W. Carter, and J. T. Boyd. Optoelectronic interconnect simulation using a mixed mode simulator. In SPIE, editor, *Optoelectronic Interconnects III*, volume 2400, pages 244–251, San Jose, Feb. 1995.

[11] B. S. Landman and R. L. Russo. On a pin versus block relationship for partitions of logic graphs. *IEEE Transactions on Computers*, C-21:1469–1479, Feb. 1971.

[12] S. P. Levitan, P. J. Marchand, M. A. Rempel, D. M. Chiarulli, and F. B. McCormick. Computer-aided design if free-space optoelectronic interconnection (FSOI) systems. In *2nd International Conference on Massively Parallel Processing Using Optical Interconnections (MPPOI'95)*, pages 239–245, San Antonio, Oct. 1995.

[13] T. Maurin and F. Devos. Optical approaches to overcome present limitations for intercommunication and control in parallel electronic architectures. In P. SPIE, editor, *Optics for computers: architectures and technologies*, volume 1505, pages 158–165, 1991.

[14] L. L. Moresco. Electronic system packaging: the search for manufacturing the optimum in a sea of constraints. *IEEE Transactions on Components, Packaging, and Manufacturing Technology*, 13(3):494–508, 1990.

[15] K. Obermann, S. Kindt, and K. Petermann. Turn-on jitter in zero-biased single-mode semicondutor lasers. *IEEE Photonics Technology Letters*, 8(1):31–33, Jan. 1996.

[16] J. W. Parker. Optical interconnection for advanced processor systems: a review of the ESPRIT II OLIVES program. *Journal of Lightwave Technology*, 9(12):1764–1773, 1991.

[17] K. Preston. The Abingdon cross benchmark survey. *IEEE Computer*, July 1989.

[18] T. Sakano, T. Matsumoto, and K. Noguchi. Three-dimensional board-to-board free-space optical interconnects and their application to the prototype multiprocessor system: COSINE-III. *Applied Optics*, 34(11):1815–1822, 1995.

[19] P. Scheer, T. Collette, D. Juvin, A. Chenevas-Paule, J.-P. Bouzinac, P. Churoux, and M. Fracès. A massively parallel SIMD multi-processor system using optical interconnects: SYNOPTIQUE. In *Optical Computing Conference (OC'96)*, pages 120–121, Sendai, Japan, Apr. 1996.

[20] J.-M. Wang, E. Kanterakis, A. Katz, Y. Zhang, Y. Li, and N. Murray. High-speed free-space interconnect based on optical ring topology: experimental demonstration. *Applied Optics*, 33(26):6181–6187, 1994.

[21] C. Weems. The next generation image understanding architecture. In *ARPA IUW*, pages 1133–1140, Monterey, CA, 1994.

[22] D. S. Wills, W. S. Lacy, C. Camperi-Ginestet, B. Buchanan, H. H. Cat, S. Wilkinson, M. Lee, N. M. Jokerst, and M. A. Brooke. A three-dimensional high-throughput architecture using though-wafer optical interconnect. *Journal of Lightwave Technology*, 13(6):1085–1092, 1995.

7

OPTICAL ARRAY LOGIC NETWORK COMPUTING: CONCEPT AND IMPLEMENTATION

Jun Tanida and Yoshiki Ichioka

Department of Material and Life Science, Faculty of Engineering, Osaka University, Suita, Osaka 565, JAPAN

ABSTRACT

An architecture of optoelectronic computing system is presented, which is called *Optical Array Logic Network Computing* (OAL-NC). Based on the idea of *optical compunection*, which means merge of optical interconnection and computation, the concept and implementation methods of the OAL-NC are described. Finally, one possible system construction of the OAL-NC is presented.

1 INTRODUCTION

Optical interconnection is considered as one of the most promising techniques for future high-performance information processing systems. Maturity of very large-scale integration (VLSI) and optoelectronic device technologies supports realization of such processing systems. In the conventional sense of optical interconnection, it is sufficient to transfer information with light waves. However, optics can provide much more functionality than just information transfer. Namely, if spatial encoding is introduced, various kinds of logical operations can be realized with simple optical procedures [4, 5]. For example, global pattern search based on template matching can be achieved effectively by the technique. Unfortunately, bare application of such an optical computing technique seems difficult to outperform electronic competitors. Instead, the technique is expected to be useful if it is applied to conventional optical interconnection scheme. Therefore, *optical compunection*, or merge of optical interconnection and computing, should be considered as an important scheme for optoelectronic hybrid systems.

P. Berthomé and A. Ferreira (eds.), Optical Interconnections and Parallel Processing: Trends at the Interface, 195-207.

In this chapter, an architecture of optoelectronic computing system based on the idea of *optical compunection* is presented. The architecture is called *Optical Array Logic Network Computing* (OAL-NC)after its fundamental processing, optical array logic (OAL), developed by the authors for digital optical computing [5]. After explanation of the OAL, the concept and implementation methods of the OAL-NC are described. Finally, one possible system construction of the OAL-NC is presented.

2 OPTICAL ARRAY LOGIC

The OAL-NC is considered as an extension of the OAL [5]. The OAL is a computational paradigm of 2-D discrete image suitable for optical implementation. Figure 1 shows a schematic diagram of the OAL. Two discrete images are the target of the OAL processing. Every couple of the corresponding pixels in the images is converted into a spatial coded pattern according to the coding rule shown in Figure 1. After the coding process, the coded image is correlated with a set of grid-like kernels called *operation kernels*. Each operation kernel consists of a set of offset delta functions. The correlated images are spatially sampled every two structuring cells, then all of the sampled images are inverted and logically 'OR'ed. The final image provides a result of logical neighborhood operation.

The operation of the OAL is expressed by the following equation.

$$c_{i,j} = \sum_{k=1}^{K} \prod_{m=-L}^{L} \prod_{n=-L}^{L} f_{m,n,k}(a_{i+m,j+n}, b_{i+m,j+n}), \tag{7.1}$$

where a, b, and c are the pixels in the two input and the output images. (i, j) is the address of the pixel, which covers from 0 to $N-1$. $f_{m,n,k}(\cdot)$ is any one of two-variable binary logic functions for the pixels at (m, n) in the local coordinate system of the neighborhood area. L and K are the size of the neighborhood area and the number of the operation kernels, respectively. Note that a set of $f_{m,n,k}$'s determine contents of the processing. Therefore, programming the OAL is accomplished by configuring the operation kernels used in the discrete correlation.

Each process of the OAL is easily implemented by optical techniques. Optical shadow casting [4] is a good example of the technique. In this case, all pixels in the images are processed in parallel. The OAL provides SIMD (single in-

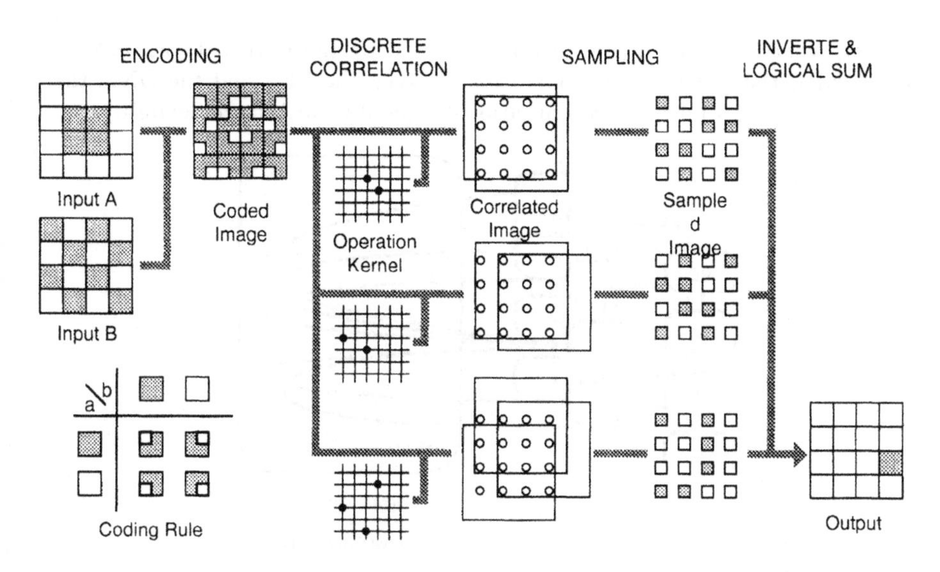

Figure 1 Schematic diagram of optical array logic. Each square frame in correlated image indicates a coded image.

struction multiple data) type of parallel processing. As a result, the OAL is an effective computational paradigm for digital optical computing.

3 OAL-NC ARCHITECTURE

Although the OAL is an attractive paradigm for digital optical computing, the authors have recognized processing inefficiencies of the OAL. As the OAL requires encoding and decoding processes for every operation, this overhead decreases processing efficiency especially for simple local operations such as $a_{i,j}b_{i,j}$. Current technologies for optoelectronic devices are still immature, thus bare application of the OAL can not provide considerable advantage over electronic equivalent. To break through this situation, the authors have presented the OAL-NC as an extension of the OAL based on more practical strategy [7].

The OAL-NC is a computational architecture in which computing property of optics is fully utilized as well as interconnection capability. Figure 2 shows a conceptual diagram of the OAL-NC. The essence of the OAL-NC is *optical compunection* for the data passed from multiple electronic processors. In the OAL-NC, local operations are executed by individual processing element's

(PE's), whereas global operations are achieved by the OAL network processor (ONP). Using cooperative operation between the PE's and the ONP, the OAL-NC provides powerful computational capability with large flexibility.

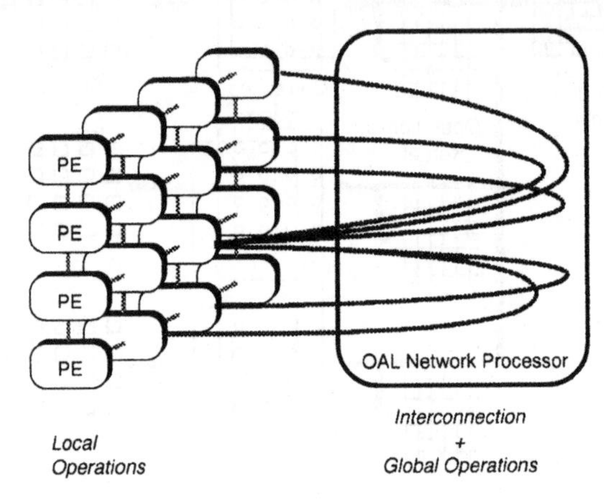

Figure 2 Conceptual diagram of OAL-NC. The ONP provides not only interconnection between the PE's but also global operations for the data from the PE's.

In the OAL-NC, various operations are executed by the ONP. Possible operations cover data transfer, data testing, data processing, system controlling, and so on. Figure 3 shows examples of operations achieved by the ONP. As the OAL takes two inputs per each pixel, two bits information from the corresponding PE are used for the operations. The results are sent back to the PE's. Therefore, the following relation exist between the data of the PE's and the pixels on the ONP.

$$a_{i,j} \leftarrow \mathrm{P}E[i,j].out1^{(p)}, \tag{7.2}$$

$$b_{i,j} \leftarrow \mathrm{P}E[i,j].out2^{(q)}, \tag{7.3}$$

$$\mathrm{P}E[i,j].in^{(r)} \leftarrow c_{i,j}, \tag{7.4}$$

where $\mathrm{P}E[i,j].x^{(k)}$ means the kth bit of the I/O register x of the PE at (i,j) position. $c_{i,j}$ is obtained from $a_{i,j}$ and $b_{i,j}$ according to Eq. (7.1). For the cases of Figure 3, the following operations are executed:

$$\overline{c_{i,j}} = \prod_{m=-L}^{L} \prod_{n=-L}^{L} \overline{a_{i+m,j+n}} \quad (Broadcasting) \tag{7.5}$$

$$\overline{c_{i,j}} = \overline{b_{i,j}} + \prod_{m=-L}^{L} \prod_{n=-L}^{L} \overline{a_{i+m,j+n}} \quad (Conditional\ Broadcasting) \tag{7.6}$$

$$c_{i,j} = a_{i,j-1} \quad (Data\, shift) \tag{7.7}$$
$$c_{i,j} = a_{i,j-2} + b_{i,j+2} \quad (Bidirectional\, shift) \tag{7.8}$$

where $\overline{c_{i,j}}$ means that the operation result must be negated after discrete correlation. This is a useful technique in the OAL to reduce the number of discrete correlations.

Note that these operations are controlled by a program of the ONP, which is an excellent feature of the OAL. As a result, a wide range of operations can be configured and achieved by the ONP.

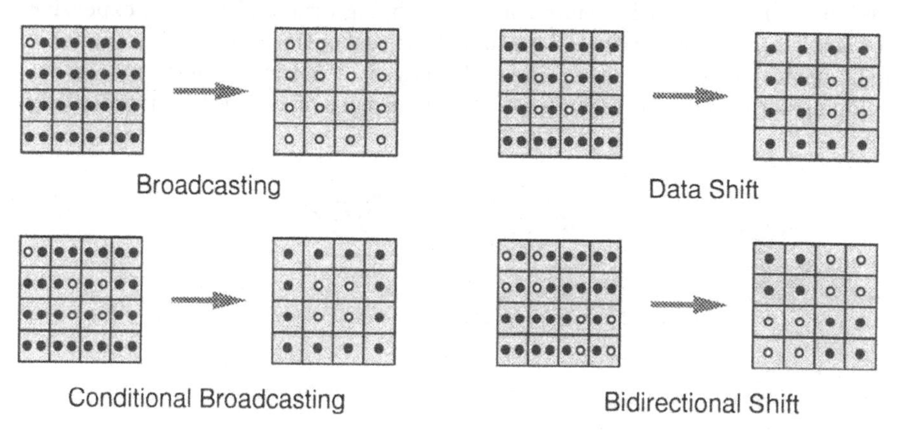

Figure 3 Example operations on the ONP. Each box indicates data associated with individual PE. Two dots in the left-hand-side box are the two bit data from the PE. One dot in the right-hand-side box is the output result by the ONP, which is sent to the corresponding PE.

The most important feature of the OAL-NC is that advantageous characteristics of both optics and electronics can be used in their suitable places. In the OAL-NC, the optics, or the ONP, is dedicated to simple and global operations, whereas the electronics, or the PE's, serves complex and local operations. Cooperative operation of the both processors enables us to extend processing capabilities of the OAL-NC. From the point of the view of system architecture, such a system is regarded as a form of heterogeneous computing system [1], which is expected to provide extremely flexible processing capabilities.

4 IMPLEMENTATION OF OAL-NC

To implement the OAL-NC, various options can be considered. To extract potential capabilities of the OAL-NC, the PE should provide considerable computational power. However, we should take account of available construction techniques. In addition, developing term and fabrication cost are also crucial issues.

For the PE implementation, there are several options related to device technologies as shown in Figure 4. VLSI implementation is expected to provide the best performance, but it requires long developing term and is much expensive for small number fabrication. As a more practical option, PLD (programmable logic device) method is promising. Although it can tremendously reduce developing term and fabrication cost, circuit density of the system is relatively low.

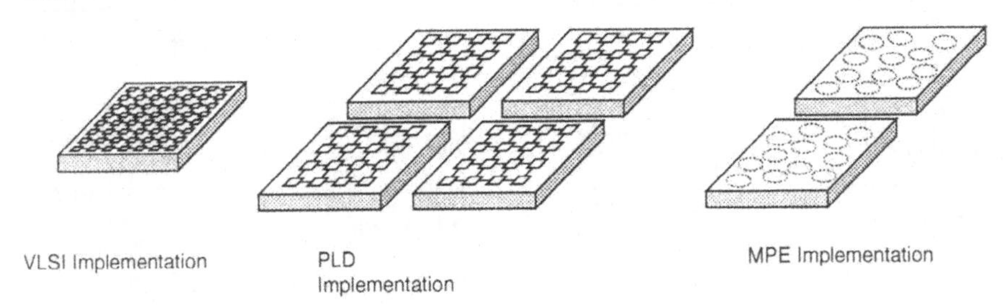

Figure 4 Implementation methods for the PE's . In MPE implementation, virtual PE's are implemented by multiple processes running on reduced number of micro processors.

Considering the situation, the authors propose a method in which multiple processes on reduced number of micro processors emulate functionalities of the PE's. We call this approach MPE (multiple process embodiment) implementation. The MPE implementation seems not to provide effective computational power, but reconfigurability of the constructed system is especially attractive compared with other methods. Table 1 summarizes features of the proposed implementation methods.

In the ONP implementation, selection of optical discrete correlator is the key issue. As described in Sec. 2, the OAL is based on discrete correlation for a coded image and an operation kernel. Thus effective implementation of the correlation is important to achieve high-performance operation of the ONP. Various kinds of configurations have been presented for the optical discrete

Table 1 Features of VLSI, PLD, and MPE implementations.

	VLSI	PLD	MPE
Using device	VLSI	PLD	micro processor
Design target	mask pattern	gate connection	program code
Circuit density	high	low	high
Functionality per PE	low	low	arbitrary
Performance	high	high	middle
Develop term	long	short	short
Modification	difficult	easy	easy
Cost	extr high	low	low

correlator, which include optical shadow casting [4], multiple imaging [6], and so on.

Figure 5 shows an example optical system for discrete correlation using multiple imaging [6]. The image on the input plane is duplicated on the modulator plane by the first half of the system. The duplicated images are overlapped with slight lateral shifts caused by the displacement of the lenses in the second half optics. Configuration of the kernel pattern is achieved by switching of the modulator. Although total length of the optical system is relatively long, it can be reduced by application of diffractive optical elements.

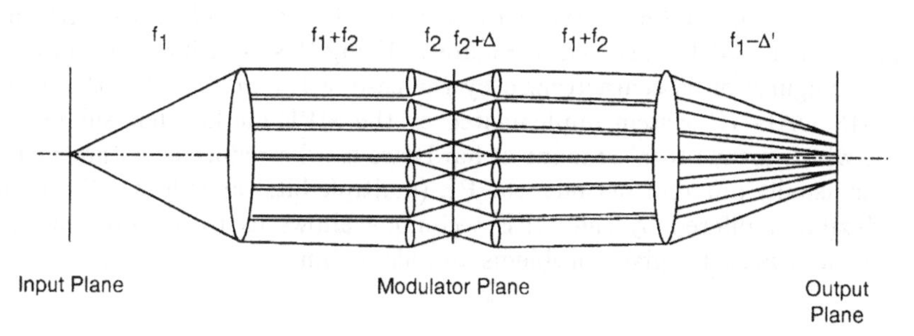

Figure 5 Optical discrete correlator based on multiple imaging.

5 MULTIPLE PROCESS EMBODIMENT

5.1 Description

The MPE implementation is an interesting idea for practical construction of an optoelectronic parallel processing system. In the conventional sense, PE's are implicitly assumed to be made by individual components. On the other hand, the MPE implementation intends to construct PE's not by *real hardware components* but by *virtual software components.* The MPE implementation has large flexibility in develop and use of the system. We can shorten developing term and easily reconfigure the system even after system construction. Note that this way of realizing high-functionality completely matches to a current trend of the computer science. In addition, reduction in hardware amount eases system packaging, which increases flexibility in system construction.

Figure 6 shows a conceptual diagram of the OAL-NC implemented by the MPE method. Most of the functionalities of the OAL-NC are realized by processes running on the micro processors. Two kinds of processes are designed to emulate function of the PE's (*PE process*) and to manage the other processes (*CTRL process*). The number of processes per processor can be determined by the specification of the OAL-NC system (pixel number, clock frequency, performance), available device, technologies, and economic factor.

For the MPE implementation, the OAL-NC is configured by three numbers: (l, m, n). They are the numbers of the *virtual* PE's, the *real* processors, and the pixels of the ONP as shown in Figure 7. Using this notation, conventional SIMD configuration is characterized by the case of $l = m = n$. On the other hand, the OAL-NC system implemented by the MPE method has smaller m than l, which means total amount of hardware can be reduced. Although bit width is usually different between the PE (multiple bits) and the ONP (single bit), flexibility offered by the MPE technique allows us to achieve suitable conversion formats for given problems and algorithms.

5.2 Performance Evaluation

In terms of processing performance, the MPE implementation is inferior to complete parallel hardware implementations, *i. e.*',VLSI and PLD methods. However, the authors presume that appropriate design of the system based on accurate analysis of the system operation can minimize such power reduction.

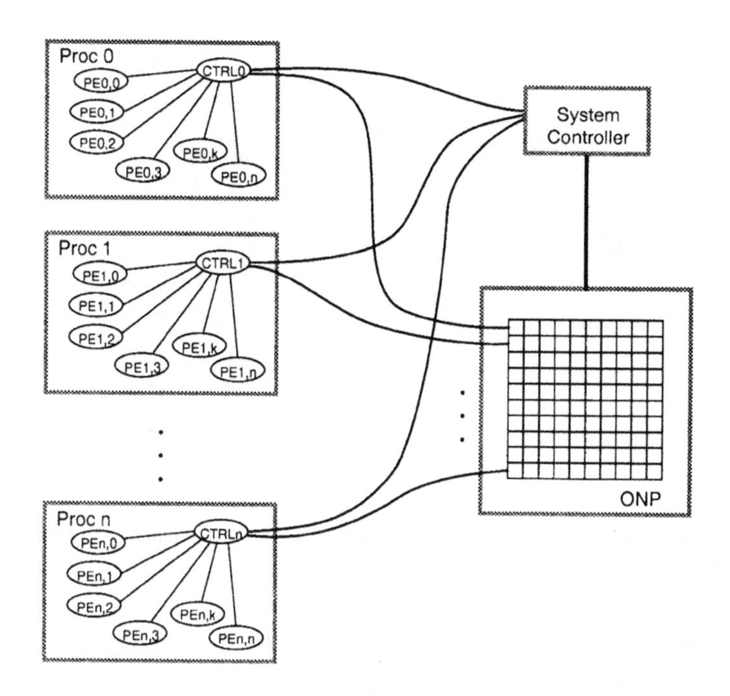

Figure 6 OAL-NC implemented by MPE method. In this example, each processor executes processes for virtual PE's located at a horizontal row.

Namely, difference between processing cycle of the PE's and the ONP can be used to accommodate the power reduction in the MPE implementation.

Now we consider power reduction by the MPE implementation comparing with the PLD implementation. We assume that single instruction of the PE requires a_m (a_p) clocks in the MPE (PLD) implementation. The clock frequency of the processor used in the MPE (PLD) implementation is f_m (f_p). The average number of instructions in the PE process is β and the required clock number for the process switching is c_{sw}. On the processing times of the both implementations, the following equation is obtained.

$$\frac{a_p\beta}{f_p} : N\frac{(a_m\beta + c_{sw})}{f_m} = \gamma : 1, \tag{7.9}$$

where γ is the ratio of the processing speed of the MPE implementation to that of the PLD implementation. N is the number of the PE processes running on a processor. For sufficient large β and for the break-even condition ($\gamma = 1$), we

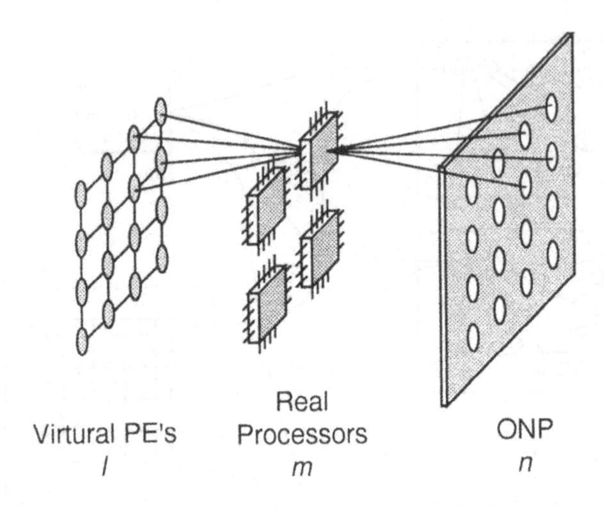

Figure 7 Configuration parameters of the OAL-NC implemented by MPE method.

obtain appropriate number of the PE processes as

$$N = \frac{a_p f_m}{a_m f_p}. \tag{7.10}$$

Figure 8 shows the relation between a_m/a_p and N for different f_m/f_p. As high-performance micro processors can be used in the MPE implementation, a_m/a_p smaller than unity and f_m/f_p larger than unity are expected. This fact suggests that processing inefficiency of the MPE implementation can be reduced as performance of micro processors increases.

6 POSSIBLE OAL-NC SYSTEM CONSTRUCTION

System construction largely depends on available technologies for device fabrication, device packaging, and optical system packaging. At this stage, systems composed of smart pixels and free-space optics are considered as a promising option. A lot of optoelectronic parallel processing systems are designed and fabricated based on the smart pixels and the free-space optics [2, 3, 8].

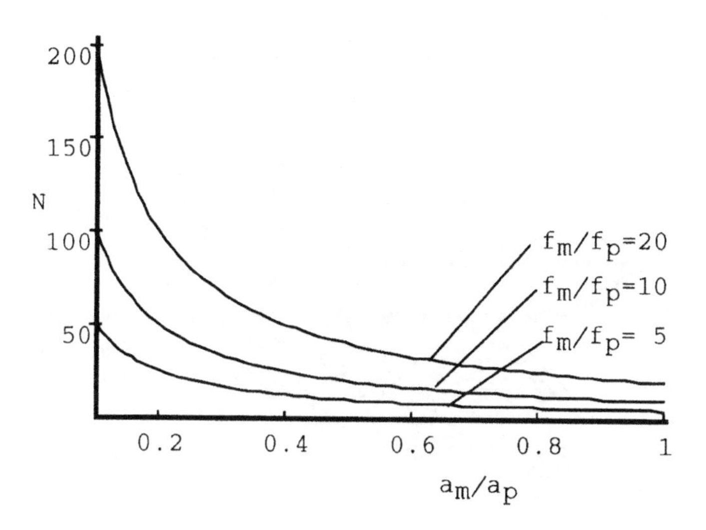

Figure 8 Appropriate PE process number on each processor for given a_m/a_p and f_m/f_p.

Following these technologies, we could draw a possible image of the packaged OAL-NC system as shown in Figure 9. The system consists of an electronic processing array (lower layer) and a folded optical correlator (upper layer). The lower layer is implemented by a smart pixel bonded to vertically stacked CMOS devices for the PE's. The smart pixel executes encoding process, which required in the OAL, and emits the coded image toward the optical correlator. The optical system is equivalent to that shown in Figure 5 except that the optical path is folded at the modulator plane. Cooperative operation of the smart pixel and the modulator accomplish parallel processing as well as data transfer according to the OAL. The bottom of the system is the system controller and I/O processor for administration tasks.

Although a lot of difficulties exist for construction of the proposed system, we hope that the abstract image of the OAL-NC system help researchers in device and packaging fields to have concrete targets in their researches.

7 SUMMARY

In this article, an architecture of optoelectronic computing system has been presented, which is called Optical Array Logic Network Computing. Based

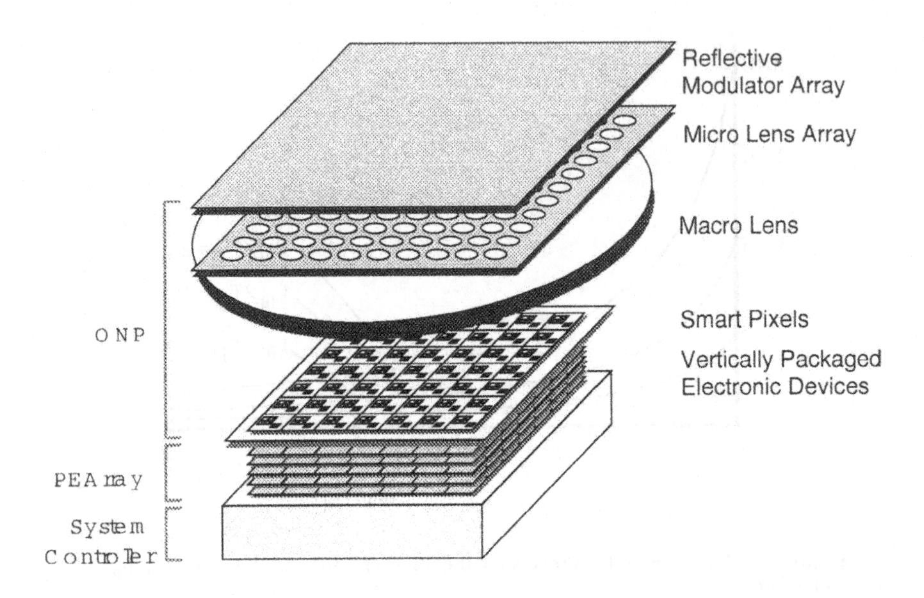

Figure 9 A possible OAL-NC construction based on smart pixel and free-space optics.

on the idea of *optical compunection*, the concept and implementation methods of the architecture are described. As an interesting implementation method, multiple process embodiment method has been explained. Finally, one possible system construction has been presented.

REFERENCES

[1] R. F. Freund and H. J. Siegel. Heterogeneous processing. *Computer*, 26(6):13–17, 1993.

[2] C. B. Kuznia and A. A. Sawchuk. Time multiplexing and control for optical cellular-hypercube arrays. *Appl. Opt.*, 35:1836–1847, 1996.

[3] Y. Li, T. Wang, and R. A. Linke. Vcsel-array-based angle-multiplexed optoelectronic crossbar interconnects. *Appl. Opt.*, 35:1282–1295, 1996.

[4] J. Tanida and Y. Ichioka. Optical logic array processor using shadowgrams. *J. Opt. Soc. Am.*, 73:800–809, 1983.

[5] J. Tanida and Y. Ichioka. A paradigm for digital optical computing based on coded pattern processing. *Intl. J. Opt. Comput.*, 1:113–128, 1990.

[6] J. Tanida and Y. Ichioka. Discrete correlators using multiple imaging for digital optical computing. *Opt. Lett.*, 16:599–601, 1991.

[7] J. Tanida and Y. Ichioka. Optical array logic network architecture. In *Optical Computing*, number 139 in Institute of Physics Conference, pages 83–86. Institute of Physics Publishing, 1994.
[8] H. Ted H. Szymanski and S. Hinton. Reconfigurable intelligent optical backplane for parallel computing and communications. *Appl. Opt.*, 35:1253–1268, 1996.

8

TOWARDS EFFECTIVE MODELS FOR OPTICAL PASSIVE STAR BASED LIGHTWAVE NETWORKS

Afonso Ferreira

CNRS LIP-ENS Lyon
Lyon, France

ABSTRACT

Distributed memory computing systems are composed of processor/memory modules that communicate by the exchange of messages through an underlying communication network. Among the many technologies used to implement such networks, the Optical Passive Star (OPS) coupler is a very efficient medium for transmitting information. It offers multiple access channels that allow a substantial reduction in the latencies for one-to-many communications, since every processor can access all its neighbors in a single step through an OPS. In this chapter, we give an overview of effective models for OPS-based lightwave networks, which can capture most aspects related to the required resources (single or multiple OPS's), technology used (single or multiple wavelengths) and diameter (single or multiple hops in pairwise communications).

1 INTRODUCTION

Distributed memory computing systems are composed of processor/memory modules that communicate by the exchange of messages through an underlying communication network. The links in such a network define for each processor a set of neighbors, to which it can send messages directly. If the message is destined to a non-neighboring processor, it then has to *hop* through intermediate processors. Hence, a basic concern in the design of communication networks is to produce feasible topologies (i.e., topologies that do not induce hardware problems) which minimize the *diameter* (the maximum number of hops in any pairwise communication).

P. Berthomé and A. Ferreira (eds.), Optical Interconnections and Parallel Processing: Trends at the Interface, 209-233.

Among the main goals of using optical technologies to implement communication networks are to achieve high *speed*, *connectivity*, and *bandwidth*, without the wiring complexity of an electronic equivalent ([13, 25]). A number of technologies have been considered for implementing optical interconnection networks [13, 28, 30, 31]. Among them, Optical Passive Stars (OPS) offer multiple access channels that allow a substantial reduction in the latencies for one-to-many communications, since every processor can access one another in a single step through an OPS [30]. Implemented either through guided or free-space optical technologies, such parallel systems would have several advantages, including simplicity, low cost, and robustness [9].

Many networks (or topologies) based on OPS couplers have been proposed in the literature, such as the optical multi-mesh hypercube ([27]), the WDM hypercube ([16]), the shuffle-exchange ([1]), the supercube ([32]), the WDM de Bruijn ([33]), POPS ([13, 21]), OTIS ([38]), and stack-rings and tori ([10, 11]).

Unfortunately, some of the problems that limited the use of most of these topologies in computers with electronic interconnections also affect computers with optical interconnections. Many of these topologies either use too many channels, have too many transceivers, or suffer from a large diameter. Moreover, in any communication step, only pairs of nodes are involved, while an intrinsic feature of optical communications is that each channel can span a large number of nodes in the network. Actually, point-to-point logical topologies do not use optical technology efficiently, and topologies in which one communication can span several nodes become very attractive ([30]). One possible reason for the use of OPS-based point-to-point networks is the lack of effective models for OPS-based one-to-many topologies. Therefore, in this chapter, we propose an overview of effective models for OPS-based lightwave networks, which can capture most aspects related to the required resources (single or multiple OPS's), technology used (single or multiple wavelengths) and diameter (single or multiple hops in pairwise communications).

Models for point-to-point topologies are based on graphs, while one-to-many topologies are best represented by hypergraphs ([2]), which can be seen as a generalization of graphs in which edges join *sets* of nodes, instead of only two nodes. As in graph models, a hypergraph edge represents a communication medium (a bus or an optical channel, for instance); a message sent on it can be read by all nodes it contains. In [3], hypergraphs were used to model bus networks, perhaps for the first time. Many papers followed, which studied structural properties of hypergraphs (*e.g.* [4, 5, 15]), algorithms for point-to-point architectures enhanced with bus systems (*e.g.* [8, 34]), and algorithms for bus interconnection networks (*e.g.* [19, 23]). Finally, proposals for new, regular topologies for one-to-many architectures appeared, as well as some communication algorithms for these topologies ([10, 13, 20, 21, 35, 36]).

Optical topologies can be designed in many ways. One or several OPS couplers can be used in the construction of single- or multi-hop, single- or multi-wavelength networks. For instance, single-hop networks can be implemented either with one multi-wavelength OPS (*e.g.* the Broadcast & Select network [30]), or with many single-wavelength OPS (*e.g.* the POPS network [13, 21]). Most of the multi-hop networks mentioned in the beginning of this chapter were also implemented with one multi-wavelength OPS, despite the fact that they do not support one-to-many communications.

This chapter focuses on topologies for one-to-many OPS systems and the hypergraph models upon which they are based. The models use the concept of *stack-graphs*, which can be briefly described as a hypergraph formed by the stacking of multiple copies of a given graph G. The stack-graph edges arise then in a natural way from the original ones: two stacks of vertices corresponding to neighboring nodes in G are connected by a stack-graph edge. It is interesting to note that the stack-graphs stem from an old and very important problem in parallel computing, namely the *embedding* problem, as will be shown later in this chapter.

In the next section, we recall the basic definitions of OPS–based systems. Then, we will show how embedding functions yield stack-graphs, giving an alternative definition of the latter. In Section 4, we show how directed stack-edges (stack-arcs, for short) can represent a shared optical medium, in this case an OPS. We will then show how directed stack-graphs can model OPS–based systems, by modeling both multi-wavelength single-OPS and single-wavelength multi-OPS systems, such as Broadcast & Select and POPS. Applications of such modeling to solve embedding problems in POPS and to design regular and modular single-OPS multi-wavelength topologies are shown in Section 5. We close the chapter with some suggestions for further research in this challenging and interesting field of one-to-many networks.

2 OPS–BASED NETWORKS

Optical networks implemented with OPS's differ with respect to the required resources (single or multiple OPS's), technology used (single or multiple wavelengths) and diameter (single or multiple hops in pairwise communications). In the following we discuss these differences in more detail.

2.1 Single-hop single-OPS multi-wavelength networks

Single-hop systems can be built using a single OPS multi-wavelength coupler, so that every processor is able to communicate directly with every other processor with no intermediate nodes (see Figure 1). In order to implement such a system, the processors' transceivers have to be tuned dynamically to the channels through which the communication takes place.

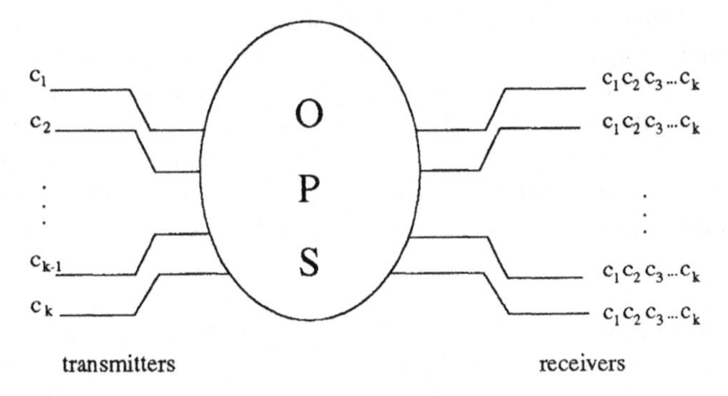

Figure 1 A k-processor single-hop multi-wavelength OPS-based lightwave network, with fixed transmitters and tunable receivers. The c_i's represent the channels which can be accessed by each processor's transceivers: processor p_j can transmit on channel c_j and receive on one out of $c_1, \ldots, c_k$.

The tuning time of the tunable transceivers may vary, depending on the number and range of the wavelengths, from a few milliseconds to a few microseconds ([12]). Unfortunately, this is considered to be very slow in comparison to a typical packet transmission time. Therefore, tunable transceivers may represent a severe drawback when building very large networks ([30]).

2.2 Multi-hop single-OPS multi-wavelength networks

One way to takle the drawback of slow tuning time is to use a single multi-wavelength OPS coupler to build a multi-hop network. In this case, a node is assigned a small static set of predefined channels (see Figure 2). Pairwise communications may then need to hop through intermediate nodes. Thus, in single-OPS multi-hop systems, communications take longer because of the multiple hops, but nodes are simpler, cheaper, and more reliable than in single-hop systems.

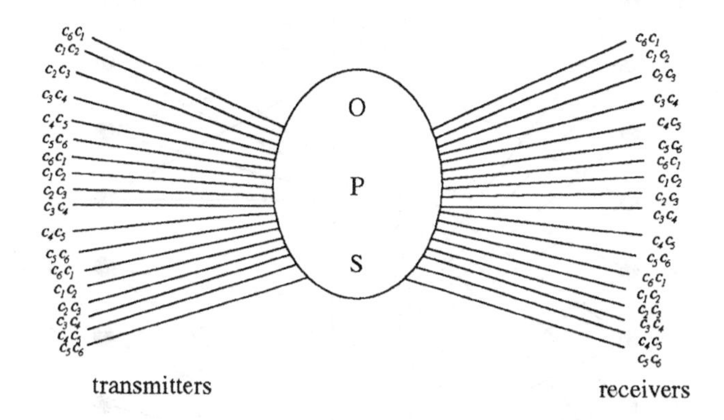

Figure 2 A multi-hop multi-wavelength single OPS lightwave network with 18 processors. The c_i's represent the channels which can be accessed by each processor's transceivers.

2.3 Single-hop multi-OPS single-wavelength networks

Another alternative to many tunable transceivers was introduced in [13]. The Partitioned Optical Passive Star network (POPS) architecture uses many single-wavelength OPS couplers and can be defined as follows.

Definition 1 *The* ***Partitioned Optical Passive Star*** POPS(d,g) *is composed of* $N = dg$ *processors and* g^2 *OPS couplers of degree* d. *The processors (respectively, the couplers) are divided into* g *groups of size* d *(respectively, of size* g*). Each OPS coupler is labeled by a pair of integers* (i,j), $0 \leq i,j < g$. *The* i*-th group of processors is connected as an output to every OPS coupler labeled* (i,j), $0 \leq j < g$ *and as an input to every OPS coupler labeled* (j,i), $0 \leq j < g$.

We shall denote by $\mathcal{I}_{(i,j)}$ the set of processors connected as input to OPS (i,j), and by $\mathcal{O}_{(i,j)}$ the set of processors connected as output to OPS (i,j). For an example of the POPS topology, see Figure 3 below.

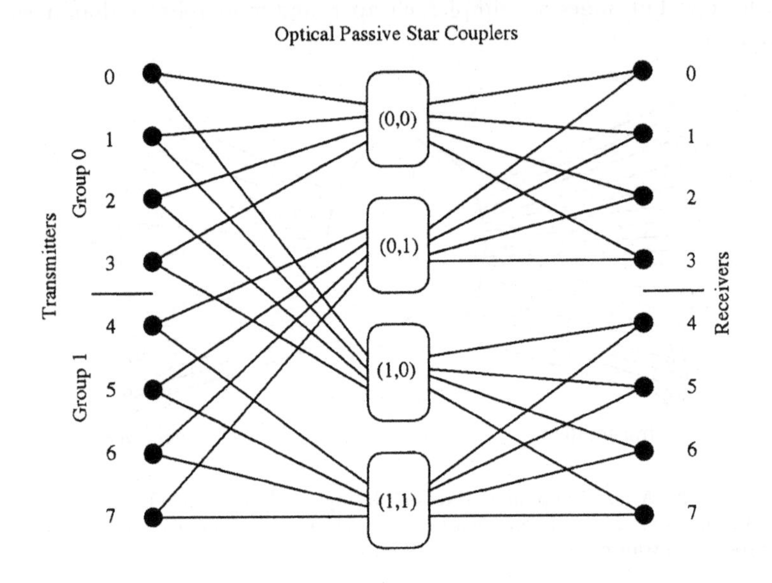

Figure 3 Partitioned Optical Passive Star POPS $(4,2)$ with 8 nodes.

3 STACK-GRAPHS

In order to introduce stack-graphs, we will first discuss the embedding problem, for reasons that will become clear at the end of this section.

3.1 About the embedding problem

One of the most important problems in parallel computing is how to ensure portability of algorithms. Many parallel algorithms were designed for specific models of computation or machine architectures. In order to translate such algorithms from one architecture to another, emulation results are required. One of the main techniques used is the embedding of the architecture for which the algorithm was originally designed into the architecture in which the algorithm has to be implemented [17, 22, 24].

The formal definition of embedding is as follows [22].

Definition 2 *Let $G = (V, E)$ and $G' = (V', E')$ be two graphs. An* **embedding** *of G into G' is an injective mapping f of V into V' combined with a mapping P_f which assigns to each edge (a, b) in E a path in G' between $f(a)$ and $f(b)$.*

Often in embeddings as defined above, the graph G is called the *guest graph* and the graph G' the *host graph*, and the quality of the embedding is measured in terms of minimizing one or several of the following parameters.

Definition 3 *Given an embedding f as above, the* **dilation** *of f is defined to be:*

$$\mathrm{dil}\,(f) = \max_{e \in E}\{\mathrm{length}\,(P_f(e))\}.$$

The **vertex-load** *of f is:*

$$\mathrm{vload}\,(f) = \max_{v' \in V'}\{|f^{-1}(v')\}\,.$$

The **edge-load** *of f is:*

$$\mathrm{eload}\,(f) = \max_{e' \in E'}|\{e \in E \mid e' \in P_f(e)\}|\,.$$

3.2 Defining stack-graphs

Notice that, once the embedding is applied, it induces a hypergraph $\varsigma\,(G, G', f)$ $(= (V_\varsigma, E_\varsigma))$ such that

1. The set of nodes of $\varsigma\,(G, G', f)$ is $V_\varsigma = V(G)$.
2. The set of hyperedges of $\varsigma\,(G, G', f)$ is $E_\varsigma = \{e_\varsigma = f^{-1}(a) \cup f^{-1}(b) \mid (a, b) \in E(G')\}$.

This kind of hypergraph is called a *stack-graph* ([7, 10, 11]), because it can be seen as a graph in which each vertex is replaced by a stack of new nodes, and the edges are, accordingly, "fattened" into hyperedges. As a matter of fact, stack-graphs can be constructed independently of existing embeddings: it suffices to build stacks of new nodes "over" each vertex of a given graph, and define the new respective hyperedges. Figure 5 shows a stack-ring based on a ring with six nodes.

Finally, we point out that the case where all the stacks have the same cardinality m is of some interest due to its regularity. In this case, m is called the *stacking-factor* of the stack-graph. The stack-graph corresponding to a graph G with stacking-factor m shall be denoted $\varsigma(G, m)$.

4 MODELING OPS–BASED NETWORKS

A wavelength in an OPS is a one-to-many communication medium that can carry one message per time slot. Thus, it can be modeled as a stack-arc, of capacity one (i.e., it can carry only one message per time slot), with input and output sets composed of d nodes each. As a stack-arc, this would be represented as in Figure 4.

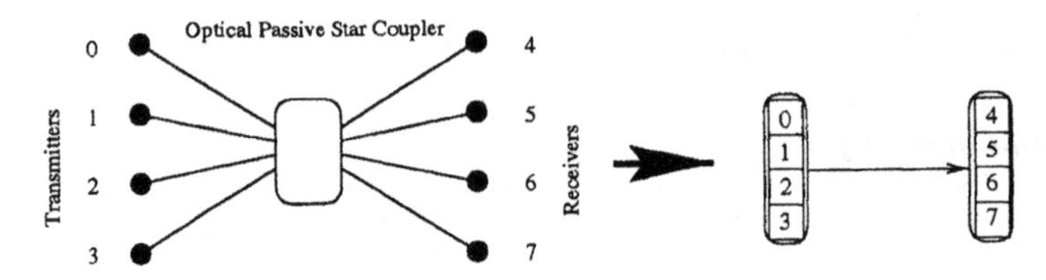

Figure 4 Modeling a single-wavelength OPS by a directed stack-edge.

Because of the directed nature of optical communications, when using stack-graphs to model optical networks we will use directed stack-graphs, unless otherwise stated. In case no orientation is given to a stack-edge, it should be assumed that it represents two stack-arcs in opposite directions.

4.1 Single-OPS

We saw how to model a single-wavelength OPS by a stack-arc. In the case where the OPS coupler is multi-wavelength, the resulting stack-graph is such that each stack-arc represents a different wavelength, and the whole stack-graph represents the single-OPS multi-wavelength network. Figure 5 shows the stack-graph representation of the multi-hop single-OPS multi-wavelength network depicted in Figure 2. Notice that the arcs are represented without direction, meaning that each processor can send and receive on that particular channel. For instance, processor (00), below, represents the processor which has the first transmitter and the first receiver (from top to bottom) in Figure 2, since it can connect to channels c_6 and c_1.

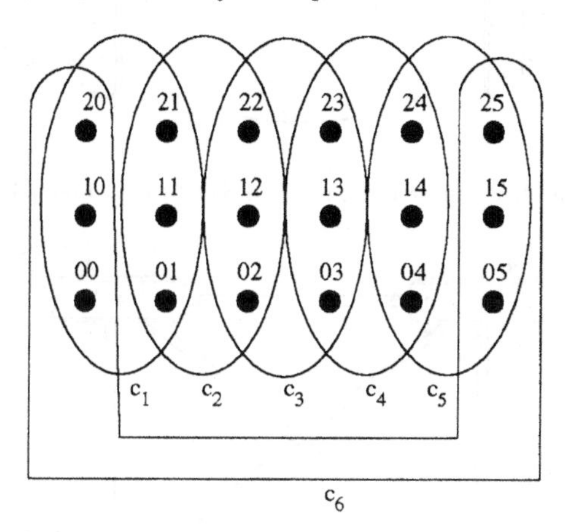

Figure 5 Stack-ring of size 6 and order 3 ($\mathcal{R}_{6,3}$). Stack-arcs are depicted encircling the vertices they contain, and are labeled with $c_{i's}$. The $c_{i's}$ represent the channels which can be accessed by each processor's transceivers.

4.2 Multi-OPS

The couplers used in POPS are single-wavelength and each OPS has input and output sets composed of groups of d nodes each. In a directed stack-graph that models POPS, each stack-arc will represent an OPS coupler and each stack-node will represent a group of d nodes. Take as example the POPS $(4, 2)$ shown in Figure 3, Section 2. There are 4 OPS divided into two groups, with $\mathcal{I}_{(0,0)} = \{0, 1, 2, 3\}, \mathcal{O}_{(0,0)} = \{0, 1, 2, 3\}, \mathcal{I}_{(0,1)} = \{4, 5, 6, 7\}, \mathcal{O}_{(0,1)} = \{0, 1, 2, 3\}$, $\mathcal{I}_{(1,0)} = \{0, 1, 2, 3\}$, $\mathcal{O}_{(1,0)} = \{4, 5, 6, 7\}$, $\mathcal{I}_{(1,1)} = \{4, 5, 6, 7\}$, and $\mathcal{O}_{(1,1)} = \{4, 5, 6, 7\}$. The directed stack-graph representing POPS $(4, 2)$ is depicted in Figure 6, where the stack-arcs are such that A represents OPS (0,0), B represents OPS (0,1), C represents OPS (1,0), and D represents OPS (1,1).

A simple exercise shows that a POPS network is perfectly represented by a directed stack-complete-graph-with-loops (stack-K_n^+, for short). A POPS (d, g) is a $\varsigma\,(K_g^+, d)$, with g stack-nodes, g^2 stack-arcs, and stacking-factor d. Figure 7 shows POPS $(3, 4)$ as a $\varsigma\,(K_4^+, 3)$.

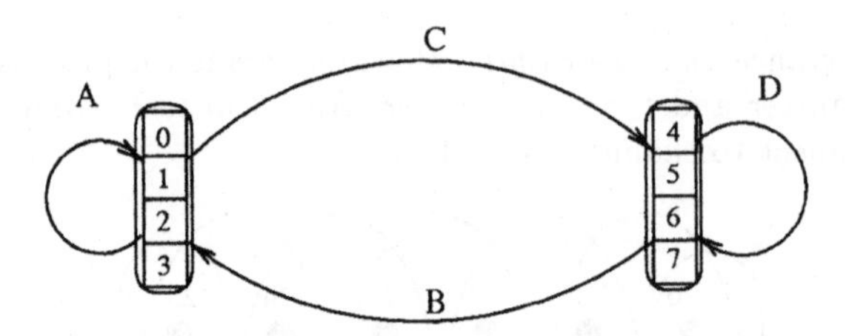

Figure 6 Another representation of POPS $(4, 2)$ as a directed stack-graph.

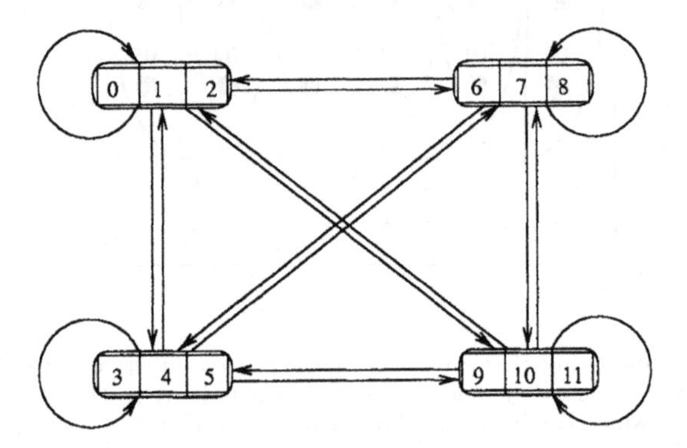

Figure 7 Stack-graph representation of POPS $(3, 4) = \varsigma\left(K_4^+, 3\right)$.

5 APPLYING THE MODELS

In this section we show how the stack-graph modeling of OPS networks can be used to solve embedding and design problems.

5.1 Optimal embeddings on POPS

Using the models presented in the previous sections, the problem of embedding a communication pattern onto POPS can be restated as follows.

Proposition 1 ([7]) *Let $G = (V, E)$ be a directed graph with $|V| = N$. Then, embedding communications which take place on G onto* POPS (d, g) *so as to minimize the completion time is equivalent to embedding G onto K_g^+, so as to minimize arc-load.*

Proof: Note that the completion time of the communication corresponds to the maximum number of point-to-point communications that would use the same OPS in the POPS (i.e., the same stack-arc in K_g^+). Thus, using Definition 2 and the equivalence of POPS and stack-graphs, the above problem is equivalent to minimizing the arc-load in K_g^+. □

Notice further that $\mathrm{POPS}\,(d,g)$ can be optimally emulated by $\mathrm{POPS}\,(d',g)$, with $d' < d$. It suffices to gather the nodes into the groups, paying attention to local communication problems only (within the same group). Consequently, embeddings onto K_g^+ can be considered with respect to the number of groups, assuming that the size of (K_g^+, d) is large enough to host the number of nodes of the guest graph, i.e., $d \geq \frac{N}{g}$, where N is the number of nodes of the guest graph.

5.2 Embedding ring communications

As an example, the problem of emulating ring algorithms by POPS was investigated in [21].

The **ring** of size n, denoted $\mathrm{R}\,(n)$, is the directed graph having n nodes labeled in $\{0, 1, \ldots, n-1\}$, and whose arcs are exactly all the pairs $(i, (i+1) \bmod n)$, $0 \leq i < n$. A preliminary solution was given in [21], where the authors showed how to optimally map communications taking place through the arcs of $\mathrm{R}\,(n)$ onto $\mathrm{POPS}\,(d,g)$, provided that n and d are powers of two.

This problem amounts to finding a way to simulate ring communications with POPS, and hence, good embeddings could help solve the problem. Indeed, with the help of stack-graph models, this problem was completely solved in [7], where it was shown how to build such embeddings for any values of n and d, such that $d \geq \sqrt{n}$. (We require that $d \geq \sqrt{n}$ in order not to have an excess of couplers with respect to processors).

Theorem 1 **([7])** *Let n, d and g be integers such that $d \geq \sqrt{n}$. There is an optimal embedding of the ring* $\mathrm{R}\,(n)$ *onto the* $\mathrm{POPS}\,(d,g)$.

Proof: Given a $\mathrm{POPS}\,(d,g)$, build its corresponding $\varsigma\left(K_g^+, d\right)$. Then, the ring node 0 can be mapped to any vertex of $\varsigma\left(K_g^+, d\right)$, since the latter is symmetric, and an Euler tour[1] in $\varsigma\left(K_g^+, d\right)$ is computed. Then, the ring node i is mapped to the i-th node of the tour. If at the end of the Euler tour there are still ring nodes remaining, another Euler tour is started to map the remaining nodes. This can be done because for any $p < g^2$, there exists an Eulerian subgraph of K_g^+ with exactly p arcs. It is very easy to see that this algorithm evenly balances the load (plus or minus 1) among the arcs of $\varsigma\left(K_g^+, d\right)$. □

In Figure 8, an example is presented: the embedding of the ring R(12) onto POPS(4, 3). Here the Euler tour places 9 of the nodes and 3 nodes remain to be mapped onto the POPS topology. A Hamiltonian cycle suffices to complete the embedding.

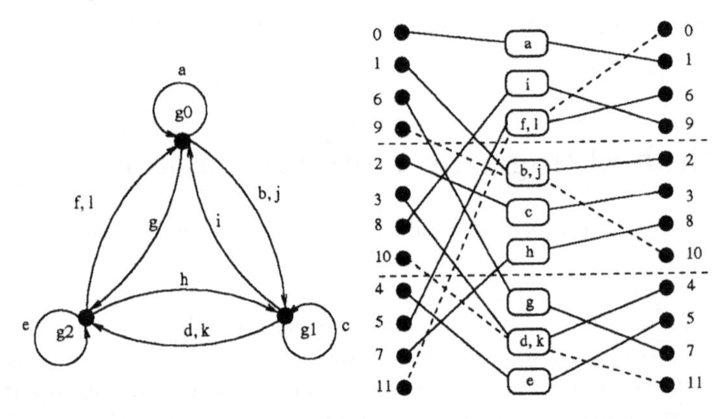

Figure 8 Ring embedding of R(12) onto POPS(4,3). The Euler tour is denoted in K_3^+ (on the left) by letters from a to i. The letters j, k and l denote the partial Euler tour required to complete the embedding. In the POPS (on the right), this partial tour is illustrated by the dotted lines.

5.3 Other embeddings

The problems of emulating tori and de Bruijn algorithms on POPS were also tackled and solved in [6, 7], using stack-graph models. In particular, in [6] optimal results were obtained for square tori, while in [7] it was shown how to embed a de Bruijn graph $B(\Delta, D)$ onto POPS(d, g), provided that $d = \Delta^{D-1}$ and $g = \Delta$.[2]

[1] An Euler tour of a graph is a closed path that traverses each edge exactly once.

[2] The de Bruijn graph $B(\Delta, D)$ is a directed graph of degree Δ and diameter D, with Δ^D nodes [26].

In order to simulate square tori communications by POPS, or in other words, embed square tori onto K_g^+, templates were found for the assignment of the tori vertices to the vertices in K_g^+. Thus, solutions for larger tori can be built from solutions for smaller ones. This yielded the following.

Theorem 2 ([6]) *Given a torus* $n \times n$ *and a* POPS (n, n), *there is an optimal embedding of the torus communications onto the POPS network.*

The main idea for the de Bruijn embedding is that stack-K_n^+ is a stack-de Bruijn $\mathrm{B}(n, 1)$, since $\mathrm{B}(n, 1)$ is exactly the complete graph with loops K_n^+. Thus, any of the automorphisms described in [37] can be used in order to embed the graph $\mathrm{B}(\Delta, D)$ into $\mathrm{B}(\Delta, D-1)$. Informally, these automorphisms guarantee that there will be Δ nodes from the guest graph mapped onto each vertex of the host graph, and that the load in the arcs is evenly distributed. Finally, $\mathrm{B}(\Delta, 1)$, with Δ nodes, is embedded onto $\mathrm{B}(\Delta', 1)$, which has Δ' nodes, through the automorphism given by the mapping $f(i) = i \bmod \Delta'$, where $\Delta = \alpha\Delta'$, $\alpha \in IN$.

The main results are

Theorem 3 ([7]) *Given a de Bruijn graph* $\mathrm{B}(\Delta, D)$ *and a* POPS (d, g) *with* $d = \Delta^{D-1}$ *and* $g = \Delta$, *there is an optimal embedding of the de Bruijn communications onto the POPS network.*

Corollary 1 ([7]) *Let be given a de Bruijn* $\mathrm{B}(\Delta, D)$ *and a* POPS (d, g) *with* $\Delta = \alpha g$ *for some integer* α *and* $d = \Delta^D/g$. *Then, there is an optimal embedding of the de Bruijn communications on the POPS network.*

5.4 Multi-hop single-OPS multi-wavelength network design

Work on multi-hop single-OPS multi-wavelength network design has concentrated mainly on point-to-point topologies, as described in the introduction. However, one of the main features of optical communications is the multicast capability, where the same message can be sent to a number of processors, at no extra cost. Hence, the need to design such networks with multicast capabilities and to study and compare their characteristics and behavior. Here again, stack-graph models were of great importance because they allowed the design of regular and modular networks ([10, 11]) as shown in Section 4.1.

It is common sense that a network topology should have properties such as the following. It should be possible to construct the network for any number of nodes; the network should be incrementally scalable, that is, it should be possible to add a new node to the existing network without a major reconfiguration; its connectivity should be high, so that in case of a failure, alternate paths can be established between the communication nodes without seriously degrading the network performance; and the network diameter — maximum hop distance — should be small, as this will enable the messages to arrive at their destinations without a major delay.

In the following, we will first discuss some topological properties of two one-to-many networks, namely with stack-ring and stack-torus topologies (see Figures 2, 5 and 9). Then, in Section 5.5, results about their stochastic behavior under one-to-one routing are recalled from [10]. These results show that even in this case, these architectures compare very well against point-to-point ones.

Design characteristics

Table 1 gives the number of nodes, total number of required wavelengths, transceivers per node, and the number of nodes per wavelength in the stack-ring, stack-torus, hypercube, and torus networks.

Model	Number of nodes	Number of wavelengths	Transceivers per node	Nodes per wavelength
Stack-ring ($n \times m$)	$n \times m$	n	2	2m
Stack-torus ($d_1 \times d_2 \times m$)	$d_1 \times d_2 \times m$	$2\ d_1 \times d_2$	4	$2m$
Hypercube	N	$N \log_2 N$	$\log_2 N$	2
Torus	$d_1 \times d_2$	$2d_1 \times d_2$	4	2

Table 1 Characteristics of the architectures.

Notice first that both stack-graph topologies have fixed numbers of transceivers per node, independent of the numbers of nodes and the dimensions of the

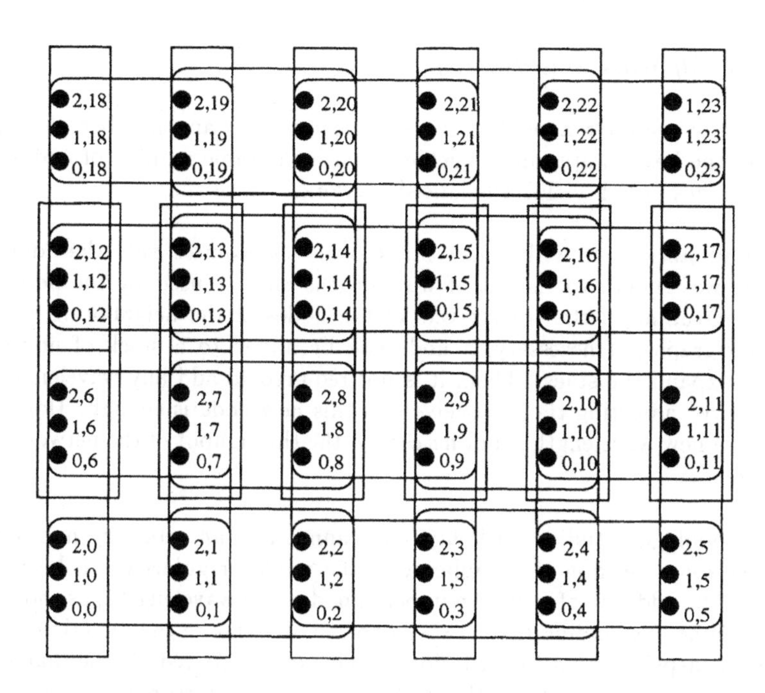

Figure 9 A stack-torus of size 4×6 and order 3 ($\varsigma(T,3)$). Stack-arcs are depicted encircling the vertices they contain. For the sake of clarity, wrap-around channels are not represented.

topologies. This is very useful when designing a topology without the *a priori* knowledge of how large the final network will be (because of insertion or deletion of nodes). In contrast, the hypercube uses a variable number of transceivers, depending on the number of nodes.

The stack-ring has a smaller number of wavelengths compared to the other networks, with respect to the same number of nodes. It is clear that this parameter can increase depending on the choice of n and m. It is important to observe that in both stack-graph topologies the number of wavelengths is smaller than the number of nodes, and that this does not occur for the graph-based topologies.

Flexibility and scalability

It is easy to see that a stack-graph can be built for any number of nodes, although the best compromise is achieved with the same number of nodes per stack of nodes, which yields a balanced network.

Expanding stack-rings with a new stack of nodes requires only the addition of a new wavelength to the set of wavelengths already used by the network, and the re-tuning of the transceivers of the nodes of the neighboring stacks. For stack-graphs, if necessary, a node can be added to a stack of nodes in order to expand the stack. Then, it is not required to add any wavelength to the network: adjusting the transceivers of this new node is enough. However, although a new wavelength is not necessary, the throughput of the network will decrease.

For the topologies studied, although one cannot add an arbitrary number of nodes without breaking the regularity of the model, regularity can be maintained by the addition of a small number of nodes and wavelengths, as shown in Table 2. When expanding the network one can also choose the direction of the expansion, depending on the criterion of the expansion: reduce the diameter, or increase the connectivity, or the number of added processors or wavelengths. Tables 1, 2 and 3 summarize the characteristics of these networks.

Network	Direction of expansion	Number of added processors	Number of added wavelengths
Stack-ring	row	m	1
$(n \times m)$	stack	n	0
Stack-torus	row	$d_1 \times m$	$2d_1$
$(d_1 \times d_2 \times m)$	column	$d_2 \times m$	$2d_2$
	stack	$d_1 \times d_2$	0
Hypercube (N)	dimension	N	$N(2 + \log N)$
Torus	row	d_1	$2d_1$
$(d_1 \times d_2)$	column	d_2	$2d_2$

Table 2 Scalability of the topologies.

Fault tolerance

In Table 3 the connectivity of the four studied architectures can be compared. Note that the stack-graphs' connectivity depends directly on the size of the stack. So, the choice of this parameter will strongly influence the connectivity of the chosen topology. The connectivity of the stack-graph based networks is larger than that of the graph-based networks, and this is one of the main advantages of those topologies.

Maximum hop distance

Table 3 also presents the maximum hop distance of the studied topologies. It is easy to notice that the hypercube has the smallest diameter, while in the stack-graph topologies it does not depend on the stack size, but on their dimensions. Notice, however, that the hypercube requires a super-linear number of wavelengths and a non-constant number of transceivers, facts which can overshadow the advantages of its small diameter.

Network	Connectivity	Maximum hop distance
Stack-ring ($n \times m$)	$2m$	$\lfloor \frac{n}{2} \rfloor$
Stack-torus ($d_1 \times d_2 \times m$)	$4m$	$\lfloor \frac{d_1}{2} \rfloor + \lfloor \frac{d_2}{2} \rfloor$
Hypercube (N)	$\log_2 N$	$\log_2 N$
Torus ($d_1 \times d_2$)	2	$2\sqrt{d_1 \times d_2}$

Table 3 Connectivity and diameter of the architectures.

5.5 Stochastic properties of routing in multi-hop single-OPS multi-wavelength networks

A simulator to evaluate the dynamic quality of stack-graph based lightwave networks was implemented, with the results described in [10]. Point-to-point routing was extensively studied. The main results and conclusions are summarized in the following.

Simulation parameters

Two of the three sets of tests (Table 4) performed are discussed here. The first was done with a varying number of nodes, the second with a varying load. The channel speed may be considered conservative, but it gives us a good picture of the behavior of the studied networks.

Set-1	**Varying nodes**
	Offered load in number of packets per second: 100
	Channel speed: 100 M-bit/s
	Number of nodes: 36, 54, 72, 90, 120, 150, 192, 256, 320
Set-2	**Varying load**
	Offered load in number of packets per second:
	10, 20, 40, 60, 80, 100, 150, 200, 250, 300, 350, 400, 450, 500
	Channel speed: 100 M-bit/s
	Number of nodes: 120

Table 4 Control sets.

When testing varying load, fixed (stack-graph) networks were used. Thus, whenever possible, (stack-graph) networks with 120 nodes were simulated. Furthermore, both stack-graph networks have diameter 3. Table 5 shows the simulated networks, giving the chosen configuration with number of wavelengths, number of fixed transceivers per node, and number of processors per wavelength.

As expected, because of their good graph-theoretic properties discussed in [20, 29], the stack-graph networks outperform graph–based networks in almost all aspects. If we further recall that stack-graph networks use only a constant number of transceivers per node, and a sub-linear number of multiplexed wavelengths, then it seems that they represent a reasonable alternative to graph–based networks. In the following, figures plotting the collected data are shown.

Varying number of nodes

Figure 10 shows that the performance of the point-to-point torus with respect to the packet delivery time and average number of hops is much worse than the other topologies. This is easily explained by the fact that the diameter of the torus increases more quickly than the others when increasing the number of nodes. The hypercube has a rather constant performance, but worse than the stack-graph networks, also because of its larger diameter.

topology	configuration	nodes	number of wavelengths	number of fixed transceivers per node	number of processors per wavelength
Stack-ring	$\mathcal{R}_{6,20}$		6	2	40
Stack-torus	$\varsigma(T(5,3),8)$	120	30	4	16
Torus	$T(10,12)$		240	4	2
Hypercube	$H(7)$	128	448	7	2

Table 5 Stack-topologies with 120 nodes and a diameter of 3. The torus has diameter 20 and the hypercube, 7.

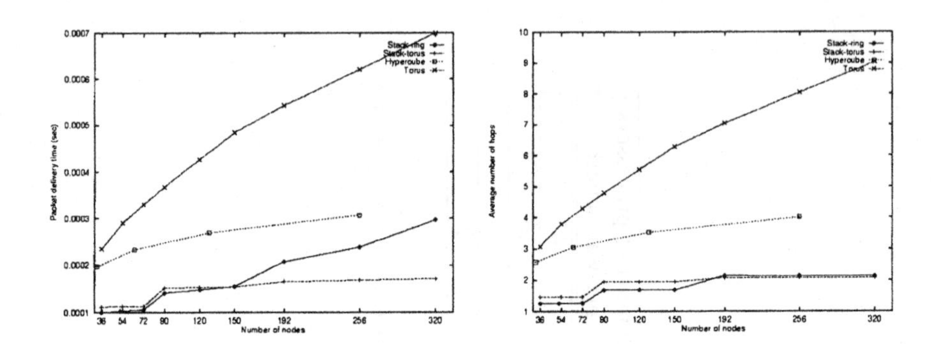

Figure 10 Packet delivery time and average number of hop versus number of nodes.

Varying load

In Figure 11 one can see that the small number of channels of the stack-ring hampers its performance. With respect to the packet delivery time, the stack-torus behaves better than both the torus and the hypercube. In contrast, the average ratio wait time to delivery time (Figure 11) shows that the number of channels in the network plays an important role, as it can be confirmed by Table 5. Hence, graph–based networks had better performance because of their larger number of edges.

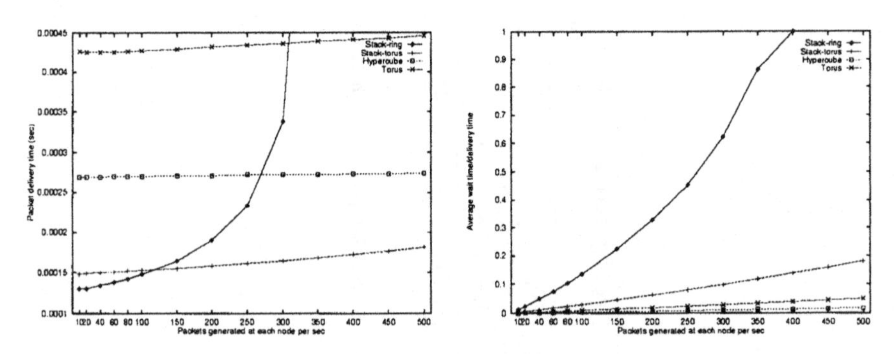

Figure 11 Packet delivery time and wait / delivery time versus load.

5.6 Remarks

Clearly, stack-graph topologies are best suited for one-to-many communications. Nevertheless, the results above show that, in general, these topologies have better performance than graph topologies even in the case of one-to-one communications. However, it is absolutely necessary to find a good tradeoff between their number of channels and diameter. For instance, the stack-torus used in the simulations had 30 channels and diameter 3, showing a good average performance, while the stack-ring that has diameter 3, but only 6 channels, behaves quite poorly with respect to experiments where the number of channels is an important factor. Therefore, an adequate balance among channels, degree, and diameter should be obtained for stack-graph networks. For instance, simulating a stack-ring with a smaller number of processors per channel could lead to better results.

In conclusion, the stack-torus behaves quite well, and requires a small number of wavelengths. Thus, if this resource is a restriction when designing the network, the stack-torus offers a good tradeoff between performance and number of channels.

6 CONCLUSIONS AND DIRECTIONS FOR FURTHER RESEARCH

Lightwave networks based on Optical Passive Stars can be designed using many combinations of resources, technology and diameter (see Table 6, below). For instance, point-to-point architectures based on graphs and implemented with Wavelength Division Multiplexing (e.g. the WDM de Bruijn [33]) are multi-hop single-OPS multi-wavelength networks.

In addition to making it easier to visualize OPS-based networks, the stack-graph model allows also the design of new one-to-many multi-hop multi-OPS single-wavelength architectures based on graph-theoretic topologies. Indeed, since the wiring complexity is no longer an issue, architectures based on small diameter, constant degree, regular stack-graphs become very interesting ([14]). It is still an open question whether stack-graphs will be of help to model multi-hop multi-OPS multi-wavelength networks. (The question marks in the table mean that those entries have yet to be studied.)

		Single-Hop	Multi-Hop
Single-OPS	Mono-λ	1 broadcast per time slot	X X
	Multi-λ	Broadcast & Select (fixed or tunable transceivers)	WDM graphs (point-to-point)
			Stack-graphs (many-to-many)
Multi-OPS	Mono-λ	OTIS POPS (K_n^+)	OTIS Stack-graphs
	Multi-λ	POPS? K_n^+?	Stack-graphs?

Table 6 Some OPS-networks classified by resources, technology and diameter.

In general, much work is to be done with respect to the feasibility, control, and efficiency of multi-OPS networks, as single-hop multi-wavelength architectures, and even more so for multi-hop ones ([18]).

Acknowledgements

This work has been partially supported by the PRC PRS of the French CNRS, by the European Human Capital and Mobility project MAP, and by AFIRST.

This work has benefitted from very fruitful discussions with the participants of the French Working Group ROI - Rencontres Optique-Informatique, and of the First Workshop in Optics and Computer Science, Metz, December 1995. The author wishes to thank particularly Pascal Berthomé, Hervé Bourdin, Pierre Chavel, Marie-Claude Heydemann, Philippe Marchand, Karina Marcus, and Ted Szymanski. He is also grateful to Joe Peters for a careful reading of the manuscript, which helped to improve it a great deal.

REFERENCES

[1] A. S. Acampora. A multichannel multihop local lightwave network. In *Proceedings of the IEEE GLOBECOM'87*, pages 459–467, Nov. 1987. Tokio, Japan.

[2] C. Berge. *Hypergraphs.* North Holland, 1989.

[3] J.-C. Bermond. Le problème des "ouvroirs" (hypergraph gossip problem). In *Colloques Internacionaux du Centre National de la Recherche Scientifique*, number 260, pages 31–34, 1976.

[4] J.-C. Bermond, J. Bond, and C. Peyrat. Interconnection networks with each node on two busses. In M. Cosnard, Y. Robert, P. Quinton, and M. Tchuente, editors, *Parallel Algorithms and Architectures*, pages 155–167. North Holland, 1984.

[5] J.-C. Bermond and F. O. Ergincan. Bus interconnection networks. Technical Report 93–56, Université Nice - Sophia Antipolis, Sept. 1993.

[6] P. Berthomé, J. Cohen, and A. Ferreira. Tori embeddings in POPS networks through stack-graph models. Technical report, LIP-ENS Lyon, 1997.

[7] P. Berthomé and A. Ferreira. Improved embeddings in POPS networks through stack-graph models. In *Proceedings of the 3rd IEEE International Conference on Massively Parallel Processing using Optical Interconnections - MPPOI'96*, pages 130–136. IEEE CS Press, Oct. 1996.

[8] L. N. Bhuyan and D. P. Agrawal. Generalized hypercube and hyperbus structures for a computer network. *IEEE Transactions on Computers*, 33:323–333, 1984.

[9] Y. Birk, N. Linial, and R. Meshulam. On the uniform-traffic capacity of single-hop interconnections employing shared directional multichannels. *IEEE Transactions on Information Theory*, 39(1):186–191, 1993.

[10] H. Bourdin, A. Ferreira, and K. Marcus. A comparative study of one-to-many WDM lightwave interconnection networks for multiprocessors. In E. Schenfeld, editor, *Proceedings of the 2nd IEEE International Workshop on Massively Parallel Processing using Optical Interconnections*, pages 257–264, San Antonio (USA), Oct. 1995. IEEE Press.

[11] H. Bourdin, A. Ferreira, and K. Marcus. On stack-graphs OPS-based lightwave networks. In *Proceedings of EuroPar'96*, Lecture Notes in Computer Science, Lyon (F), Oct. 1996. Springer-Verlag.

[12] C. A. Brackett. Dense wavelength division multiplexing networks: Principles and applications. *IEEE Journal on Selected Areas in Communications*, pages 948–964, Aug. 1990.

[13] D. Chiarulli, S. Levitan, R. G. Melhem, J. Teza, and G. Gravenstreter. Multiprocessor interconnection networks using partitioned optical passive star (POPS) topologies and distributed control. In E. Schenfeld, editor, *Proceedings of the Second IEEE International Conference on Massively Parallel Processing using Optical Interconnections (MPPOI'94)*, pages 70–80, Cancun, Mx, Apr. 1994. IEEE Press.

[14] D. Coudert, A. Ferreira, and X. Muñoz. Efficient multi-hop multi-OPS lightwave networks. Technical report, LIP - ENS Lyon, 1997.

[15] C. Delorme. Graphes et hypergraphes sommet-transitifs. Research Report 383, LRI – University Paris-Sud, Bat. 490, 91450 Orsay, France, Nov. 1987.

[16] P. W. Dowd. Wavelength division multiple access channel hypercube processor interconnection. *IEEE Transactions on Computers*, 41:1223–1241, 1992.

[17] A. Ferreira. *Handbook of Parallel and Distributed Computing*, chapter Hypercubes. McGraw-Hill, New York (USA), 1995.

[18] A. Ferreira, E. Fleury, and M. Grammatikakis. Multicasting control and communications on multihop stack-ring ops networks. In *Proceedings of the 4th IEEE International Conference on Massively Parallel Processing using Optical Interconnections - MPPOI'97*, 1997.

[19] A. Ferreira, A. Goldman vel Lejbman, and S. W. Song. Bus based parallel computers: A viable way for massive parallelism. In Springer-Verlag, editor, *PARLE'94 Parallel Architectures and Languages Europe*, volume 817, pages 553–564, July 1994. Lecture Notes in Computer Science.

[20] A. Ferreira and K. Marcus. Modular multihop WDM-based lightwave networks, and routing. In S. I. Najafi and H. Porte, editors, *Proceedings of The European Symposium on Advanced Networks and Services, Conference on Receivers, transmitters, and WDMs for fibre optic networks*, volume 2449, pages 78–86, Amsterdam, Mar. 1995. SPIE - The International Society for Optical Engineering.

[21] G. Gravenstreter and R. G. Melhem. Embedding rings and meshes in partitioned optical passive stars networks. In E. Schenfeld, editor, *Proceedings of the Second IEEE International Conference on Massively Parallel Processing using Optical Interconnections (MPPOI'95)*, pages 220–227, San Antonio, Tx, Oct. 1995. IEEE Press.

[22] M.-C. Heydemann, J. Opatrny, and D. Sotteau. Embeddings of hypercubes and grids into de Bruijn graphs. *Journal of Parallel and Distributed Computing*, 23:104–111, 1994.

[23] A. Hily and D. Sotteau. Communications in bus networks. In M. Cosnard, A. Ferreira, and J. Peters, editors, *Parallel and Distributed Computing*, volume 805 of *Lecture Notes in Computer Science*, pages 197–206. Springer-Verlag, May 1994.

[24] J. S. Jwo, S. Lackshmivarahan, and S. K. Dhall. Embedding cycles and grids in star-graphs. In *Second Symposium on Parallel and Distributed Processing*, pages 540–547, Dallas, Texas, Dec. 1990.

[25] P. Lalanne and P. Chavel, editors. *Perspectives for parallel optical interconnects.* Basic Research Series. Springer-Verlag, 1993.

[26] F. T. Leighton. *Introduction to algorithms and architectures: Arrays, Trees, Hypercubes.* Morgan Kaufmann Publishers, 1992.

[27] A. Louri and H. Sung. A hypercube-based optical interconnection network: a solution to the scalability requirements for massively parallel computers. In E. Schenfeld, editor, *Proceedings of the First International Workshop on Massively Parallel Processing Using Optical Interconnections*, pages 81–93, Cancun (Mx), Apr. 1994. IEEE Press.

[28] P. Marchand, A. Krishnamoorthy, S. Esener, and U. Efron. Optically augmented 3-D computer: Technology and architecture. In *Proceedings of the 1st IEEE International Conference on Massively Parallel Processing using Optical Interconnections - MPPOI'94*, pages 133–139, Cancun, Mx, 1994. IEEE Press.

[29] K. Marcus. A comparative study of topological properties of some WDM lightwave networks. In E. H. D'Hollander, G. R. Joubert, F. J. Peters, and D. Trystram, editors, *Proceedings of the International Conference on Parallel Computing PARCO'95.* North Holland, 1995.

[30] B. Mukherjee. WDM-based local lightwave networks Part I: Single-hop systems. *IEEE Networks*, 6(3):12–27, May 1992.

[31] B. Mukherjee. WDM-based local lightwave networks Part II: Multi-hop systems. *IEEE Networks*, pages 20–32, July 1992.

[32] A. Sen, A. Sengupta, and S. Bandyopadhyay. Generalized supercube: An incrementally expandable interconnection network. *Journal of Parallel and Distributed Computing*, 13:338–344, 1991.

[33] K. N. Sivarajan and R. Ramaswami. Lightwave networks based on de Bruijn graphs. *IEEE/ACM Transactions on Networking*, 2(1):70–79, Apr. 1994.

[34] Q. F. Stout. Meshes with multiple busses. In *Proceedings of the 27th IEEE Symposium on the Foundations of*, pages 264–273, 1986.

[35] T. Szymanski. A fiber-optic "hypermesh" for SIMD/MIMD machines. In *Supercomputing-90 Conference*, pages 710–719, New York, Nov. 1990.

[36] T. Szymanski. Hypermeshes: Optical interconnection networks for parallel computing. *Journal of Parallel and Distributed Computing*, 26:1–23, 1995.

[37] P. Tvrdik, R. Harbane, and M.-C. Heydemann. Uniform homomorphisms of de Bruijn and Kautz networks. Technical Report 986, LRI- Université Paris-Sud, June 1995.

[38] F. Zane, P. Marchand, R. Paturi, and S. Esener. Scalable network architectures using the optical transpose interconnection system (OTIS). In *Proceedings of the 3rd IEEE International Conference on Massively Parallel Processing using Optical Interconnections – MPPOI'96*, pages 114–121. IEEE CS Press, Oct. 1996.

9

EMBEDDING PROPERTIES OF RECONFIGURABLE PARTITIONABLE OPTICAL NETWORKS

Ted H. Szymanski and S. Thomas Obenaus

Departments of Electrical Engineering and Computer Science
McGill University,
Montreal, Quebec, Canada, H3A 2A7

ABSTRACT

This chapter describes the embedding properties of some emerging reconfigurable and partitionable optical networks, and motivates and formalizes several combinatorial optimization problems associated with embeddings in these networks. In particular, the embedding properties of a reconfigurable multichannel free-space optical backplane called the "HyperPlane" will be described. The networks to be embedded can be conventional point-to-point networks which are modeled as graphs $G(V, E)$, or bus-based networks which are modeled as hypergraphs $H(V, E)$. By partitioning the backplane optical channels appropriately, the optical backplane can be dynamically reconfigured to embed arbitrary networks in real time. The optical backplane can thus provide terabits of low latency bandwidth for message-passing multiprocessors based upon graphs, and shared memory multiprocessors based upon broadcast busses. It is also shown that partitionable optical networks exhibit a significant improvement in performance over non-partitionable optical networks.

1 INTRODUCTION

Optics offers a fundamental "bandwidth advantage" when compared with electronics. The bandwidth advantage can be exploited in three fundamental ways: By exploiting the "spatial parallelism" of optics, an optical network can support more optical channels than electrical channels, each occupying a different region of space, as shown in Figure 1. By exploiting the "temporal parallelism" of optics, the optical channels can be clocked at much faster rates than the electrical channels. Each channels occupies a different time slot, as shown in Figure 1. Hence, a single very high bandwidth optical channel is functionally

P. Berthomé and A. Ferreira (eds.), Optical Interconnections and Parallel Processing: Trends at the Interface, 235-257.

equivalent to several lower bandwidth electrical channels. By exploiting the "wavelength parallelism" of optics, the optical channels can use distinct wavelengths, as shown in Figure 1. Combinations of these three approaches can also be used to exploit the bandwidth advantage [37]. Hence, many emerging optical networks naturally support a large number of distinct high bandwidth channels. In this chapter, the embedding properties of a reconfigurable free-space optical backplane based upon multiple partitionable optical channels will be described. The networks to be embedded can be described as graphs or hypergraphs. Graph-based models are fundamental in the design, modeling and performance analysis of computing and communications systems. The conventional graph consists of a set of vertices V and a set of edges E, where each edge represents a logical relationship between precisely two vertices. A fundamental characteristic with graph-based models is the restriction that each edge joins at most two vertices [2], and many of the theorems in the field of graph theory rely on this property.

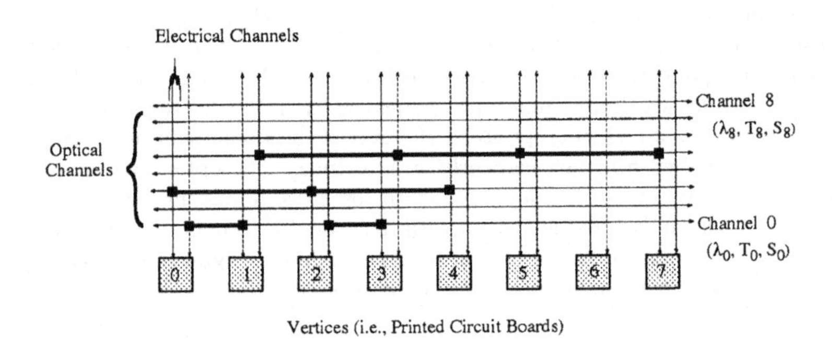

Figure 1 Multichannel optical network based upon Space, Wavelength or Time multiplexing.

Emerging optical technologies allow for the construction of distributed optical channels which interconnect many nodes. Let the channels in Figure 1 represent communication channels over distinct wavelengths. A single wavelength division multiplexed (WDM) transmission can be broadcasted over a single channel to multiple receivers simultaneously, without the intervention of intermediate printed circuit boards (i.e., vertices). Figure 1 illustrates a point-to-point connection connecting two vertices 0 and 1, and a bus (or channel) connecting three vertices 0, 2 and 4. It is not convenient to model such distributed optical channels as edges in a point-to-point graph, since the channels span a large number of vertices. For the same reasons, large optical systems built with distributed optical channels arranged into various topologies cannot be properly model-led as graphs. Without an accepted modeling formalism, it is difficult to recognize and characterize essentially identical optical architectures which differ only in implementation details.

Researchers have recognized the limitations of the graph model and have considered hypergraphs for modeling bus-based networks, including Bermond, Bond, Scale and Djelloul [3, 4], Szymanski [35, 36], Bourdin, Ferreira and Marcus [8], Mackenzie, Ould-Khaoua, Sutherland and Kelly [27]. A "hypergraph" model is a generalization of the conventional graph model in which edges can contain multiple vertices [2]. A hypergraph consists of a set of vertices V and a set of hypergraph edges E, where each hyper-edge consists of an arbitrary set of vertices. (We often use the phrase "hyper-edge" to distinguish these from regular graph edges.) Hypergraphs are well suited to formally model any network in which sets of vertices can share logical attributes, such as membership to a distributed optical bus or channel, or membership to a distributed optical crossbar switch.

In this chapter, the embedding capabilities of a reconfigurable and partitionable free-space optical backplane based upon multiple optical channels will be described. The unique features of this network include its dynamic reconfigurability, where the network can be reconfigured in fractions of a microsecond, and partitionability, i.e., the ability to partition an optical channel into multiple smaller segments [37]. The networks to be embedded include conventional point-to-point networks which are modeled as graphs $G(V, E)$, or bus-based networks which are modeled as hypergraphs $H(V, E)$. This chapter will motivate and formalize some interesting combinatorial optimization problems based upon hypergraphs. The chapter is written for a broad audience of researchers working in three multi-disciplinary areas, including *optics, engineering and computer science*. Hence, the chapter will focus upon identifying and unifying similar basic concepts from each field, towards the description of reconfigurable optical networks. In the process, we hope to highlight certain basic similar concepts from these fields which differ primarily in terminology.

2 THE "HYPERPLANE" RECONFIGURABLE OPTICAL BACKPLANE

The architecture of a general purpose free-space optical backplane for parallel computing and telecommunications is first summarized [20, 37]. The optical backplane consists of a number of Printed Circuit Boards (PCBs) or MultiChip Modules (MCMs) which are arranged in a opto-mechanical packaging assembly, as shown in Figure 2. Each PCB typically contains multiple opto-electronic "*Smart Pixel Arrays*" which manage the optical I/O between PCBs in the backplane. Smart pixel arrays are opto-electronic integrated circuits with optical I/O, electrical I/O and electronic processing capabilities. The opto-electronic smart pixel arrays essentially provide an interface between the electrical and optical domains.

The optical backplane typically supports between 1,000 and 10,000 parallel optical bit-wide channels (i.e., "bit-channels") spaced a few hundred microns apart. Each optical bit-channel operates at between 100 Mbit/sec and perhaps 1 Gbit/sec, and the peak capacity of the backplane is thus between 1 Terabit/sec and perhaps 10 Terabit/sec. The optical bit-channels are organized into "optical channels", i.e., a collection of w optical bits are switched together as an indivisible entity. An optical channel is typically between 8 and 64 bits wide, corresponding to popular datapath widths in electronic systems.

Each PCB has access to some or all of the Z optical channels through X electronic **Injector** and Y electronic **Extractor** Channels labeled $\{I_1, I_2, \ldots, I_X\}$ and $\{E_1, E_2, \ldots, E_Y\}$, where in general $X \leq Z$ and $Y \leq Z$. The injectors provide the capability of injecting electronic signals into a selected subset $\mathcal{I}$ of optical channels while the extractors are used to extract information from another subset $\mathcal{E}$ of optical channels. The connectivity provided with the unique combination of electronic channels and partitionable optical channels creates a three-dimensional reconfigurable partitionable optical backplane called the "*HyperPlane*".

This architecture uses a dynamically programmable smart pixel array which provides a programmable interface between the electrical and optical domains. Due to the bandwidth advantage of optics over electronics, the interface typically consists of a programmable switch between a relatively small number of electrical channels and a large number of optical channels. The *HyperPlane* smart pixel array transports and filters of the order of Terabits/second (Tb/s)

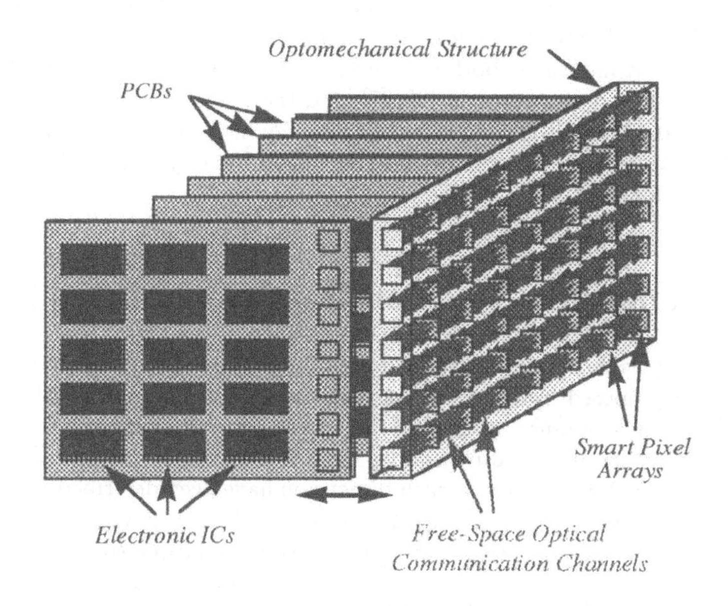

Figure 2 Free-space optical backplane [37].

of optical data, and inserts or extracts of the order of tens or hundreds of Gigabits/second (Gb/s) of electrical data for each PCB or MCM. A number of different opto-electronic technologies can be used to implement smart pixel arrays [37].

In this architecture, each of the optical channels can be partitioned as a single channel spanning the entire backplane (end-to-end), or it can be partitioned into multiple channel "segments" spanning smaller physically distinct sections of the backplane. Due to this partitionability, many conventional interconnection networks can be efficiently embedded into the *HyperPlane*. Many networks, including 2D and 3D meshes, shuffles, hypercubes and hypermeshes can be embedded into the backplane and the embeddings can be changed in fractions of a microsecond.

The *HyperPlane* intelligent optical backplane is under development at McGill University and at the University of Colorado at Boulder. The Canadian Institute for Telecommunications Research (CITR), financed by the Canadian government and members of the Canadian telecommunications industry, is funding the development in Canada. The research program is multi-disciplinary and includes the collaboration of physicists, engineers and computer scientists affiliated with many institutes, both in Canada and abroad. The principles behind the architecture are described in [20, 37], and the design of the smart pixel arrays is described in [31]. The first generation demonstrator become operational in 1996. A demonstration of a second generation intelligent optical backplane is scheduled for 1997-98 [10].

Partitionability vs. Non-Partitionability

Most reconfigurable optical networks do not support partitionable optical channels. Many proposed optical networks multiplex multiple wavelengths over a single fiber, in a manner such that the wavelength channels are **not** partitionable. For example, 2 dimensional mesh-like systems with multiple non-partitionable WDM channels in each dimension have been described in Kostuk et. al. [22], Dowd [14], Li et. al. [25], Liu et. al. [26], Szymanski [36] and others. Some proposed optical networks multiplex multiple spatially distinct optical channels in a region of free-space, such that the spatial channels are not partitionable. A 1 dimensional ring-like system with multiple non-partitionable space division multiplexed (SDM) channels has been described by Redmond and Schenfeld, called the OPAM architecture [30]. These non-partitionable networks will be shown to be limited in their embedding capabilities, since they cannot achieve any optimization by partitioning the optical channels to embed more effectively. In this chapter, it is shown that partitionable optical networks typically exhibit a significant improvement in performance over non-partitionable networks. In other words, a partitionable network can typically have a significantly fewer number of optical channels than a non-partitionable network, and still have the same embedding capability.

We point out that the electrical engineering community has been exploiting the concept of partitionability for many years. The concept of partitionability of conventional electrical interconnection networks has been formalized by Siegel in [33]. The concept of partitionability of an individual channel has been referred to as a "segmented channel" in the VLSI community, and this scheme is used universally in field programmable gate arrays [9].

3 GRAPH AND HYPERGRAPH MODELS FOR BUS-BASED OPTICAL NETWORKS

Due to the multi-disciplinary nature of the intended audience, some basic concepts from the field of graph theory will be summarized. By conventional definition, a graph $G(V, E)$ consists of non-empty set of vertices V and a set of edges E which is disjoint from V, such that if e is an edge and u and v are vertices where $e = \{u, v\}$, then edge is said to "join" vertices u and v, and these vertices are called the "ends" of e. The power of the graph model to capture the key features of an electrical point-to-point interconnect is one reason for the success of the graph model. Once a problem can be formulated in terms of a conventional graph, many algorithms and theorems from the field of graph-theory which can be brought to bear upon the problem. Identical graphs can be detected, and the fundamental characteristics of different graphs can then be systematically identified.

The hypergraph can be defined as a generalization of the conventional graph model to allow edges to "span" multiple vertices. A hypergraph $H(V, E)$ consists of a non-empty set of vertices V and a set of edges (or "hyper-edges") E which is disjoint from V, and an incidence mapping that associates with each hyper-edge an unordered set of vertices of V [2]. For example, if e is an hyper-edge and u, v and w are vertices such that $e = \{u, v, w\}$, then the hyper-edge e is said to "join" vertices u, v and w, and these vertices are called "members" of e. These vertices are said to be "adjacent" or "neighbors", reflecting the fact that they are members of the same hyper-edge. The "degree" of a vertex is defined as the number of hyper-edges incident to that vertex, and the degree of a hypergraph is defined as the maximum degree taken over all of its vertices. The reader is referred to [2–4] for a treatment on hypergraphs.

The topology of a hypergraph can be formally characterized in the same manners as a conventional graph, including the "incident matrix" representation, the "adjacency matrix" representation, and the "edge-set" representation. In this chapter we will consider only the edge set representation, where each hyper-edge is represented by explicitly enumerating its vertices. We point out that the electrical engineering and computer science communities have been using different notations to denote essentially the same concept for many years. The engineering community has developed a formalism for electrical "net-lists" which is routinely used in computer aided design packages [9]. A typical net-list consists of an enumeration of all electrical connections (or "nets"), and associated with each net a list of all "contact-points" associated with that net. The net-list description of the connections in Figure 1 is essentially the same as the hypergraph description shown in Table 1;

Hypergraph edge	Vertices
0	0, 1
1	2, 3
2	0, 2, 4
3	1, 3, 5, 7

Table 1 A Hypergraph and its edge-set representation.

Hypergraphs are well suited to formally model optical networks in which vertices can **share common attributes**, such as membership to a distributed optical bus, or membership to a distributed optical crossbar switch. If a problem can be formulated within the hypergraph model, many algorithms and theorems from the field of graph-theory can then be brought to bear upon the problem.

4 A MODEL FOR RECONFIGURABLE PARTITIONABLE OPTICAL INTERCONNECTS

The optical backplane in Figure 2 has two main variations, the *Linear HyperPlane* which does not have "wrap-around" optical channels, and the *Circular HyperPlane* which has wrap-around optical channels between the ends in Figure 2. An "embedding template" for the 1 dimensional *Circular HyperPlane* is shown in Figure 3. The template uses a **box** to denote each backplane PCB (or MCM). Each PCB has a number of **vertical lines** which represent electrical **Injector** and **Extractor** channels to and from the PCB. The template has a large number of **horizontal lines** which in this chapter represent uncommitted "**optical channels**". In this chapter the phrase "optical channel" will refer to a 32 bit-wide channel. The horizontal lines in the 2D template denote optical channels without specifying their precise physical location in 3D free-space. Optics provides a large 3D spatial bandwidth advantage which the 2D template does not fully reflect.

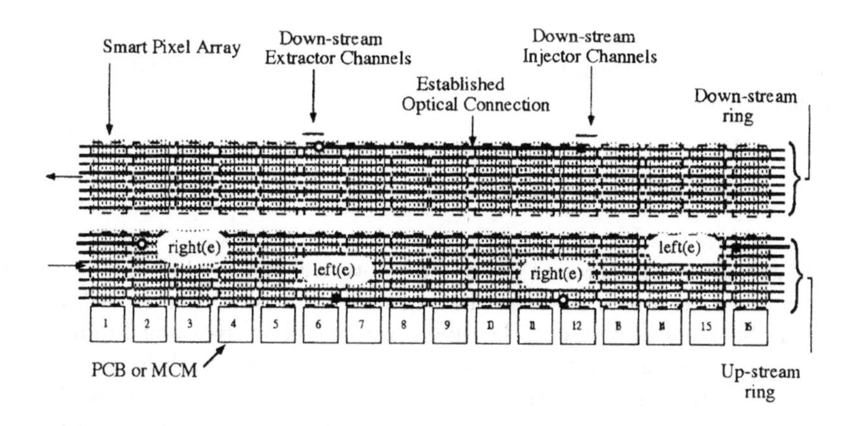

Figure 3 1 Dimensional multichannel circular array (i.e., *Circular Hyper-Plane*) with multiple partitionable optical channels.

Boxes, circles and bold lines represent the connections in the *Circular Hyper-Plane*. The solid boxes represent injection points, i.e., connections between vertical electronic injector channels and horizontal optical channels. The circles represent extraction points, i.e., connections between vertical electronic extractor channels and horizontal optical channels. The bold lines represent established point-to-point or multi-point optical connections. The K vertical injector and extractor channels emanating from a PCB represents the PCB "degree", i.e., the maximum number of optical channels which can be accessed by a PCB. When a connection between PCBs is established, a bold horizontal line is drawn between the optical channel end-points. The embeddings can be changed in real-time by down-loading appropriate control bits to the smart pixel arrays.

The 1 dimensional optical interconnect in Figure 1 can be generalized to higher dimensions, as shown in Figure 4. Consider a regular n-dimensional network with d vertices aligned in each dimension. This multi-dimensional optical interconnect can be formally modeled as a hypergraph. An n dimensional multichannel optical backplane can be modeled as an orthogonal hypergraph with $|E|$ hyper-edges, where each hyper-edge represents one or more partitionable optical channels spanning all vertices aligned along a dimension. This mesh-like network with multiple optical channels in each row or column has been described as "HyperMesh" in [36].

There are d^n vertices, each with a unique identifier called its "coordinate vector", where the elements correspond to the vertices' coordinates in the n-dimensional space:

$$\text{vertex coordinate vector} = (x_{n-1}, x_{n-2}, \ldots, x_1, x_0).$$

Each hyper-edge can represent one or more optical channels which interconnects a set of d vertices which are aligned along some dimension. Hence, each hyper-edge can be assigned a unique identifier consisting of a vertex and a dimension; by moving along the dimension a set of $d-1$ other vertices belonging to the same hyper-edge can be identified. This hyper-edge can thus be represented by a vector of n elements, corresponding to the vertex coordinates, along with a symbol Θ which denotes the "free-variable". The vertices belonging to a hyper-edge can be identified by letting the symbol assume a value $\Theta \in (0, \ldots, d-1)$. The hyper-edge with label

$$z(x_{n-1}, \ldots, x_{m+1}, \Theta, x_{m-1}, \ldots, x_0)$$

consists of the set of d vertices as follows:

$$z(x_{n-1}, \ldots, x_{m+1}, \Theta, x_{m-1}, \ldots, x_0) = \bigcup_{i=0}^{d-1} \{(x_{n-1}, \ldots, x_{m+1}, i, x_{m-1}, \ldots, x_0)\}.$$

With this convention, the n-dimensional optical interconnect shown in Figure 4(b) can be modeled as an orthogonal hypergraph, where each hyper-edge identifies a row or column of vertices which share one or more optical channels as follows:

$$Z = \left\{ \begin{array}{c} z(x_{n-1}, x_{n-2}, \ldots, x_1, \Theta) \\ z(x_{n-1}, x_{n-2}, \ldots, \Theta, x_0) \\ \vdots \\ z(\Theta, x_{n-2}, \ldots, x_1, x_0) \end{array} \right\},$$

$\forall x_i \in \{d-1, \ldots, 0\}$ and $\forall 0 \leq i < n$. In Figure 4(a), there are multiple optical channels in each row or column. If we wish to label these optical channels individually, we may append an identifier to the labels

$$z(x_{n-1}, \ldots, x_{m+1}, \Theta, x_{m-1}, \ldots, x_0),$$

so that the resulting labels identify unique optical channels in each row or column. By adding attributes to the hyper-edges, this model captures the essence of many proposed electrical or optical networks, including those described in [5, 8, 14, 15, 17–19, 21, 22, 25–27, 30, 34–37].

5 COMBINATORIAL OPTIMIZATION PROBLEMS ENCOUNTERED IN THE OPTICAL EMBEDDINGS

In this section, we describe some interesting combinatorial optimization problems associated with embeddings in reconfigurable optical interconnects. (For the benefit of the multi-disciplinary readership, the basic concepts are summarized.) According to Garey and Johnson [16], a "combinatorial optimization problem" Π can be either a minimization problem or maximization problem consisting of three parts:

1. A set D_Π of *problem instances*;
2. for each instance $I \in D_\Pi$, a finite set $S_\Pi(I)$ of candidate solutions; and
3. a function m_Π which assigns to each instance $I \in D_\Pi$ and each candidate solution $\sigma \in S_\Pi(I)$ a positive rational number $m_\Pi(I, \sigma)$ called the *solution value* for σ.

An "*optimal solution*" is a candidate solution such that its solution value is minimized or maximized respectively. Previous sections have summarized the use of the hypergraph formalism for modeling optical networks. In this section, we will formalize the optimization problem of embedding a guest hypergraph $G(V, E)$ into the host with multiple partitionable optical channels, representing the multi-dimensional optical interconnect described in section 4. An NP-complete problem is a problem which essentially requires an exponential amount of time to solve optimally. To establish the NP-completeness of a given problem Π, it suffices to establish that Π contains a known NP-complete problem as a special case [16]. In other words, one may specify additional restrictions

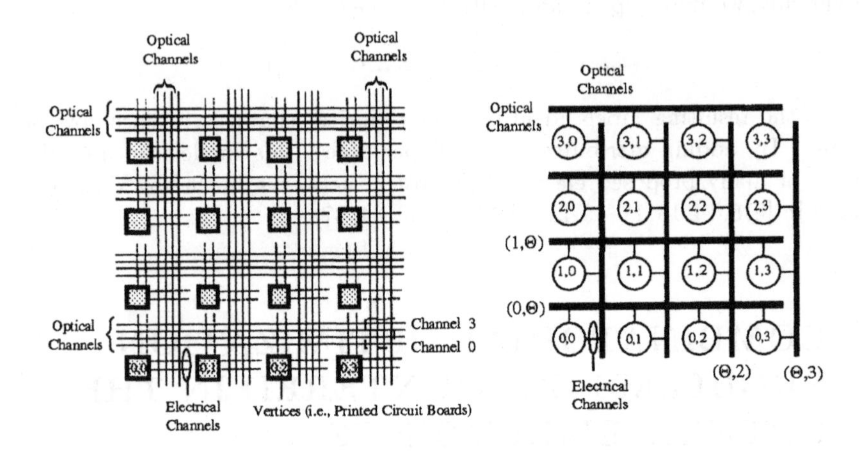

Figure 4 2 Dimensional multi-channel mesh ("*HyperMesh*") with multiple optical channels in each dimension. (a) Hyper-edges shown explicitly. (b) Abstract model.

or constraints on the original problem so that the constrained problem will be identical to a known NP-complete problem, in a technique is called "*proof by restriction*". Many of the following optimization problems can be easily seen to be NP-complete, by containing known NP-complete problems as restricted cases.

Definition: Let E^+ and E^- denote the sets of edges which do and do not cross the "wrap-around" region once embedded, respectively. A "hyperplane mapping" of a hypergraph $H(V, E)$ onto the hyperplane is a one-to-one mapping of vertices $f : V \longrightarrow \{1, 2, \ldots, |V|\}$ and a two-way (reversible) mapping of edges onto channel segments such that all members of an edge e are in between the left and the right end nodes (denoted respectively $L(e)$ and $R(e)$), as shown in Figure 3, i.e.,

$$\forall e \in E^+, \forall u \in e, L(e) \leq f(u) \leq R(e);$$

$$\text{and } \forall e \in E^-, \forall u \in e, f(u) \leq R(e) \text{ or } \leq f(u).$$

Optimization Problem Class I - *"HyperGraph Minimum Cut n-Dimensional Linear (Circular) Arrangement"*

The objective here is to maximize the bisection bandwidth of a hypergraph which has been embedded into an n-dimensional optical multi-channel linear or circular array (i.e., a multichannel mesh or *HyperMesh* for short). More formally, determine a one-to-one mapping of guest nodes into the host nodes, and a mapping of guest edges onto unique segments of partitionable optical channels, such that the number of optical channels required for the embedding is minimized. Once the number of optical channels is minimized, the bisection bandwidth of the guest can be maximized by replicating the embedding, i.e., embedding multiple copies of the hypergraph into the backplane. The embedded hypergraph is replicated until either the node degree limits further replication, or the number of optical channels in the backplane limits further replication. Consider all vertical *"bisectors"* of the host graph, where a bisector divides the host graph into two halves of arbitrary sizes. The decision problem is, for each vertical bisector located at any index i, is the cardinality of the set of embedded edges crossing the bisector (i.e., the "cut-width") less than K? If so, then the guest graph can be embedded using no more than K partitionable optical channels. The one dimensional problem can be stated as:

Instance: Hypergraph $H(V, E)$, positive integer K.

Question: Is there a hyperplane mapping such that for all i, $1 \leq i \leq |V|$,

$$|\{e \in E^+ : L(e) \leq i < R(e)\} \cup \{e \in E^- : not(R(e) \leq i < L(e))\}| \leq K?$$

In this formulation, the problem instances consists of a given hypergraph and all integers K. For each instance, the node and edge mappings define a set of candidate solutions. Each candidate solution is associated with a number representing its "cut-width", and an optimal solution is one with the minimum cut-width.

This optimization problem is stated in the context of an optical backplane with multiple spatially distinct channels which are partitionable, as shown in Figure 1. However, the problem is equivalent to minimizing the number of time slots used to embed a network when TDM is used for the bandwidth advantage, and minimizing the number of wavelengths when WDM is used for the bandwidth advantage, provided that the channels are partitionable.

The problem of embedding hypergraphs into a linear array has been considered by Bhasker and Sahni [6], who presented heuristic algorithms to minimize the number of wires crossing a bisector in an electrical backplane by ordering the boards. Our problem is a generalization to include the circular multichannel array, a generalization made possible with optical technology. When the general problem is constrained such that (i) each hyper-edge contains at most two vertices so the hypergraph degenerates into a graph, (ii) the "wrap-around" capability of the optical channels is removed, and (iii) the dimensionality is reduced to one, the problem degenerates to the NP-complete problem "*Minimum-Cut Linear Arrangement*" [16].

Optimization Problem Class II - "*HyperGraph Optimal n-Dimensional Linear (Circular) Arrangement*"

The objective here is to minimize the propagation delays of a HyperGraph which has been embedded into an n-dimensional optical multi-channel linear or circular array. More formally, determine a one-to-one mapping of guest nodes into the host nodes, and a one-to-one mapping of guest edges onto unique subsections of partitionable optical channels, such that the sum of the lengths of embedded edges is minimized. It follows that the sum of the propagation delays over the embedded hyper-edges is minimized. The one dimensional problem can be stated as:

Instance: Hypergraph $H(V, E)$, positive integer K.

Question: Is there a hyperplane mapping such that

$$\sum_{e \in E^+} (R(e) - L(e)) + \sum_{e \in E^-} (|V| + R(e) - L(e)) \leq K?$$

When the general problem is constrained such that (i) each hyper-edge contains at most two vertices so the hypergraph degenerates into a graph, (ii) the "wrap-around" capability of the optical channels is removed, and (iii) the dimensionality is reduced to 1, the problem degenerates to the known NP-complete problem "*Optimal Linear Arrangement*" [16].

Optimization Problem Class III - "*Hypergraph Minimum Cut Optimal n-Dimensional Linear (Circular) Arrangement*"

The objective here is to maximize the bisection bandwidth while simultaneously minimizing the propagation delays of a hypergraph which has been embedded into an n-dimensional optical multi-channel linear or circular array. More formally, determine a one-to-one mapping of guest nodes into the host nodes, and a one-to-one mapping of guest edges onto unique subsections of partitionable optical channels, such that the number of optical channels required for the embedding is minimized and the sum of the lengths of embedded hyper-edges is minimized. The problem can be stated as:

Instance: Hypergraph $H(V, E)$, positive integers K_1, K_2.

Question: Is there a hyperplane mapping such that for all i, $1 < i < |V|$,

$$|\{e \in E^+ : L(e) \leq i < R(e)\} \cup \{e \in E^- : not(R(e) \leq i < L(e)\}| \leq K_1$$
$$\text{and} \sum_{e \in E+} (R(e) - L(e)) + \sum_{e \in E-} (|V| + R(e) - L(e)) \leq K_2?$$

The above three classes of optimization problems have been motivated by the practical need to maximize bandwidth or minimize delays in embeddings, and apply to optical interconnects which use SDM, WDM or TDM with partitionable channels, as shown in Figures 1 and 4. These optimization problems lead to the question "how does one identify an optimal solution to problem class III", if an optimal solution indeed exists? The following simple theorem may help.

Theorem: Any embedding into the multi-channel linear or circular array which minimizes the sum of horizontal distances and simultaneously utilizes all segments of partitionable optical channels between vertices for embedding edges or hyperedges is an optimal solution to the "*Hypergraph Minimum Cut Optimal 1D Linear (Circular) Arrangement*" problem.

Proof: It is given that the sum of distances is minimized. Hence, it only remains to establish that the number of optical channels used in minimized. Since sum of distances is minimized the nodes cannot be rearranged to yield a smaller sum of distances. The nodes can potentially be rearranged to use a smaller number of optical channels. However, since every channel segment is used, reducing the number of optical channels would reduce the sum of distances, contradicting the minimum sum of distances hypothesis. QED.

Optimization Problem Class IV - "*Hypergraph Contractability and Homomorphism*"

In practice, the host hypergraph may have many more nodes than the guest hypergraph, and it is necessary to embed multiple guest nodes into every host node. This technique can be accomplished by performing a "contraction" operation on the original hypergraph. In this case, it is useful that the degree of the contracted nodes be minimized so that once they are embedding into a PCB, the I/O bandwidth of a PCB is not exceeded. This leads to two useful optimization problems called *Hypergraph Contractability and Homomorphism*, which can be stated as follows;

Instance: Hypergraphs $G(V_1, E_1)$, $H(V_2, E_2)$

Contractability Question: Can a hypergraph isomorphic to H be obtained from G by a sequence of edge contractions, i.e., a sequence in which each step replaces all vertices belonging to the same hyper-edge by a single vertex, adjacent to exactly those vertices that were previously adjacent to at least one of the vertices of the hyper-edge ?

Homomorphism Question: Can a hypergraph isomorphic to H be obtained from G by a sequence of identifications of non-adjacent vertices, i.e., a sequence in which each step replaces two non-adjacent vertices u, v by a single vertex w adjacent to exactly those vertices that were previously adjacent to at least one of u and v?

When the general problems are constrained such that (i) each hyper-edge contains at most two vertices so the hypergraph degenerates into a graph, these problems degenerate to two well known NP-complete problems "Graph Contractability" and "Graph Homomorphism" [16].

Optimization Problem Class V - *"Hypergraph Line-of-Sight Embedding in an n-Dimensional Grid"*

Another important problem is determining whether a guest hypergraph can be embedded into a host such that each hyper-edge is embedded in one dimension only, thereby providing a "line-of-sight" optical path between the vertices it interconnects. Equivalently, every hyper-edge has no bends. The 2 dimensional problem can be stated as follows;

Instance: Hypergraph $G(V, E)$, positive integers M, N.

Question: Is there a two-way function $f : V \longrightarrow \{1, 2, \ldots, M\} \times \{1, 2, \ldots, N\}$ such that if $\{u, v\} \in E$, $f(u) = (x_1, y_1)$ and $f(v) = (x_2, y_2)$, then either $x_1 = x_2$ or $y_1 = y_2$, i.e., $f(u)$ and $f(v)$ are both on the same "line" of a grid?

When the original problem is constrained such that each hyper-edge contains at most two vertices, and $M \cdot N = |V|$ and f is one-to-one, then the hypergraph degenerates into a graph and the problem degenerates to a well known NP-complete problem "Edge Embedding on a Grid" [16].

6 EMBEDDINGS OF MESHES

In this section, we summarize the embedding process by illustrating the embedding of various meshes (with wrap-around) into reconfigurable optical interconnects. Meshes (i.e., d-ary n-cubes) are popular interconnects in large scale multiprocessor systems due to their scalability, modularity and suitability to handle multi-dimensional arrays which occur in scientific algorithms.

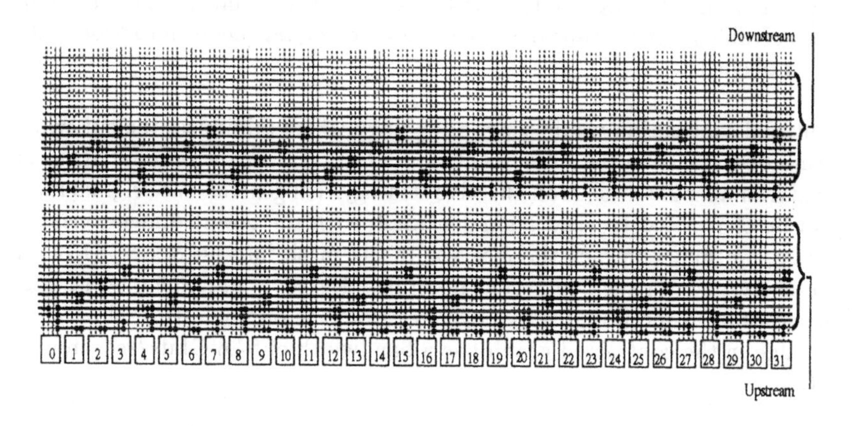

Figure 5 Embedding of a 3D mesh for a Cray Research T3D supercomputer into a bidirectional Circular *HyperPlane*. (Each bold line represents a 57.6 Gigabit/sec. channel.)

The embedding of a 3D mesh of a Cray T3D supercomputer [12] is shown in Figure 5. The embedding template represents an optical backplane consisting of multiple "upstream" channels and multiple "downstream" channels, where each stream supports 1,024 optical bits. Assuming an optical clock rate of 600 Mhz, each stream supports an aggregate bandwidth of 600 Gigabits/sec and the optical backplane supports an aggregate (or bisection) bandwidth of 1.2 Terabits/sec. Each box represents a PCB which contains 16 processors.

These processors and their associated logic can be incorporated into multiple Multi-Chip Modules (MCMs) which reside on each PCB.

A large Cray T3D supercomputer can be organized as a 3D $8 \times 8 \times 16$ mesh, where each edge has a bandwidth of approx. 3.6 Gbits/sec. in each direction (24 wires $\times$ 150 Mhz) [12]. Since this system has 1,024 processing nodes and assuming the backplane has 64 PCBs, it is necessary to embed 16 processing nodes into each PCB. (The embedding in Figure 5 can be extended by adding another 32 PCBs).

The first step is to perform a graph contraction operation so that the guest graph and host graph have the same number of nodes (optimization problem class IV). The second step is to determine an mapping of guest nodes into host nodes which minimizes the cut-width, minimizes the delay, or simultaneously achieves both (optimization problem classes I, II, and III). A final step would involve mapping the edges into partitionable optical channels, and this final step is solvable in linear time once the mapping of guest nodes into host nodes is accomplished.

An optimal contraction would yield a topology which minimizes the cut-width and sum of distance in the subsequent steps. Contracting the original 3D mesh to yield a smaller 2D mesh is an effective strategy. (In general, the original mesh should be contracted to yield a smaller mesh with the fewest dimensions). In this case, we may contract the $8 \times 8 \times 16$ 3D mesh into an 8×8 2D mesh, where every node in the contracted mesh has degree 4, and where every edge in the contracted mesh represents 16 edges from the original 3D mesh. Hence, every edge in the contracted graph has a bandwidth of 57.6 Gbits/sec. in each direction.

In the embedding of the 8×8 mesh, there are 8 rings in dimension 1, spanning nodes (0,1,2,3,4,5,6,7), ..., (57,58,59,60,61,62,63). These 8 rings can be embedded with 2 partitionable optical channels spanning nodes $0 \ldots 63$, since the rings are non-overlapping, as shown in Figure 5. (In Figure 5, these rings span 4 nodes and will be extended to span 8 nodes in system with 64 PCBs.) There are 8 rings in dimension 2 spanning nodes (0,8,16,24,3 2,40,48,56), ..., (7,15,23,31,39,47,55,63). These 8 rings are overlapping and thus each rings requires 2 partitionable optical channels. Hence, the total number of optical channels is $2 + 2 \times 8 = 10$, as shown in Figure 5. The embedding process can be generalized to meshes with varying number of nodes in each dimension, and to a varying number of dimensions.

Note that if the optical channels are not partitionable, the embeddings are much less efficient. Consider the OPAM architecture proposed by Schenfeld et. al. [1,30]. The OPAM and Circular HyperPlane architectures have some similarities, reinforcing basic design decisions; they both exploit clusters of electronic processors, i.e., the Printed Circuit Boards in Figures 1 and 5. They both exploit ring-like free-space optical data paths with hundreds or thousands of spatially distinct micro-optical channels. However, they differ significantly in areas of dynamic reconfiguration, partitionability, and ability to process optical data. The OPAM architecture is essentially a multi-channel circular array similar to that shown in Figure 3, without partitionability or dynamic reconfiguration of optical channels. Since the optical channels are not partitionable, every ring will require 16 optical channels in order to be embedded (8 channels for edges to the right neighbors, and 8 channels for edges to the left neighbors). There are 16 rings in the 2D mesh, and hence 256 optical channels will be required. With this contraction, each channel requires 57.6 Gbits/sec, and hence the mesh embedding in the non-partitionable OPAM architecture will require 256×57.6 Gbps = 14,745 Gbps. In other words, the OPAM architecture will require an aggregate bandwidth of 14.75 Terabits/sec in order to embed the Cray T3D mesh, resulting in a low efficiency. With this contraction the Cray mesh has a bisection bandwidth of roughly $8 \times 2 \times 57.6 = 900$ Gigabits/sec (there are 8 rings crossing the bisector, and each ring supports 57.6 Gbps in each direction). In contrast, the Circular Hyperplane will require an aggregate bandwidth of $10 \times 2 \times 57.6 = 1.15$ Terabits/sec to embed the mesh, resulting in a high efficiency. The same arguments apply to other optical networks which are not partitionable - the lack of partitionability of optical channels can result inefficient embeddings.

Extension to Higher Dimensions

Consider the problem of embedding an n dimensional mesh into a 2 dimensional optical backplane. (Suppose that n is even.) The first step is to perform a graph contraction operation so that the guest graph and host graph have the same number of nodes (optimization problem class IV). The second step is to determine an embedding of the contracted graph which maintains line of sight connectivity (optimization problem class V), since line of sight connectivity eliminates the need to "bend". The next step is to determine an embedding which retains the line of sight connectivity, and which also minimizes cut-width, minimizes delay, or simultaneously achieves both (problem classes I, II, and III). In other words, the ideal embedding would be an optimal solution to optimization problem classes I, II, III, IV and V. Fortunately, when one is embedding the graphs usually found in scientific algorithms, systolic algorithms and data flow graphs such as the FFT, efficient solutions are easily found.

Finally, we point out that similar problems appear in the field of "Field Programmable Gate Arrays" (FPGA). An FPGA is typically a 2 dimensional array of "logic cells" or "logic blocks", with a multiplicity of partitionable electrical channels in each of the rows and columns [9]. A typical FPGA may have a $32 times 32$ array of logic blocks, with perhaps 32 partitionable electrical channels in each row and column. Programming the FPGAs involves the embedding of a "data flow graph" into the grid of logic blocks. The data flow graph specifies a task to be performed in the logic blocks. Due to the combinatorial nature of the optimization problems, heuristic algorithms are used to perform the mapping of guest nodes into host nodes and the embedding of edges, in a process called "place-and-route". Hence, many heuristic algorithms used in the field of FPGAs can be applied to the embedding in two dimensional optical networks.

7 CONCLUSIONS

This chapter has described the embedding properties of some emerging reconfigurable and partitionable optical networks. In particular, the embedding properties of a reconfigurable multichannel free-space optical backplane called the "HyperPlane" have been described. The chapter has motivated and formalized many interesting combinatorial optimization problems associated with embeddings in optical networks. The networks to be embedded include conventional point-to-point networks which are modeled as graphs $G(V, E)$, or bus-based networks which are modeled as hypergraphs $H(V, E)$. By partitioning the backplane optical channels appropriately, the optical backplane can be dynamically reconfigured to embed arbitrary networks in real time. It was also shown that partitionable optical networks exhibit a significant improvement in performance over non-partitionable optical networks.

Acknowledgements

The assistance of many graduate students in the Microelectronics and Computer Systems (MACS) Laboratory at McGill University is acknowledged, in particular Palash Desai (Lucent Technologies), Manoj Verghese (NorTel), Victor Tyan and Boonchuay Supmonchai. Thanks to Prof. H. Scott Hinton of the University of Colorado at Boulder, who co-developed the backplane architecture, and to Prof. D. Plant and Prof. F. Tooley of McGill University. Funding from NSERC Canada Grant OGP0121601 and the Canadian Institute of Telecommunications Research (CITR) Grant 96-3-4 is acknowledged. (For additional information on this project, please see http: //www.macs.ee.mcgill.ca)

REFERENCES

[1] A. Barak and E. Schenfeld. Embedding classical communication topologies in the scalable OPAM architecture. *IEEE Transactions on Parallel and Distributed Systems*, 7(9):962–978, Sept. 1996.

[2] C. Berge. *Hypergraphs.* North-Holland Mathematical Library, Amsterdam, 1989.

[3] J.-C. Bermond, J. Bond, and S. Djelloul. Dense bus networks of diameter 2. Research Report 94-46, CNRS, Université de Nice Sophia-Antipolis, Aug. 1994.

[4] J.-C. Bermond, J. Bond, and J.-F. Saclé. Large hypergraphs of diameter 1. In Bollobas, editor, *Graph Theory and Combinatorics.* Academic Press, 1984.

[5] P. Berthomé and A. Ferreira. Improved embeddings in POPS networks through stack-graph models. In *Third International Workshop on Massively Parallel Processing using Optical Interconnections*, pages 130–136. IEEE CS Press, Oct. 1996.

[6] J. Bhasker and S. Sahni. Optimal linear arrangement of circuit components. *Journal of VLSI and Computer Systems*, pages 87–109, 1987.

[7] J. A. Bondy and U. S. R. Murphy. *Graph Theory with Apllications.* North Holland, 1984.

[8] H. Bourdin, A. Ferreira, and K. Marcus. A comparative study of one-to-many WDM lightwave interconmnection networks for multiprocessors. In *Second International Workshop on Massively Parallel Processing using Optical Interconnections*, pages 257–253, San Antonio (USA), Oct. 1995. IEEE Press.

[9] S. D. Brown, R. J. Francis, J. Rose, and Z. G. Vranesic. *Field Programmable Gate Arrays.* Kluwer Academic Publishers, 1992.

[10] Canadian Institute for Telecommunications Research. Research program 1996-97, photonic devices and systems, Aug. 1996. (http://www.citr.ee.mcgill.ca).

[11] J. P. Cohoon and S. Sahni. Heuristics for backplane ordering. *Journal of VLSI and Computer Systems*, pages 37–60, 1987.

[12] Cray Research Inc. Cray T3D system architectural overview, Sept. 1993.

[13] P. Desai. Embeddings of a cray T3D supercomputer into the optical backplane. Microelectronics and Computer Systems (MACS) Laboratory, McGill University, Montreal, Quebec, Canada.

[14] P. W. Dowd. Wavelength division multiple access channel hypercube processor interconnection. *IEEE Transactions on Computers*, 41(10):1223–1241, Oct. 1992.

[15] E. E. E. Frietman. Opto-electronic processing and networking: A design study. Delft University of Technology Printing Office, 1995.

[16] M. Garey and D. Johnson. *Computers and Intractability. A guide to the theory of NP-completeness.* W. Freeman and Compagny, New York, 1979.

[17] G. Gravenstreter and R. G. Melhem. Embedding rings and meshes in partitioned optical passive star networks. In *Second International Workshop on Massively Parallel Processing using Optical Interconnections*, pages 220–227, San Antonio (USA), Oct. 1995. IEEE Press.

[18] Z. Guo, R. G. Melhem, R. W. Hall, D. M. Chiarulli, and S. P. Levitan. Pipelined communications in optically interconnected arrays. *Journal of Parallel and Distributed Computing*, 12(3):269–282, July 1991.

[19] J. H. Ha and T. M. Pinkston. The SPEED cache coherence protocol for an optical multi-access interconnect architecture. In *Second International Workshop on Massively Parallel Processing using Optical Interconnections*, pages 98–107, San Antonio (USA), Oct. 1995. IEEE Press.

[20] H. S. Hinton and T. H. Szymanski. Intelligent optical backplanes. In *Second International Workshop on Massively Parallel Processing using Optical Interconnections*, pages 133–143, San Antonio (USA), Oct. 1995. IEEE Press.

[21] J. Kilian, S. Kipnis, and C. E. Leiserson. The organization of permutation architectures with bused interconnections. *IEEE Transactions on Computers*, 39(11):1346–1358, Nov. 1990.

[22] R. K. Kostuck, T. J. Kim, D. Ramsey, T.-H. Oh, and R. Boye. Connection cube and interleaved optical backplane for a multiprocessor data bus. In *Second International Workshop on Massively Parallel Processing using Optical Interconnections*, pages 144–151, San Antonio (USA), Oct. 1995. IEEE Press.

[23] B. Krishnamurthy and M. S. Krishnamoorthy. The difficulty of funding good embeddings of program graphs onto the OPAM architecture. In *Second International Workshop on Massively Parallel Processing using Optical Interconnections*, pages 124–129, San Antonio (USA), Oct. 1995. IEEE Press.

[24] F. T. Leighton. *Introduction to Parallel Algorithms: Arrays, Trees, Hypercubes.* Morgan-Kaufmann, San Mateo, CA, 1991.

[25] Y. Li, S. B. Rao, I. Redmond, and T. Wing. Free-space WDMA optical interconnects using mesh-connected bus topology. In *Proc. Int. Conf. Optical Computing (OC'94)*, pages 153–156, Edinburgh, 1994. Institute of Physics Publishing.

[26] G. Liu, K. Y. Lee, and H. F. Jordan. n-dimensional processor arrays with optical buses. In *Second International Workshop on Massively Parallel Processing using Optical Interconnections*, pages 116–123, San Antonio (USA), Oct. 1995. IEEE Press.

[27] L. M. Mackenzie, M. Ould-Khaoua, R. J. Sutherland, and T. Kelly. Cobra: A high-performance interconnection for large multicomputers. Computing Science Research Report 1991/R19, University of Glasgow, Oct. 1991.

[28] T. S. Obenaus. Topology of a high speed free-space photonic network. Master's thesis, Depts. Elec. Eng. and Computer Science, McGill University, .

[29] T. S. Obenaus and T. H. Szymanski. Embedding star graphs into optical meshes without bends. Submitted.

[30] I. Redmond and E. Schenfeld. A distributed reconfigurable free-space optical interconnection network for massively parallel processing architectures. In *Proc. Int. Conf. Optical Computing*, pages 215–218, Edinburgh, Aug. 1994. Institute of Physics Publishing.

[31] D. R. Rolston, D. V. Plant, T. H. Szymanski, H. S. Hinton, M. H. Ayliffe, D. N. Kabal, A. V. Krishnamoorthy, K. W. Goosen, J. A. Walker, B. Tseng, S. P. Hui, J. C. Cunningham, and W. Y. Jan. A hybrid-SEED smart pixel array for a four-stage intelligent optical backplane demonstrator. *Journal of Quantum Electronics*, pages 97–105, Apr. 1996. Special Issue on Smart Pixels.

[32] I. Scherson. Orthogonal graphs for a class of interconnection networks. *IEEE Transactions on Parallel and Distributed Systems*, 2(1):3–19, Jan. 1991.

[33] H. J. Siegel. The theory underlying the partitioning of permutation networks. *IEEE Transactions on Computers*, 29(9):791–800, 1980.

[34] Q. F. Stout. Mesh-connected computers with broadcasting. *IEEE Transactions on Computers*, 32(9):826–830, Sept. 1983.

[35] T. Szymanski. Graph-theoretic models for photonic networks. In I. Scherson, editor, *Proceedings of New Frontiers: A Workshop on Future Directions of Massively Parallel Processing*, pages 85–96. IEEE Computer Society, IEEE Press, Oct. 1992.

[36] T. H. Szymanski. Hypermeshes - optical interconnection networks for parallel computing. *Journal of Parallel and Distributed Computing*, 26:1–23, Apr. 1995.

[37] T. H. Szymanski and H. S. Hinton. Reconfigurable intelligent optical backplane for parallel computing and communications. *Applied Optics*, pages 1253–1268, Mar. 1996. Special Issue on Optical Computing.

10

TIME DIVISION MULTIPLEXED CONTROL OF ALL-OPTICAL INTERCONNECTION NETWORKS

Charles Salisbury and Rami Melhem

Department of Computer Science
University of Pittsburgh
Pittsburgh, PA 15260, USA

ABSTRACT

Circuit switching techniques are preferred in optical multiprocessor interconnection networks because they do not require any optical to electronic signal conversion to route messages. Thus, they can provide all-optical paths between message sources and destinations and the large communication bandwidth of optical signals is available to the nodes attached to the network. However, circuit switching techniques increase the complexity of managing the network, and thus may increase the communication delay. In a massively parallel processor, this added delay may affect the performance of a parallel program. In this chapter we model the performance impact of multiplexed circuit switched network management techniques. We show how program performance is affected by the choice of circuit switching techniques, by the multiplexing degree, and by the characteristics of the network.

1 INTRODUCTION

The electrical interconnection networks used in current multiprocessors often use packet switching techniques to exchange messages between processors. Packet switching networks route messages between a sender and a receiver through intermediate routing nodes that may be either switches or other processors. These intermediate nodes decode routing information in the packet header and retransmit the message along the next network link in the path toward the receiver. With current technology, packet processing is performed by electronic circuitry. Thus, a message sent as an optical signal must be converted to an electronic form by the intermediate node and stored in a data buffer while the routing information is processed. The message is then retrans-

P. Berthomé and A. Ferreira (eds.), Optical Interconnections and Parallel Processing: Trends at the Interface, 259-282.

mitted as an optical signal on the appropriate network link. The retransmission may be delayed if there is contention for the link or if the buffer at the other end of the link is full. Thus each routing step introduces a delay in message transmission that increases the network *latency*, which is the time it takes a message to cross the network. The processing time at each intermediate node may also limit the network *bandwidth*, which is the quantity of data that can pass through the network in any period of time. The impact on performance is especially significant in optical networks, where the potential bandwidth can be very high and the potential latency very low. As a result, packet switching is not currently capable of attaining the full performance possible from an optical interconnection network.

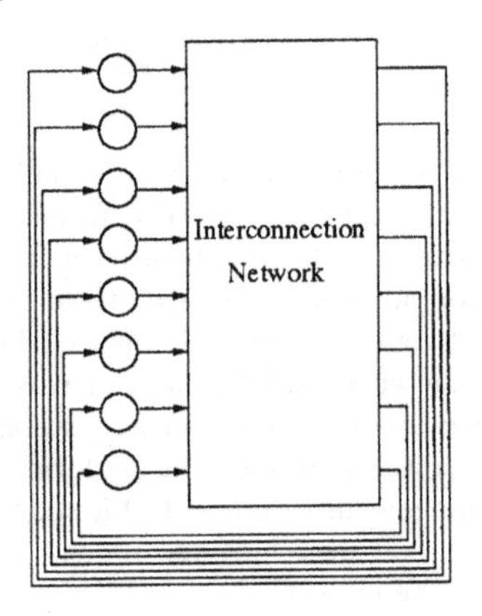

Figure 1 An 8 x 8 interconnection network.

An alternative to packet switching is called circuit switching. Circuit switching techniques do not process messages at intermediate nodes. Rather, they provide a direct connection or *path* between the source of a message and its destination. Since the entire path is constructed from optical components, a circuit switched network can therefore exploit more of the optical potential. If the network has enough resources to provide a dedicated circuit between each source and each destination, as in a cross-bar switch, network management is simplified and does not limit the communication performance. However cross-bar switches require expensive, complex hardware and are impractical for a large number of processors. A less expensive circuit switched network can be built by sharing network resources. However, the network management techniques that implement sharing may have an impact on communication performance. A simplified diagram of a network that interconnects eight processors is shown in Figure 1. At any point in time, only a subset of all possible connections are actually provided by the network. The network management technique controls the sharing to ensure that all communication requests are satisfied. The shared components of an optical network might include switches, registers, optical wavelengths, and optical links. Each path using a shared component may be assigned a different optical wavelength, as in *wavelength division multiplexing* (WDM) [1], or a different time slice, as in *time division multiplexing* (TDM) [4].

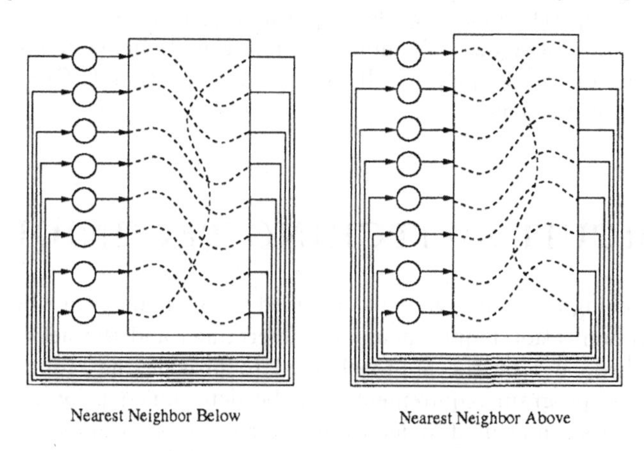

Figure 2 Network states for a circular communication pattern.

The set of paths provided by the network is determined by the state of the various network components. The collective state of these components is called the *network state*. For our purposes, a network state can be represented by the set of connections it provides. Figure 2 shows how the network must interconnect the processors to provide a circular communication pattern. Many algorithms logically view processors as being connected in a ring, where processors communicate with their nearest neighbors around the ring. The figure shows the two network states that provide these connections. When a network cannot provide all the connections required for the execution of a program in a single state, the state will need to change while the program is executing. The network management activities that change the network state may introduce delays that affect the performance of a parallel program.

In this chapter, we will consider a network which is not able to provide all the paths required by a program in a single network state. We will describe several circuit switching techniques. Our focus is on techniques that change the network state in response to communication requests from the executing program, and on the use of time division multiplexing. We discuss the nature of the delays a parallel program will encounter when communicating, and model the performance of the program as the sum of these delays. We refer to the communications impact on performance as the cost of network management. We show how the costs are related to the network characteristics, and how these characteristics suggest the choice of the appropriate management technique.

The rest of the chapter is organized as follows. In Section 2 we describe several circuit switching techniques. We develop a model of the cost of network management for an arbitrary set of communication requests in Section 3, and refine the model to describe looping parallel programs in Section 4. Our conclusions are contained in Section 5.

2 CIRCUIT SWITCHING TECHNIQUES.

Circuit switching techniques may be classified by the manner in which the program's communication requirements are determined and the network's states are established, and by the approach used to provide the required network states. When program requirements can be determined prior to execution, *static* techniques can be used to determine the required network states. Alternatively, *dynamic* techniques use information gathered during program execution to modify the current network state in order to satisfy program requirements. Both kinds of technique can be used with multiplexing or without multiplexing.

2.1 Static and dynamic techniques.

It is sometimes possible for a compiler to use sophisticated techniques to determine the communication requirements of a program for each phase of execution. This type of static technique is called *compiled communication* [2, 9]. Depending on the size of the network, it may not be possible to provide all the required paths in a single network state. Since this activity is performed at compile time, complex optimization algorithms can be used to determine the minimum number of network states that are required [12, 14]. During execution, the program directs the network to establish the required paths by loading the next network state from a predetermined sequence of states. To ensure that all processors have a consistent view of the network, all communication activity must cease and the processors must be synchronized while the state is changed.

Dynamic techniques use information gathered at program execution time to determine which paths the program needs. These techniques respond to individual requests generated by the application. This is a natural technique to use in a packet switched network, where establishment of one path is independent in time and space from establishment of any other path. In a circuit switched network, a dynamic technique alters the network state in an incremental fashion to provide the required paths. Each request is delayed while control activity occurs to insert the path into the network state. Communication along existing paths can continue while the new path is being added. Since resources are reserved for each path, existing connections are not affected by a request for a new path. Thus, dynamic techniques do not need to halt all communication to coordinate changes to the network state.

Static techniques have a potential performance advantage over dynamic techniques. When a static technique can accurately determine the required set of paths, it can minimize the number of times the network state must be altered. Further, the overhead of creating network states is incurred at compilation time and does not impact program execution. While the communication pattern is often set at time of compilation, at other times it is not [8]. This may occur when the communication pattern is determined by the data, as when indirect addressing is used. For example, it may be used explicitly (e.g. `C[i] = B[A[i]]`), or implicitly by the **`scatter`** and **`gather`** operations commonly used to manipulate sparse arrays. In these cases, the compiler may need to invoke a sequence of network states that provides all possible paths. This requires the maximum possible number of network states, is extremely wasteful of communication bandwidth, and can introduce large delays in communication.

2.2 Time Division Multiplexing

A non-multiplexed approach to circuit switching is to establish a network state which provides as many of the paths required by the program as possible, up to the maximum that the network can provide in a single state. When a new path is required, a new network state is established which includes the new path. Resources are dedicated for the duration of a given network state. Resources are shared over time by changing the network state. Resources can also be shared by using multiplexing techniques. In optical networking, wavelength division multiplexing can be used to provide multiple communication paths through each network component using different wavelengths of light. Time division multiplexing can be used to allocate resources for a short time interval, and to share the resources by automatically changing the allocation for subsequent time intervals. This requires additional hardware and new control techniques to provide a more complex network state called the *multiplexed network state*. These two forms of multiplexing can be used independently or together.

Time division multiplexing can be used in electronic networks as well as optical networks. However, it is especially attractive in optical networks for several reasons. First, optical networks are currently more expensive than electronic networks and additional resource sharing can help to amortize the costs. Second, an interconnection network built from optical components can provide communication bandwidth of several hundred gigabits per second [11]. This greatly exceeds the communication requirements of any single processor, so that an optical network has a capacity that is available to be shared. Third, network control operations are electronic and relatively slow compared to optical data transmission. Multiplexing increases the number of circuits that can be provided by the network without the need for a control operation, thus reducing delay and increasing performance.

Communication networks have long used time as a dimension over which to multiplex data. Wide area networks use statistical multiplexing where packets traveling different routes are temporally interleaved along the links of the network. Modem multiplexers use time multiplexing to combine signals from several low speed communication links for transmission over a high speed link. Time division multiplexing more closely resembles the time-slicing technique used in scheduling the execution of tasks on a processor. The network establishes a particular state for a small time interval, called a *time slot*. This state provides one set of interconnections required by the program. The length of the time slot is sufficient for one message or one packet of a message to be transmitted. The multiplexed state consists of a sequence of these sets of interconnections. The network automatically cycles through this sequence, providing a set of connections for one time slot before establishing the next state with another set of paths.

Referring back to Figure 2, a pattern of nearest neighbor communication around a ring could be provided by establishing each of the two network states in alternating time slots. Each processor may communicate with another processor during the time slot when the network provides the communication path between them. When all states in the sequence have been established, the sequence is repeated. Depending on the length of a time slot and the number of states in the sequence, the cycle can be short enough so that the path is available again when the processor has additional data to transmit. The number of states in the cycle is referred to as the *multiplexing degree*. The degree of multiplexing may be constant for the entire program execution, or it may change as the program requires and the network allows. These alternatives correspond to a *fixed* and a *variable* degree of multiplexing.

TDM requires the use of a global clock to synchronize the time slots for all the transmitters and receivers connected to the network. The actual data transmission does not require a global clock, as self-clocking can be used at the bit level [2]. The length of a time slot is chosen to allow for data transmission and for the components of the network to change state to provide the required set of paths. For example, a sequence of switch settings may be specified by the bits in a register, and the switch may cycle through the settings by rotating the bits. A time slot of 30 nanoseconds can be used to transmit four words of 64 bits at 10 gigabits per second and also allow enough time for the network components to change state. Time multiplexing may improve performance when the delays incurred by the use of a cycle of states are less than the delays incurred to create new network states and synchronize the processors while the new states are being established.

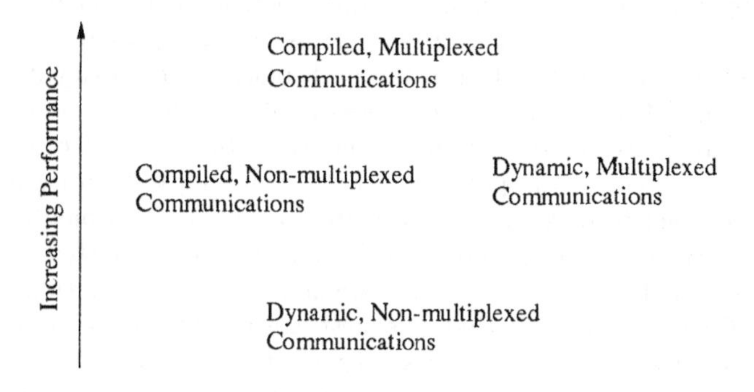

Figure 3 Expected relative performance of circuit switching techniques.

2.3 Using TDM with static and dynamic techniques.

The most straight-forward network management approach is to use dynamic state establishment without multiplexing. It requires neither multiplexing hardware nor sophisticated compilation techniques. In a circuit switched optical network, we can attempt to improve performance through the use of compiled communication or by the addition of multiplexing. We can also do both. We would expect that under normal circumstances, determined by network characteristics and program requirements, these techniques would improve performance. The expected relative performance of these techniques in circuit switched networks as shown in Figure 3.

The process of establishing a path in a circuit-switched network requires two steps. First, the required network components must be identified and a new network state prepared. Then the processors must be synchronized and the state loaded. The network topology determines the required network components, and the network's blocking characteristics and current state can be used to determine resources for which there is contention.

In a multiplexed network, contention could be resolved simply by increasing the multiplexing degree until, potentially, all possible paths were provided in a single multiplexed state. However, each increase in the degree of multiplexing increases the potential to delay a communication request along an established path. If we want to bound the maximum degree of multiplexing or if we do not multiplex, then we must resolve path conflicts due to the blocking characteristics of the network. Current dynamic control techniques may establish a connection separately for each message or use *explicit path release*, in which the blocked request is delayed until the path blocking it is explicitly released by the program. The effectiveness of these approaches depends on the number of paths used relative to the size of the network, the speed of control operations, and the ability of the programmer and compiler to identify paths that should be released to make network resources available. Optionally, a control protocol may be used to negotiate with the holder of the contested resource for the release of the blocking path. There are additional considerations with this approach. First, it is possible that a request can be blocked by more than one existing connection. Second, networks such as Clos and WDM star networks can provide more than one route between a source and destination. In these networks, it is possible that all alternate routes are blocked and a choice must be made about which blocking path to release.

Given the complexity of parallel programs, it seems desirable to automate the allocation and de-allocation of network resources as much as possible. One automated approach is to allow the network to detect when the resources provided to a path should be used to satisfy another communication request. The paths provided by a circuit switching network can therefore be managed similar to the way operating systems manage virtual and physical memory[3]. Analogous to page stealing in memory operations, the network could locate and remove paths that hold needed resources but are unlikely to be used. We call this *replacement* mode of operation. Multiplexing increases the network's capacity, therefore decreasing the number of times path replacement is needed. To implement replacement mode, a processor must be able to recognize that a path has been stolen so it can request re-establishment of the path when it is again required.

The decision to replace a path should be based on network and program characteristics such as the predictability of the communication pattern, the time to process requests, the duration of a time slot, the variability of message lengths, the frequency or recency of path use, and similar factors. For example, the existence of large, fixed length messages, frequently changing communication patterns, and a fast network management algorithm suggests the use of a long time slot in which to transmit an entire message, and that each message will require the establishment of a new path. In such a case, it would be reasonable to provide a path for a single time slot only, and to make any path that has been used eligible to be replaced. Alternatively, a long request processing time, highly variable message length, and stable or slowly changing communication patterns might suggest the use of small time slots for the transmission of packets and the retention of paths to allow for potential reuse. In this case we could allow paths to be held until they are *explicitly released* by the requester. Replacement mode could be used as an adjunct to these explicit path releases in order to optimize the use of network resources.

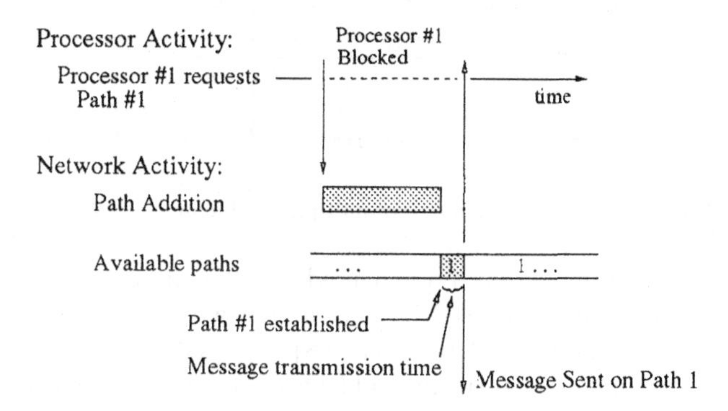

Figure 4 Operation of a dynamic algorithm.

Figure 4 shows how a path is added to a network state when network management is performed using a dynamic approach. At the top of the figure, processor 1 makes a request for communication path 1 and is blocked while the network establishes the path to satisfy the request. The bottom of the figure depicts the operation of the network. When the path request is received, network management determines how to insert the requested path into the network state. We assume preemption mode operation, but the time the request is blocked could just as easily represent the time to wait for or to negotiate for the release of a needed resource. Note that other communications can proceed in parallel with this activity on paths that are already established. When the new network state has been determined, the results are communicated back to the affected processors and the new state is loaded. The newly inserted paths are then available and the replaced paths are not. With path 1 established, the message is sent and control is returned to the program. If processor 2 requests path 2, this process is repeated and path 2 is added to the network state. If path 2 and path 1 cannot be provided simultaneously (i.e. due to network blocking characteristics), then processor 1 must be notified that path 1 is no longer available. Alternating requests for path 1 and path 2 will result in frequent changes to the network state and large communication delays. This situation is depicted in Figure 5.

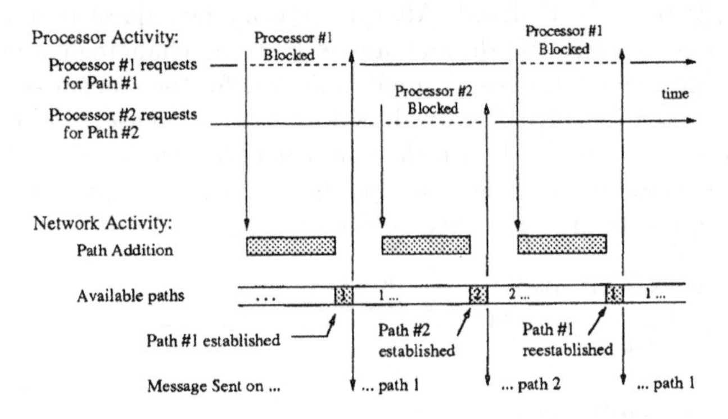

Figure 5 Operation of a non-multiplexed dynamic algorithm.

The dynamic algorithm can be implemented in different ways. For example, a centralized approach may be used in conjunction with a parallel control network (not shown in the figures) to accept requests and return the results. Alternatively, a distributed algorithm could be used where processors exchange requests and negotiate new network states using normal communication paths through the network. Some implementations may be able to process several requests at once. However, the basic steps in network management are the same regardless of the implementation.

In a non-multiplexed network, all paths are added following the same basic procedure. The number of paths that can be established at any instant of time is determined by the size of the network and its blocking characteristics. When a path is established, a message can be sent at any time. In a multiplexed network, paths are provided during discrete, synchronized time slots, as shown in Figure 6. The figure depicts a network with a multiplexing degree of two. At the top of the figure, we see two processors submitting requests for paths that cannot be provided simultaneously in a single network state. Path 1 has been added to the network state following the process shown in Figure 4. The

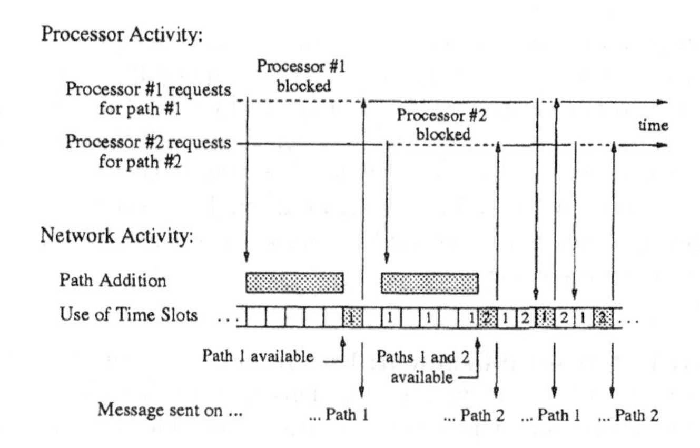

Figure 6 Operation of a multiplexed dynamic algorithm.

only difference from Figure 4 is that the path is established during only one of the two time slots. Subsequent requests for path 1 can be satisfied without network management activity. Note that while processors can send messages at any time, the physical transmission of messages must now be synchronized to the start of the time slot that provides the required path. There may be some delay between the time when a message is ready and when the path is available. However, this delay will be much shorter than the time needed to process a request and create a new network state. As in the non-multiplexed case, the first request for path 2 requires the same processing as the first request for path 1. The requesting processor is blocked while the path is added. When the request is processed, processor 2 is notified of the establishment of path 2 and the message is sent. Processor 1 may or may not be notified of the existence of path 2, depending on the network management algorithm. Subsequent requests for either path 1 or path 2 do not require network management activity. While these simple examples use only one path in each network state, the actual number of paths in a network state is determined by the blocking characteristics of the network.

3 COST OF COMMUNICATION IN TDM NETWORKS

The cost of communicating in parallel processing systems is a significant factor that affects the design of many parallel algorithms. In this section, we present a general approach for evaluating the costs of communication in a circuit switched network. This approach is used to quantify the circumstances under which communication cost can be reduced by using time division multiplexing. To model the impact of circuit switched network control we will only consider the network component of communication latency. The time for a message to be transmitted and to propagate through the network is unaffected by multiplexing. Thus, there are two key components of network latency which are affected by multiplexing.

- ▷ **Network state establishment:** Each request for a new connection must be inserted into the network state. This is done either by the compiler or dynamically during program execution. Loading the new state into the network components must be done during program execution. Multiplexing reduces the number of times a network state must be created and loaded.
- ▷ **Path access:** When TDM is used, a program will encounter a delay when it requires a path that is in the multiplexed state, but not in the state provided at the time of the request. The delay occurs while the network cycles through the sequence of states until it reaches the state providing the desired path.

Note that the size of the delay for establishing a network state is determined by electronic processing speeds, while the size of the path access delay is related to optical transmission rates.

We model the cost of processing a request and establishing a network state with the following two components.

- ▷ A fixed component, denoted by E_f. It includes the time to synchronize processors and load the new network state. With a static technique, this is the only establishment cost incurred at program execution time. A dynamic technique may also have a fixed component, such as the time between communication of the new state to the processors and the time when the state is loaded.
- ▷ A component proportional to some characteristic of the network or the communication pattern. For example, a dynamic algorithm might require a certain amount of processing for each path present in the network state, or for each request being handled. The constant that reflects proportional

processing time is E_p. The network management algorithm determines the network or program characteristic to which this is applied.

In a non-multiplexed circuit switched network, a path that has been established may be used by the program without any further delay. However, with TDM a program may encounter an access delay for a path. While the length of the delay will depend upon the timing of the request relative to timing of the multiplexing cycle, on average the delay is proportional to the multiplexing degree. We denote the proportionality constant as A_m, the cost to access a path in a multiplexed network.

While time division multiplexing introduces a cost to access a path, it also offers the potential to reduce establishment costs. With TDM, we set up a large multiplexed state and incur the overhead to cycle between the states. If we can contain all paths required by the program in a single multiplexed state, the establishment cost will be incurred only once. However, if we provide infrequently used paths in the multiplexed state, communication requests may be delayed while the network provides connections that are not needed during a particular phase of program execution. In a non-multiplexed network we incur only the overhead to set up a single network state, but state establishment will be performed much more often. In the rest of this section we develop a framework for evaluating this cost tradeoff for an arbitrary sequence of requests. In the following section, we study the effect of multiplexing on the communication cost of looping programs. While the form of the solutions for looping will apply only to specific communication patterns, the insight into the cost tradeoffs in a TDM network should be generally applicable.

In the discussion that follows, we will assume that the network can provide a fixed number of any arbitrarily chosen paths simultaneously. The number of paths will be denoted by m. For example, networks such as the Benes, Clos, cross-bar, and WDM star can provide connections for any arbitrary permutation [5–7]. We will represent the multiplexing degree of the network as K. Since the network can provide a maximum of m paths in one state, the multiplexed state can contain up to Km paths. Note that $K = 1$ corresponds to a non-multiplexed network.

We represent the communication requests of a parallel program as a sequence of requests for paths, which we call $\mathcal{T}$. We assume messages are of fixed size and can be completely transmitted in a single time slot in a multiplexed network. We assume the parallelism in the program is such that the network control handles a group of m requests simultaneously. Each group of m requests must be satisfied before the subsequent group of m requests arrives. In this way, the model corresponds to the use of blocking sends and non-blocking receives. We make the pessimistic assumption that requests for new paths must be processed before any of the m messages can be sent. Thus, the total communication delay encountered by a program will be the sum of the delays encountered by all groups of communication requests.

When a static technique can be used to establish network states, $\mathcal{T}$ can be determined in advance. We can create network states by partitioning $\mathcal{T}$ into the minimum number of sub-sequences that use exactly m different paths. We call this *m-path partitioning*. The ith set of m distinct paths forms the ith network state, and the sequence of states for $i = 1, 2, \ldots$ provides paths that satisfy the program's sequence of communication requests. Given a knowledge of the network, the optimal multiplexing degree can also be determined. A more detailed explanation of the creation of network states and costs for static techniques is found in [13].

When a dynamic approach is used, m path partitioning cannot be applied because the network does not know the sequence of requests in advance. It must construct each new network state from the current state by inserting any new paths required by the group of m requests being processed. We call this *m-request partitioning*. The states created from m-request partitioning will in general be different from the states created by m-path partitioning, since the two methods handle multiple requests for the same path in different ways. We will let N_i be the network state that exists after processing the ith group of m requests. With this approach, successive network states may be identical ($N_i = N_{i-1}$) if they contain requests for the same set of paths. Let s be the total number of states. Since there are $|\mathcal{T}|$ requests, $s = \frac{|\mathcal{T}|}{m}$.

The choice of multiplexing degree is particularly difficult in a dynamically controlled network, since the network only sees a few requests at a time. The multiplexing degree may be set using the following approaches.

- ▷ The network may provide only a fixed multiplexing degree which is predetermined by the network hardware.
- ▷ The network may provide a fixed multiplexing degree which can be selected for a program at the start of its execution.
- ▷ The network may provide a variable multiplexing degree which can be changed arbitrarily during program execution.

Each approach suggests a different amount of complexity in the network and its control algorithms. We will assume that the network provides a variable multiplexing degree, so that it can adapt to changes in the program's communication requirements.

We will investigate the costs of dynamic network management under the assumption that the time to insert new paths into a network state is proportional to the number of paths that must be inserted. Intuitively this represents the minimum amount of work that must be done. We also assume that there is a fixed delay to coordinate the processors and load the network state. The number of paths that must be added to satisfy the ith set of m requests is $|N_i - N_{i-1}|$, where $N_0 = \emptyset$. Since each network state provides m paths, this number is no greater than m.

To determine the cost of providing s states in a non-multiplexed network we need to know how many times changes to the network state are required. For each network state N_i, define $C_i = 1$ whenever $1 \leq |N_i - N_{i-1}|$ and $C_i = 0$ otherwise. Thus, $\sum_{i=1}^{s} C_i$ is the number of times the network state must be changed to provide the paths required by $\mathcal{T}$. Further, $\sum_{i=1}^{s} |N_i - N_{i-1}|$ is the total number of path insertions that must be processed to create these network states. The costs of a dynamic, non-multiplexed approach for an arbitrary sequence of requests is:

$$\text{Cost}_{\text{ non-multiplexed}}(\mathcal{T}) = E_f \sum_{i=1}^{s} C_i + E_p \sum_{i=1}^{s} |N_i - N_{i-1}| \tag{10.1}$$

In a multiplexed network, the multiplexed state provides Km paths. We construct a state containing more than m paths by multiplexing u consecutive network states. We assume that u divides s so that we create $\frac{s}{u}$ multiplexed states, denoted N'_j for $j = 1, \ldots, \frac{s}{u}$. Each multiplexed state provides the paths required by um consecutive requests. We call this sub-sequence of requests T_j. The paths provided by the multiplexed state N'_j are the paths in $\cup_{i=u(j-1)+1}^{ju} N_i$. The degree of multiplexing, K_j, required by this multiplexed state is $K_j = \lceil \frac{|N'_j|}{m} \rceil$. Since the number of paths in N'_{j-1} can be different from the number of paths in N'_j, the multiplexing degree of these two states can be different. It is also possible that $N'_{j-1} = N'_j$, and that no changes are required to the multiplexed state.

We consider the cost of establishing a new multiplexed state in two parts. Similar to the non-multiplexed case, it seems reasonable that the cost of determining multiplexed network state j is proportional to the number of paths that must be inserted, $|N'_j - N'_{j-1}|$. However, the actual cost may be more than this because the network does not receive all um requests at once. A network management algorithm using replacement mode might insert the new paths $(N'_j - N'_{j-1})$ into the multiplexed state by first removing paths common to both states $(N'_j \cap N'_{j-1})$. In the worst case, a multiplexed network management algorithm will do no better (and no worse) than the non-multiplexed algorithm, and costs could be proportional to $|N'_j|$. We will model the cost of determining a multiplexed state as proportional to the number of new paths and recognize that this may underestimate the actual cost. Where we earlier used a pessimistic assumption about parallelism in request processing, here we use an optimistic assumption about the effectiveness of the network management algorithm.

Although the multiplexed network handles only m requests at a time, it may require fewer changes to the network state than a non-multiplexed network. Instead of providing m paths in the state, it provides Km paths which are a superset of the m paths provided in the non-multiplexed network. A group of requests that require the insertion of a new path into a network state of m paths may not require the insertion of a new path into a multiplexed state of Km paths.

Let C'_j be the number of groups of m requests that require a new path to be added to create the multiplexed state N'_j. The cost to establish the multiplexed state is the cost of C'_j changes to the network state plus the cost of processing the paths to be added. Thus, the cost to establish the multiplexed state is $E_f C'_j + E_p|N'_j - N'_{j-1}|$. From the above discussion and the fact that N'_j is formed from u batches of requests, it follows that the number of changes to the multiplexed and non-multiplexed states are related by the following inequality:

$$0 \leq C'_j \leq \sum_{i=u(j-1)+1}^{ju} C_i \leq u \tag{10.2}$$

Also from the above discussion and as shown in detail in [10], the total number of paths inserted into the multiplexed state cannot be greater than the total number of paths inserted into the non-multiplexed states. This means that

$$0 \leq |N'_j - N'_{j-1}| \leq \sum_{i=u(j-1)+1}^{ju} |N_i - N_{i-1}| \leq um \tag{10.3}$$

When multiplexing is used, a communication request will be delayed on average for one half of the length of a multiplexing cycle. Thus, the proportionality constant A_m is just the length of a time slot and the delay is $A_m \frac{K_j}{2}$. A network state N'_j provides paths for $um = |T_j|$ requests. Since m requests are processed in parallel, this path access delay will be encountered $\frac{|T_j|}{m} = u$ times. Thus, the total path access delay for the requests satisfied by network state N'_j is $A_m u \frac{K_j}{2}$.

The total establishment and access cost for the paths in state N'_j is:

$$E_f C'_j + E_p |N'_j - N'_{j-1}| + A_m u \frac{K_j}{2}$$

Summing the costs for all $\frac{s}{u}$ multiplexed states, the total cost of communication for the program is:

$$\text{Cost}_{\text{multiplexed}}(\mathcal{T}) = E_f \sum_{j=1}^{s/u} C'_j + E_p \sum_{j=1}^{s/u} |N'_j - N'_{j-1}| + A_m u \sum_{j=1}^{s/u} \frac{K_j}{2} \qquad (10.4)$$

It follows from inequalities 10.2 and 10.3 that, with dynamic control, the establishment costs in a multiplexed network can not exceed the establishment costs in a non-multiplexed network. However, the introduction of path access costs with multiplexing creates a performance tradeoff that must be evaluated in order to determine overall the value of multiplexing. We cannot evaluate this tradeoff further without knowing more about the structure of $\mathcal{T}$. Since parallel programs often use looping structures to perform computations and communicate results, we will use a loop as the basis for further analysis. We will consider only loops which have distinct paths, meaning that each path is used only once in each iteration of the loop.

4 COMMUNICATION COST IN LOOPING PROGRAMS.

We use the following notation to describe the key communication characteristics of a loop in a parallel program. The loop has r iterations, and the sequence of requests issued in each iteration of the loop is L. We refer to the entire sequence of communication requests from the loop as $\mathcal{L}$. The sequence is assumed to be the same in each loop iteration. The number of different paths used in the loop is l. Since the loop has distinct paths, every path in L is different and $l = |L|$. The network provides m paths simultaneously, where $m < l$.

In the non-multiplexed network where $m < l < 2m$, the first group of requests requires m paths to be inserted into the network state. Every subsequent group of m requests requires $l - m$ paths to be inserted into the network state so that $|N_i - N_{i-1}| = l - m$ for $i > 1$. The total number of path insertions is $(s-1)(l-m) + m$. We will approximate this as $s(l-m)$, which should introduce only a small error for loops with a large number of iterations. When $2m \leq l$, m paths must be inserted into every new state. In both cases a new state must be established for each group of m requests, and $C_i = 1$ for all i. Equation 10.1 can now be expressed as:

$$\text{Cost}_{\text{ non-multiplexed}}(\mathcal{L}) = sE_f + \begin{cases} E_p sm & \text{for } 2m \leq l \\ E_p s(l-m) & \text{for } m < l < 2m \end{cases} \tag{10.5}$$

In the multiplexed network, we again group u adjacent states. When $l \leq um$, all l paths will be added to build the initial multiplexed network state, and no further changes to the multiplexed state are needed. Thus $|N'_j - N'_{j-1}| = 0$ and $C'_j = 0$ for all $j > 1$. The multiplexing degree K is equal to $\lceil \frac{l}{m} \rceil$. After the initial state has been established, the only recurring cost is the path access cost. Note that this model describes the cost incurred by a preemptive replacement policy. Such a policy will not make changes to the multiplexed network state once it has been established. Depending on the implementation, an explicit release policy may force changes to the network state and hence incur a greater cost.

When $um < l$, we again have $\frac{s}{u}$ multiplexed states. Since the paths are distinct, each multiplexed state contains exactly um paths. Thus the multiplexing degree for all states must be u, so that $K = K_j = u$ for all j. To determine the cost of establishing these multiplexed states, we must consider two cases.

When $u < \frac{1}{2}\frac{l}{m}$, the loop uses a very large number of paths and the paths in adjacent multiplexed states will be completely different. Hence $|N'_j - N'_{j-1}| = um$. Each group of m requests will require the insertion of m paths into the network state. Since there are u such groups in the multiplexed state, $C'_j = u$ for all j. Since every request is for a path not present in the previous multiplexed state, every request requires insertion processing and the path replacement strategy will not affect the performance of the network.

When $\frac{1}{2}\frac{l}{m} \leq u < \frac{l}{m}$, adjacent multiplexed states will have some paths in common. To build N_1', um paths must be inserted into the multiplexed state in groups of m. Thus, $C_1' = u$. The succeeding states require the insertion of only $l-um$ paths, so that $|N_j'-N_{j-1}'| = l-um$ for $j > 1$. The minimum number of groups of requests that cause a path to be added to the multiplexed state is $C_j' = \lceil \frac{l-um}{m} \rceil$. For clarity we will assume m divides l and $l-um$ evenly, and we will again use the approximation that the cost of the first state is the same as the cost of the remaining states. In this case, the replacement algorithm has a large effect on network performance. For example, an algorithm that replaces the least recently used paths would perform poorly, as the paths that are used least recently are most likely to be used again soon.

We can combine these observations with the cost given by Equation 10.4 to determine the costs of the multiplexed network for this loop, Cost $_{\text{multiplexed}}(\mathcal{L})$:

$$\begin{cases} sE_f + smE_p + A_m s\frac{K}{2} & \text{for } 1 < K \leq \frac{1}{2}\frac{l}{m} \\ E_f[\frac{s}{K}(\frac{l-Km}{m})] + E_p m[\frac{s}{K}(\frac{l-Km}{m})] + A_m s\frac{K}{2} & \text{for } \frac{1}{2}\frac{l}{m} < K < \frac{l}{m} \\ E_f\lceil \frac{l}{m} \rceil + E_p l + A_m s\frac{K}{2} & \text{for } \frac{l}{m} \leq K \end{cases} \tag{10.6}$$

To simplify the cost equation further, we define the factor β which combines the key network characteristics that affect communication cost.

$$\beta = \frac{E_f + mE_p}{A_m/2}$$

This ratio reflects the cost of establishing a network state relative to the cost of using paths that are multiplexed. Larger values of β correspond to networks that have a relatively high cost to establish network states, or time slots with a relatively short duration. We would expect multiplexing to be more effective in networks with a large β value. Using β and the facts that $s = \frac{|\mathcal{T}|}{m}$ and $|\mathcal{T}| = rl$, we can restate the cost equations as follows:

$$\text{Cost}_{\text{multiplexed}}(\mathcal{L}) = \begin{cases} \frac{|\mathcal{T}|A_m}{2m}(\beta + K) & \text{for } 1 < K \leq \frac{1}{2}\frac{l}{m} \\ \frac{|\mathcal{T}|A_m}{2m}(\frac{\beta(l-Km)}{Km} + K) & \text{for } \frac{1}{2}\frac{l}{m} < K < \frac{l}{m} \\ \frac{|\mathcal{T}|A_m}{2m}(\frac{\beta}{r} + K) & \text{for } \frac{l}{m} \leq K \end{cases} \tag{10.7}$$

The non-multiplexed network cost can also be stated in terms of β.

$$\text{Cost}_{\text{non-multiplexed}}(\mathcal{L}) = \frac{|\mathcal{T}|A_m}{2m}\beta \quad \text{for } K = 1 \tag{10.8}$$

We can combine Equations 10.7 and 10.8 to show when multiplexing will reduce cost for networks with different values of β. The graph in Figure 7 shows the cost per communication request in units of $A_m/2m$. To see how the cost changes in each of the three regions of Equation 10.7, we arbitrarily fixed $\frac{l}{m} = 8$ and $r = 10$.

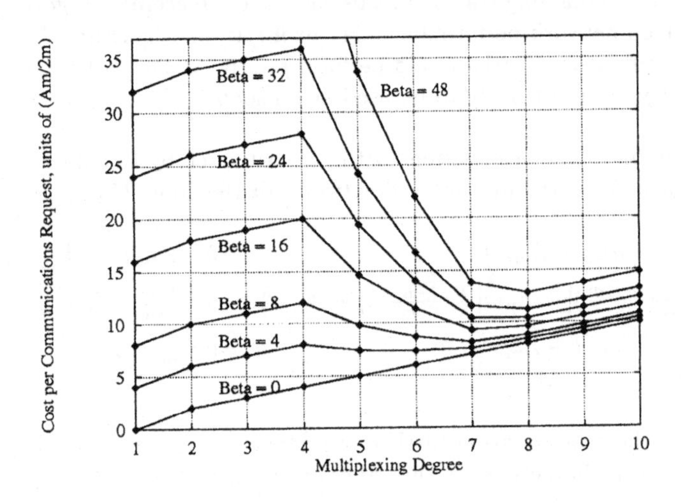

Figure 7 Predicted costs in a dynamic multiplexed network.

The graph of Figure 7 can be divided into three regions. On the left where $K \leq \frac{1}{2}\frac{l}{m}$, the degree of multiplexing is small. The replacement algorithm is not a significant factor in network performance because established paths are not being reused. In this region, multiplexing is unlikely to pay off. In the center section of the graph, adjacent multiplexed states do have paths in common, and established paths are being reused. For networks with large β, cost rapidly decreases with increased multiplexing degree. An effective replacement algorithm is crucial to obtaining the cost reductions shown. On the right side of the graph where $K \geq \frac{l}{m}$, all paths required by the loop are provided in a single multiplexed state. If β is large, this degree of multiplexing provides the lowest cost obtainable in a multiplexed network, even with the relatively small number of ten iterations. Increasing K further adds a degree of multiplexing beyond that required by the program, and cost increases as a result of the increasing delay to access paths. The replacement algorithm again is not significant, this time because path replacement is not needed.

We see that it is desirable to pick the correct degree of multiplexing for the program, and that the penalty for choosing too small a value of K is greater than the penalty for choosing too large a value. We also see that multiplexing reduces cost only when K is close to $\frac{l}{m}$ and β is large. These conclusions have been confirmed by network simulations.

Application to dynamic multiplexing control.

To further clarify when multiplexing should be used, we consider the case which generally provides the best performance in a multiplexed network. This is when $K = \frac{l}{m}$ and the entire loop is contained in a single multiplexed state. From Equations 10.7 and 10.8, we find that this reduces cost when:

$$\beta > \frac{\frac{l}{m}}{1 - \frac{1}{r}} \tag{10.9}$$

For a loop with a large number of iterations, this reduces to the requirement that $\beta > K$. Just as β is determined solely by network characteristics, the right hand side of inequality 10.9 is determined solely by the communication requirements of the loop. Thus, from the key characteristics of the loop and the network over which it will run, we can determine whether multiplexing will be effective in reducing cost. Simply stated, if $\beta < \frac{l}{m}$ do not multiplex. If $\beta > \frac{l}{m}$, multiplex to degree $K = \frac{l}{m}$

The key question is how to determine l (and hence K) when receiving requests dynamically in groups of m. One way to do this in a loop with distinct paths would be to check for the reuse of the path that was requested first, counting the number of intervening requests. If appropriate to multiplex, the multiplexed state could be built during the second iteration of the loop. Alternatively, the network can be passed information from the program to help determine l. For example, the program could signal the network at the end of the first iteration of the loop.

An approach that does not require the loop to have distinct paths would be to continuously increase the multiplexing degree until the rate of requests for new paths falls below a threshold. The resulting degree of multiplexing could be compared to β to determine if multiplexing should continue.

These techniques could be enhanced to handle programs that consist of a series of loops. The decision to multiplex a loop and the choice of multiplexing degree can be different for each loop. In this case the network needs to know when one loop has ended and the next loop has begun. One way to obtain this information is to have the program notify the network prior to entering a loop. Notification could include the number of different destinations that each processor will communicate with in each iteration of the loop, allowing the network to estimate l.

5 CONCLUSIONS

Several different techniques can be used to control a circuit switched optical interconnection network for multiprocessors. We have shown how time division multiplexing and a dynamic path management scheme can be used together to implement this control. We presented a model of the cost of communication in such a network using a general technique that can be applied to static and dynamic networks, with and without multiplexing. We applied the model to dynamic control of looping parallel programs and described when multiplexing can reduce communication cost.

With dynamic control, we found that the communication cost with multiplexing can be divided into three regions. When the degree of multiplexing is low or the network performance characteristics favor frequent reconfiguration, there may be little or no benefit from multiplexing. For other networks, as the degree of multiplexing increases the effectiveness of the path management algorithms becomes a critical determinant of performance. As the multiplexing degree becomes large enough so that a single multiplexed state contains all paths used by a loop, communication cost in these networks is minimized. A comparison of network and program characteristics can be used to determine if multiplexing to this degree will reduce cost. While it may be possible for the network to determine these program characteristics dynamically, program directives may help to improve the effectiveness of network management.

Overall, we expect that a dynamic approach to time division multiplexing can be effective under the following conditions.

1. The multiplexed state holds most of the paths required by each loop in the program so that the rate of requests for new paths is low.
2. Message transmission occurs in parallel with program computation and network processing of requests for new paths.

3. The time to multiplex states is sufficiently small compared to the time to add a path to the network state.

We note the similarity between replacing paths in a network state and the paging process used to manage a processor's physical memory. The memory management concepts of *locality of reference* and *working set* can be applied to path management in a circuit switched network, as well. We looked at a single loop with distinct paths, and found that performance was best when the network provided the program's entire working set of paths. We expect that the ability to identify and manage the working set of paths used by a parallel program will have a strong impact on communications performance in a circuit switched network .

Acknowledgements

This work is supported in part by NSF awards ASC-9318185 and MIP-9633729 and by AFOSR award F49620-93-1-0023DEF to the University of Pittsburgh

REFERENCES

[1] C. A. Brackett. Dense wavelength division multiplexing networks: Principles and applications. *IEEE Journal on Selected Areas of Communications*, 8:948–964, Aug. 1990.

[2] F. Cappello and C. Germain. Toward high communication performance through compiled communications on a circuit switched interconnection network. *Proceedings of the First IEEE Symposium on High-Performance Computer Architecture*, pages 44–53, Jan. 1995.

3] D. M. Chiarulli, S. P. Levitan, R. G. Melhem, and C. Qiao. Locality based control algorithms for reconfigurable optical interconnection networks. *Applied Optics*, 33:1528–1537, Mar. 1994.

] I. Chlamtac and A. Ganz. Channel allocation protocols in frequency-time controlled high speed networks. *IEEE Transactions on Communications*, 36(4):430–440, 1988.

C. Clos. A study of non-blocking switching networks. *Bell Systems Technical Journal*, 15(1):406–424, 1953.

P. Dowd. Random access protocols for high-speed interprocessor communications based on an optical passive star topology. *IEEE Journal of Lightwave Technology*, 9(6):799–808, 1991.

[7] K. Hwang. *Advanced Computer Architecture.* McGraw-Hill, New York, NY, 1993.

[8] D. Lahaut and C. Germain. Static communications in parallel scientific programs. In *PARLE '94 Parallel Architecture and Languages.* IEEE, July 1994.

[9] J. Li and M. Chen. Compiling communication-efficient programs for massively parallel machines. *IEEE Transactions on Parallel and Distributed Systems*, 2(3):361–375, 1991.

[10] R. Melhem. Time-multiplexing optical interconnection networks; Why does it pay off? In *Proceedings of the 1995 ICPP Workshop on Challenges for Parallel Processing*, pages 30–35. CRC Press, Aug. 1995.

[11] P. Prucnal, I. Glesk, and J. Sokoloff. Demonstration of all-optical self-clocked demultiplexing of TDM data at 250Gb/s. In *Proceedings of the First International Workshop on Massively Parallel Processing Using Optical Interconnections*, pages 106–117. IEEE, Apr. 1994.

[12] C. Qiao and R. Melhem. Reconfiguration with time division multiplexing MINs for multiprocessor communications. *IEEE Transactions on Parallel and Distributed Systems*, 5(4):337–352, 1994.

[13] C. Salisbury and R. Melhem. Modeling communication costs in multiplexed optical switching networks. In *11th International Parallel Processing Symposium*, pages 71–79, Geneva, Switzerland, Apr. 1997. IEEE.

[14] X. Yuan, R. Melhem, and R. Gupta. Compiled communication for all-optical TDM networks. In *Supercomputing '96.* IEEE, Nov. 1996.

11

BOUNDS AND ANALYSIS TECHNIQUES FOR GREEDY HOT-POTATO ROUTING

Assaf Schuster

Computer Science Department, Technion, Haifa, Israel 32000

ABSTRACT

In this chapter we consider a type of packet routing known as hot-potato routing. In hot-potato routing there is no intermediate storage for the packets (messages) that are on their way to their destinations, which is an important feature for communication networks that are based on optical hardware and for which the messages are composed of beams of light. In particular we consider a "practical" mode of routing, known as greedy routing. In greedy routing, unless some local congestion forbids it, an intermediate network node always attempts to send packets towards their destinations. We present several algorithms and analysis methods that were recently suggested by the author and his colleagues for greedy routing, along with some negative results by means of a general lowerbound.

1 INTRODUCTION

In this chapter we consider packet routing, and in particular a routing mode known as *hot-potato* or *deflection* routing [1, 16–18, 24, 28, 32, 33]. The important characteristic of algorithms which assume this mode is that they use no buffer space for storing delayed packets. Each packet, unless it has already reached its destination, must leave the processor at the step following its arrival. Packets may reach a processor from all its neighbors and then are redirected, each one on a different outgoing link. This may cause some packets to be "deflected" away from their preferred direction. Such an unfortunate situation cannot happen in the "store-and-forward" routing mode, in which a packet can be stored at a processor until it can be transmitted to its preferred direction.

P. Berthomé and A. Ferreira (eds.), Optical Interconnections and Parallel Processing: Trends at the Interface, 283-354.

Much work has recently focused on exact analysis and design of hot potato routing algorithms. One reason is that variants of hot-potato routing are used by parallel machines such as the HEP multiprocessor [31], and by high-speed communication networks [24]. In particular, hot-potato routing is very important in fine-grain massively-parallel computers, such as the Caltech Mosaic C [30]. For such machines, even the inclusion of a small sized storage buffer at each processor causes a substantial increase in the cost of the machine. An important domain in which deflection-type routing is highly desirable is optical networks [1,16,32,33]. In such networks, storage must take the electronic form. Thus, packets that should be stored must be converted from (and back to) the optical form. In the current state of technology, this conversion is very slow compared to optical transmission and switching rates.

The other reason for the growing interest in hot-potato routing, which made this field the subject of several recent works is that the hot-potato paradigm seems by its nature a 'distributed dynamic' process, where packets move at each step according to some simple set of *local* rules. As it turned out, this process proved to be hard to analyze, especially if the algorithm is somewhat 'simple', (we shall see along the sequel a possible definition of what 'simple' should be).

The first hot-potato algorithm was proposed by Baran [4]. Borodin and Hopcroft suggested an algorithm for the hypercube [10]. Prager [29] showed that the Borodin-Hopcroft algorithm terminates in n steps on the 2^n-nodes hypercube for a special class of permutations. Hajek [18] presented a simple greedy algorithm for the same network that runs in $2k+n$ steps, where k is the number of packets in the system.

The algorithm by Hajek gives priority to packets that are *closer* to their destinations [18]. It is straightforward to see that using this basic rule, any greedy packet routing algorithm terminates in $O(k \cdot diam)$ where k is the number of packets in the system and $diam$ is the network diameter. This implies, for example, $O(n^3)$ upper bound for such an algorithm on the two-dimensional $n \times n$ mesh, where each node is the origin of a single packet.

The work of Hajek was simplified and generalized in a work by Brassil and Cruz [12]. For any regular network with undirected edges (such as the mesh and the hypercube) the algorithm of Brassil and Cruz assumes some pre-specified order on the destinations, and packets are given priority according to the rank of their destination in that order. They show a bound of $diam + P + 2(k-1)$, where k is the number of packets, $diam$ is the network diameter, and P is the length of a walk connecting all destinations.

The goal in these earlier works [12, 18], which was later pursued in [8, 11, 14], was to present a "simple" algorithm for hypercubes and meshes which routes k packets with any combination of origins and destinations, in $d_{max} + 2(k-1)$ steps, where d_{max} is the maximal source-to-destination distance. The general idea (see Section 4) is that if there is no livelock, then a "chain of deflections" cannot be too long, say, not longer than the number of packets in the system.

Several works considered permutation routing. Feige and Raghavan [15] presented an algorithm for the torus that routes most permutations in $2n+O(\log n)$ steps. An optimal deterministic result was given by Newman and Schuster [26], who presented an algorithm that is based on sorting for permutation routing on the mesh. Their algorithm routes every permutation in $7n+o(n)$ steps. The reduction to sorting was later improved by Kaufmann *et al.* [21], so that the leading term was reduced to 3.5.

Bar-Noy *et al.* [3] gave a simple $O(n\sqrt{m})$ step permutation routing algorithm, where m is the maximum number of packets destined to a single column. Their result holds for any routing problem (not just partial permutations) on the mesh or torus with at most m packets destined to any column. Since at most n packets are destined for any column in a partial permutation, their algorithm runs in $O(n^{3/2})$ steps for any partial permutation. Kaklamanis *et al.* [20] presented an algorithm that routes most batch problems within $\frac{1}{2}dn + O(\log^2 n)$ steps on the d-dimensional torus, and within $2n + O(\log^2 n)$ steps on the mesh. Meyer auf der Heide and Scheideler [25] gave algorithms for all vertex-symmetric networks, terminating in $O(\log n \cdot diam \log^{1+\epsilon} diam)$ if the diameter of the network $diam = \Omega(\log n)$.

Despite the above collection of results, the problem of efficient hot-potato routing is still open. The main reason is that the routing decisions should be fast since all incoming packets are sent out by the next step. The decision speed is determined by the simplicity of the routing algorithm's exit assignment. Many of the algorithms mentioned earlier make complex routing decisions and would never be implemented in practice. For example, sorting packets based on their destinations is costly, both in time and in area on a routing chip. A second reason is that in order to support practical situations, *continuous* or *dynamic routing* (in contrast to batch routing) is desirable, where a packet may be injected into the network at any time during the algorithm. Thus the routing decisions should be independent of the time elapsed, making sorting-based algorithms undesirable. A third reason is that routing should be adaptive to the overall load. When the load is high, packets are not expected to arrive too fast, hence they may afford long detours from the shortest paths to their destinations in order to avoid congested points. When the traffic is low, packets may head directly to their destinations with no interruptions. Unfortunately, most algorithms in the literature lack this greedy-like behavior, and so they send packets on long paths even though they do not need to do so.

The above discussion suggests that practical algorithms should be "simple" and "adaptive." Several recent works tried to address these notions. Ben-Dor *et al.* [9] formally defined a notion of greedy algorithms for hot-potato routing, giving a potential function analysis for a large set of such algorithms. Feige [14] gave a collection of insights on the nature of greedy hot-potato routing algorithms, resulting in several bounds. Borodin *et al.* [11] gave algorithms for many-to-one routing that appear to be "simpler" and better suited for practical implementations. They achieve bounds that are comparable to the original works of [12, 18], using refinements of the "chain of deflections" idea.

Other recent results by Ben-Aroya *et al.* [7, 8] consider routing to a single target, and achieve optimal time (in [7] up to some additive term, see Sections 5 and 6). In contrast to the "chain of deflections" argument, these results utilize the target bandwidth (number of in-links). Such results give hope in eventually achieving the bound of $O(d_{max} + W)$ for general type routing requests, where W is the *network bandwidth lower bound* (defined as the maximum, over all node subsets S, of the number of packets destined to S divided by the number of links leading to S from nodes not in S). In fact, such a result already exists in the store-and-forward domain [23], using network flow techniques. Although the related methodology might tolerate a "technology transfer" into the more dynamic setting of the hot-potato domain (see also Newman and Schuster [27], for a general derivation of hot-potato routing algorithms using bounded-buffers packet routing algorithms), it does not exhibit other desirable features we seek, since the routing is composed of phases.

There have been several attempts to give a formal or semi-formal definition of the concepts "simple" and "adaptive" with respect to routing algorithms. Some of these features, such as independence of the time elapsed, locality, and determinism, are desirable in store-and-forward routing as well. We explicitly mention two other features that, as far as we know, are unique to the hot-potato literature.

- ▷ Hajek [18], and subsequently Feige and Raghavan [15], suggested the *one-pass* property of algorithms: The packets entering the node are considered one at a time in a pre-specified order on the incoming links, and each packet, in its turn, is assigned to an outgoing link that is not already taken by the previously considered packets. The involved hardware scans the incoming packets and assigns outgoing links to them in a single pass. The one-pass property was generalized by Borodin *et al.* [11] to *k-pass*, where k passes over the incoming links are allowed for routing decisions.

▷ Ben-Dor *et al.* [9, 19] defined *greedy* algorithms, where a packet always advances towards its target unless all its "good" outgoing links are already taken by advancing packets. The motivation is that a packet always tries to advance unless some global congestion forbids it, in which case we can afford to deflect it. This definition was later extended to the *strongly greedy* concept [8] or, equivalently, the *totally greedy* [11] or the *maximal advancing* concept [14], where not only is the algorithm greedy, but also the number of deflected packets is minimized for each node.

We focus here on presenting greedy routing algorithms and techniques that were suggested recently for their analysis. The underlying network topology is the 2-dimensional mesh, sometimes considering also higher dimensional meshes or the hypercube. The work is organized as follows. Section 2 describes the model and give precise definitions. Section 3 presents the potential function analysis method for the analysis of greedy routing algorithms. Section 4 gives the "chain of deflections" argument. Section 5 describes an algorithm for optimal routing to a single destination on the 2-dimensional mesh. Section 6 presents a randomized algorithm for routing to a single target on higher dimensional meshes. Finally, Section 7 gives a general lowerbound for permutation routing by algorithms that are "too simple" and "too greedy".

2 THE MODEL, TERMINOLOGY, DEFINITIONS AND PRELIMINARIES

We represent a network of processors by a graph in which each node corresponds to a processor and each link represents a communication link. The information sent on the communication link moves in the direction of the link. A network for which the communication links are bidirectional will be represented by an undirected graph, or by a graph in which for each link there exists an anti-parallel link. For each node there is a unique *id* and each node knows the *id* of its neighbors.

A *routing problem*, or, a *routing request*, is a set of ordered pairs of nodes. Each such pair corresponds to an origin node at which a packet originates at time

$t = 0$ and a destination node to which the packet is sent. Unless otherwise explicitly stated, we assume that no node can be the origin of more packets than its out-degree. We denote by k the size of the routing problem, i.e., the initial number of packets in the system.

In routing requests of the *general type* a node may be the destination of many packets. A routing problem is said to be of the *single-target* type if all packets are destined to the same node. A routing problem is a *(partial) permutation*, if each node is the destination of at most one packet.

The networks we consider are synchronous, which means that packets are sent in discrete time steps. The time it takes for a routing algorithm to solve a routing problem is the number of steps that elapse until the last packet reaches its destination.

In the *hot-potato* routing style there are no buffers, hence a packet cannot stay in a node (other than its destination node), and must leave every intermediate node in the step that follows its arrival. Each node in the network performs the following operations at every step:

- ▷ Receives the packets that were sent to it in the previous step on incoming links. This is not performed for $t = 1$.
- ▷ Performs a local calculation that depends on the headers of the packets stored at the node and on the identity of the links through which these packets entered the node (except for the first step). A packet that has reached its destination is considered delivered and removed from the network.
- ▷ According to the result of the calculation, the node sends all its packets on the outgoing links. At most one packet can be sent on each outgoing link.

2.1 The d-dimensional mesh

The network we are dealing with is the d-dimensional mesh which is a graph with a set of n^d nodes that correspond to all the vectors of dimension d over $\{1, \cdots, n\}$. There is an arc between the nodes $\bar{a} = (a_1, \ldots, a_d)$ and $\bar{b} = (b_1, \ldots, b_d)$ if and only if the L_1 distance between these two vectors is one. The diameter of the d-dimensional mesh is $d(n-1)$, and the degree of the nodes in the network is between $2d$ (for interior nodes) and d (for nodes in the

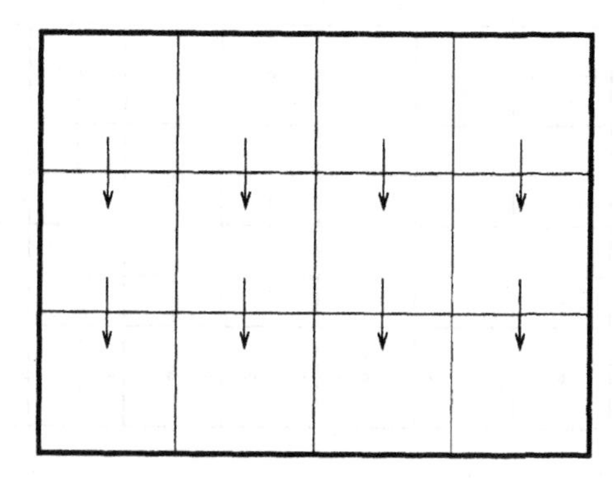

Figure 1 Direction $'-'$ in the second coordinate in a 2-dimensional mesh. The squares in the picture represent nodes, and their faces represent arcs.

corners of the mesh). Let $\bar{a} = (a_1, \ldots, a_d)$ and $\bar{b} = (b_1, \ldots, b_d)$ be two nodes in the d-dimensional mesh, then the distance between $\bar{a}$ and $\bar{b}$ (i.e. the length of the shortest path between them) is $\sum_{i=1}^{d} |a_i - b_i|$.

Let p be a packet in the mesh. The distance between p and its destination at time t, is the distance between the node that contains p in the beginning of step t and the destination node of p. We denote by d_{max} the maximal distance at the beginning of the algorithm (namely, as given in the routing request) from any source to any destination.

Since every arc in the mesh connects two nodes with id's that are different in exactly one location, we can divide the arcs into $2d$ distinct directions (e.g. arcs that increase the first coordinate, arcs that decrease the fourth coordinate, etc.) Direction $'+'$ in the i'th coordinate is the set of all arcs in the mesh of the form $(a_1, \ldots, a_{i-1},\ a_i,\ a_{i+1}, \ldots, a_d) \longrightarrow (a_1, \ldots, a_{i-1},\ a_i + 1,\ a_{i+1}, \ldots, a_d)$ Similarly, direction $'-'$ in the i'th coordinate is the set of all arcs in the mesh of the form $(a_1, \ldots, a_{i-1},\ a_i,\ a_{i+1}, \ldots, a_d) \longrightarrow (a_1, \ldots, a_{i-1},\ a_i - 1,\ a_{i+1}, \ldots, a_d)$ (See Figure 1).

We sometime say that an arc that goes out from some node is "going in direction X", if that arc belongs to direction X.

In particular, the $n \times n$ mesh, or the two-dimensional mesh, which we sometimes simply call a mesh, is a graph with n^2 nodes. Each node corresponds to the pair (row, col) where $1 \leq row, col \leq n$ (row is the row number, and col is the column number). We envisage the mesh in the plane so that the column

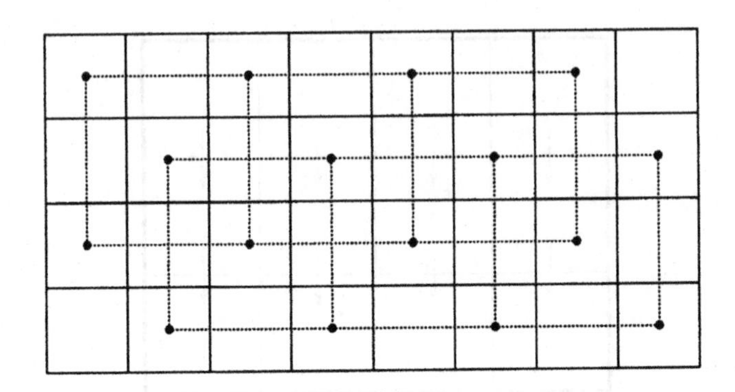

Figure 2 2-neighbors in the 2-dimensional mesh. Nodes that are connected by dashed lines are 2-neighbors

numbers increase from left to right and the row numbers increase from the bottom upwards. We also refer to outgoing links of each node in that manner: left, up, right, and down (or, alternatively, West, North, East, and South). Every link of the mesh connects two nodes that differ in either the row number or the column number. Thus, for the two-dimensional mesh we distinguish between the two dimensions: Nodes connected by dimension X links differ in their column numbers, whereas nodes connected by dimension Y links differ in their row numbers.

Definition 2.1 *We say that a node $\bar{b}$ is a* 2-neighbor *of the node $\bar{a}$ in the direction X, if there is a path of length 2 from $\bar{a}$ to $\bar{b}$ that contains only arcs in the direction X. We say that $\bar{b}$* is a 2-neighbor of $\bar{a}$ *if there is a direction X such that $\bar{b}$ is a 2-neighbor of $\bar{a}$ in the direction X (See Figure 2).*

For example, in a 2-dimensional mesh, the node $(1,2)$ is a 2-neighbor of the node $(3,2)$ (in the direction '$-$' in the first coordinate), but the node $(2,3)$ is not a 2-neighbor of the node $(3,2)$: Although there are paths of length 2 between these nodes, none of these paths contains two arcs in the same direction.

From the last definition, it follows that 2-neighbors is a symmetric relation. Therefore, the transitive closure of this relation is an equivalence relation. It is easy to see that it divides the mesh into 2^d equivalence classes, each of which is isomorphic to a d-dimensional mesh with $(\frac{1}{2}n)^d$ nodes (assuming n is even).

2.2 Greedy Hot-potato Routing Algorithms

Definition 2.2 *Let S be a node in the mesh, and let p be a packet in S. We say that an arc that goes out of S is a* good arc *for p, if it enters a node that is closer to p's destination. Similarly, we say that an arc that goes out of S is a* bad arc *for p, if it enters a node that is farther from p's destination.*

We say that a certain direction is a good direction *for p, if there is a good arc for p that goes out of S in that direction. Otherwise, we say that it is a* bad direction. *That is, a bad direction for p either contains a bad arc for p, or does not contain any arc that goes out of S (if S is a node on an edge of the mesh).*

For example, a packet p in the 5-dimensional mesh that is currently in node $(1,3,2,6,1)$, *and its destination is node* $(4,3,8,2,1)$ *has 3 good directions: '+' in the first coordinate, '+' in the third coordinate and '−' in the fourth coordinate. All the other directions are bad for p.*

We say that a packet p *advances* in step t if it gets closer to its destination in that step. Otherwise, we say that p is *being deflected.* Let p and q be two packets that are in the same node at the beginning of step t, we say that p is deflected by q (or q is deflecting p) in step t if (a) p is deflected in step t, and (b) q is advancing in that step via an arc that is good for p.

In this chapter we consider *greedy* hot-potato routing algorithms in which a packet always attempts to advance. We make this formal in the following definition.

Definition 2.3 ([9]) *A hot-potato routing algorithm is said to be* greedy *if whenever a packet p is being deflected, all its good arcs are used by other advancing packets.*

From the definition it follows that in a greedy algorithm, in order to deflect a packet that has i good directions, at least i other packets must advance from the same node.

We say that an algorithm is *strongly greedy* when, in addition to being greedy, it also maximizes the number of advancing packets from every node. We say that the algorithm is *weakly greedy* when, from every non-empty node, at least one packet advances.

2.3 Restricted Packets

Note that a greedy algorithm is not always guaranteed to terminate, as livelock may easily occur even when only four nodes are involved. One way to resolve the livelock problem is to give priority to certain packets, e.g. those that are "restricted" in a certain sense.

Definition 2.4 *We say that a packet in the 2-dimensional mesh is* restricted *if it has exactly one good direction.*

The technique that is presented in the next section shows that this approach indeed solves the livelock problem by means of a tight upperbound (up to constant factors) on the running time of greedy routing algorithms that give priority to restricted packets.

3 THE POTENTIAL FUNCTION ANALYSIS: PRIORITY TO RESTRICTED PACKETS

In this section we describe a method for routing general requests which uses potential function analysis to obtain upper bounds on the running time of greedy algorithms in meshes [9]. The set of algorithms for which the running time is bounded here is very large, paying by a relatively weak result.

3.1 Potential Function Analysis

We present here a general method for using potential function in the analysis of hot-potato routing algorithms. In Section 3.2, this method is used for obtaining upper bounds on the termination time of algorithms in the 2-dimensional mesh. This approach can also be generalized to the d-dimensional mesh [19].

Let $\mathcal{A}$ be a greedy routing algorithm, and suppose that we have a potential function $\phi_p(t)$ for every packet p, such that $0 \leq \phi_p(t) \leq M$ in every step t (where M is some positive constant), and that $\phi_p(t) = 0$ only if the packet p reached its destination by step t. Consider the "global" potential function $\Phi(t)$ that is defined by $\Phi(t) \stackrel{\text{def}}{=} \sum_p \phi_p(t)$. We give a general scheme which yields an upper bound on the running time of $\mathcal{A}$. This scheme uses some local property of the potential function, which we define next. In order to define this property, we first define the notion of a node that *loses potential.*

Definition 3.1 *A node S in the mesh* loses potential *in step t if the sum of potential of the packets that entered S at time t is greater than the sum of potential of the same packets at time $t+1$. We denote the difference between the two sums by $\Delta\Phi_S$.*

The local property of Φ that we need can be stated as follows:

Property 3.2 *Let S be a node in the d-dimensional mesh that contains ℓ packets in step t.*

- ▹ *If $\ell \leq d$ then S loses at least ℓ potential units at step t (that is $\Delta\Phi_S \geq \ell$).*
- ▹ *If $\ell > d$ then S loses at least $2d - \ell$ potential units at step t (that is $\Delta\Phi_S \geq 2d - \ell$).*

This property must be satisfied in every node of the mesh (even for nodes near the edge of the mesh) in every step.

We show now that the above property, together with a bound on ϕ_p, yields an upper bound on the running time of the algorithm $\mathcal{A}$. Later, we show some examples of routing algorithms and potential functions that satisfy this property. For the sake of analysis, we divide the nodes in the mesh into "good" nodes and "bad" nodes.

Definition 3.3 *We say that a node in the mesh is a* bad node *at a certain step if it contains more than d packets at the beginning of that step. Otherwise it is a* good node*. We use $B(t)$ to denote the total number of packets in bad nodes at time t, and $G(t)$ to denote the total number of packets in good nodes at time t.*

We can interpret Property 3.2 as follows: If S is a good node, it loses at least one potential unit for every packet in it. If S is a bad node, it loses at least one potential unit for every "missing" packet. As an immediate corollary we get

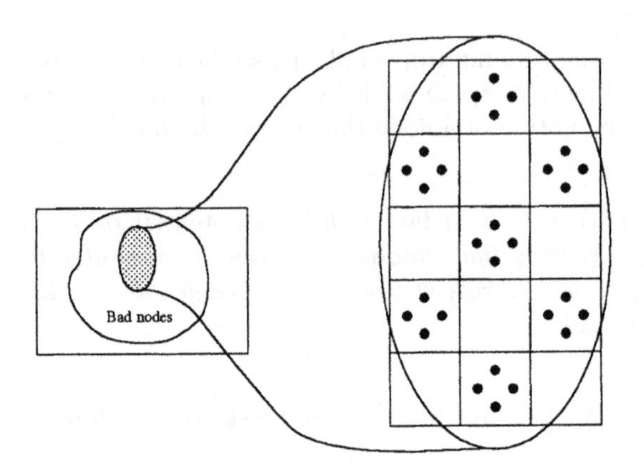

Figure 3 An area of bad nodes in the 2-dimensional mesh

Corollary 3.4 *For every step t,* $\Phi(t+1) \leq \Phi(t) - G(t)$.

In particular, from Corollary 3.4 it follows that the potential function is a monotonic non-increasing function. The main difficulty we face is that we do not have any a-priory lower bound on $G(t)$, as almost all the packets can be in bad nodes at any given time. In order to deal with situations where $G(t)$ is almost zero, we consider the d-dimensional "volume" of bad nodes (See Figure 3.), and prove that in every two successive steps, there is a loss of at least one potential unit for every arc on the surface of that volume.

Definition 3.5 *We say that an arc e, that goes out of a node S is a* surface arc *if*

- ▷ *The node S is a bad node.*
- ▷ *Either the 2-neighbor of S in the direction of e is a good node, or S does not have a 2-neighbor in this direction (i.e., if it is on an edge of the mesh).*

We also consider an arc that leads "out of the mesh" (in a bad node on the edge of the mesh) as a Surface arc(see Figure 4.). We denote the number of Surface arcs at time t by $F(t)$.

Lemma 3.6 *For every step t,* $\Phi(t+2) \leq \Phi(t) - F(t)$.

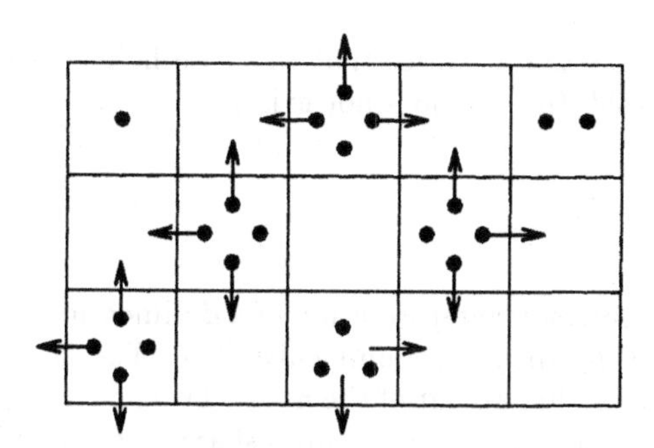

Figure 4 Surface arcs in the 2-dimensional mesh

Proof: We show that for every Surface arc we can account a loss of at least one potential unit either at step t or at step $t+1$. Let S be a bad node, and consider some Surface arc that goes out of S. Denote the node in the other side of the Surface arc (if exists) by C, and the node in the other side of C in the same direction (if exists) by N.

N is a good node	N
The arc from S to C is a Surface arc	C
S is a bad node	S

If no packet is leaving S towards C, then S contains less than $2d$ packets, and thus, by Property 3.2, S loses at least one potential unit for every "missing" packet at step t. One unit of that loss can be account to the arc from S to C. Notice that this case holds even if the node C is "out of the mesh" (i.e., C does not exist).

If there is a packet that leaves S towards C then we consider two cases:

1. A packet entered C from N at step t. Since N is a good node in step t, it loses one potential unit for every packet in it. Thus, we can account the loss of the packet that entered C to the arc from S to C.

2. No packet entered C from N at step t. In this case, C contains less than $2d$ packets. If C is a good node at time $t+1$, then it loses one potential unit for every packet in it. Thus, we can account the loss of potential caused by the packet that entered C from S, to the Surface arc from S to C.

   If C is a bad node then it loses one potential unit for every "missing" packet. Thus we can account the loss due to the "missing" packet from N

to the arc from S to C. Notice that this case holds even if the node N is "out of the mesh" (i.e., N does not exist).

∎

We make use of a geometric interpretation of a d-dimensional mesh: Each node is represented by a d-dimensional unit cube (1^d). The $2d$ faces of each cube correspond to the arcs that go out of the node. Two nodes in the d-dimensional mesh are adjacent if and only if their cubes share a common face.

Let e be a Surface arc that goes out of a bad node S, and consider the equivalence class of the node S under the transitive closure of the 2-neighbor relation. Since e is a Surface arc, the 2-neighbor of S (in the original mesh) in the direction of e is either a good node, or does not exists at all. Thus, the face that corresponds to the arc e in the equivalence class of S has a bad node in one side (S itself) and a good node (or no node at all) on the other side.

Therefore, if we consider the d-dimensional volume that is composed of the bad nodes in every equivalence class, then every face in the surface of that volume corresponds to a Surface arc in the mesh. In other words, the surface of the "bad volume" of a certain equivalence class is equal to the number of Surface arcs that go out of nodes in this class. Thus, in order to bound from below the number of Surface arcs (and thus the number of packets exiting the "bad volume"), we use an isoperimetric inequality on the surface. The inequality, which is due to Roy Meshulam (see the proof in [9, 19]), state that any d-dimensional volume V that is composed of d-dimensional unit cubes, has surface of size at least $2dV^{(d-1)/d}$.

Using Meshulam's isoperimetric inequality and the geometric interpretation of the mesh, we get that if there are many bad nodes, then there must also be many Surface arcs. Thus it can be shown that

Lemma 3.7 *If there are $B(t)$ packets in bad nodes at the beginning of step t, then the number of Surface arcs in that step is at least $(2d)^{1/d} \cdot B(t)^{(d-1)/d}$.*

Lemma 3.8 *If the potential at the beginning of step t is $\Phi(t)$, then in the following two steps the potential is decreased by at least $(2d)^{\frac{1}{d}} \left(\frac{\Phi(t)}{2M}\right)^{\frac{d-1}{d}}$ (where M is the a-priory bound on the potential of every packet).*

The proof essentially uses some algebra and convexity considerations in order to manipulate the equations $\Phi(t+2) \le \Phi(t+1) \le \Phi(t) - G(t)$ (Corollary 3.4) and $\Phi(t+2) \le \Phi(t) - (2d)^{1/d} \cdot B(t)^{(d-1)/d}$ (Lemma 3.6 and Lemma 3.7).

Theorem 3.9 *If $\mathcal{A}$ is a routing algorithm and Φ is a potential function which satisfies Property 3.2, then $\mathcal{A}$ solves every routing problem with k packets in the d-dimensional mesh within at most $(4d)^{1-\frac{1}{d}} \cdot k^{1/d} \cdot M$ steps.*

The proof uses the fact that the potential function satisfies the following conditions -

1. At the beginning of the algorithm, $\Phi(0) = \Phi_0 \le k \cdot M$, since there are k packets and each packet has at most M units of potential.
2. By Lemma 3.8, $\Phi(t+2) \le \Phi(t) - (2d)^{1/d} \cdot \left(\frac{\Phi(t)}{2M}\right)^{(d-1)/d}$.

The proof then goes by bounding the decrease of Φ in phases, then letting the length of the phase decrease to zero [9].

3.2 Potential Function in the 2-dimensional mesh

In this subsection, we show a potential function in the 2-dimensional mesh which satisfies Property 3.2, and use Theorem 3.9 to obtain an upper bound on the running time of a large class of greedy algorithms. To this end, we need to restrict somewhat the routing algorithm.

Priority to Restricted Packets

Recall that a packet in the 2-dimensional mesh is called restricted if it has exactly one good direction. We divide the restricted packets into two types (see Figure 5):

Type A: Packets that were restricted in the previous step, and advanced in it.

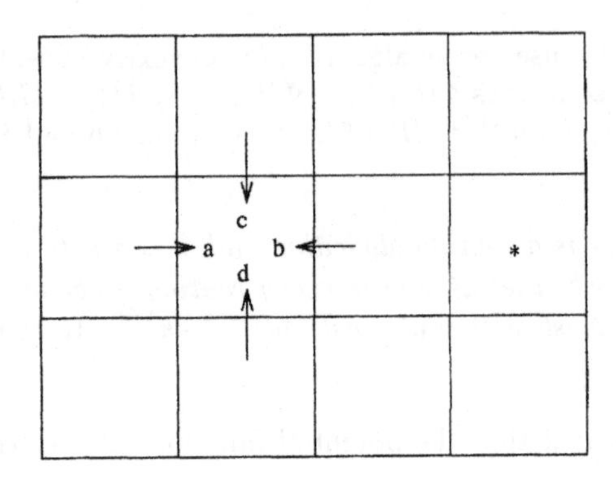

Figure 5 Restricted packets: b, c and d are of type B, and a is of type A

Type B: All the other packets, i.e., either packets that were deflected in the previous step, or packets that were not restricted in the previous step.

Definition 3.10 *We say that a greedy routing algorithm* gives priority to restricted packets *if a non-restricted packet can not deflect a restricted one. This implies that whenever a restricted packet is deflected, the packets that deflect it must also be restricted.*

For the definition of the potential function, we need the following properties of algorithms that give priority to restricted packets.

1. A restricted packet can deflect at most one type A packet at every step.
2. If a type A packet, q, is deflected by another restricted packet, p, then p must be of type B.

Definition of the Potential Function

We now show that a routing algorithm in the 2-dimensional mesh which gives priority to restricted packets routes any routing problem with k packets in at most $8\sqrt{2} \cdot n\sqrt{k}$ steps.

To this end, we define the potential function as follows: The potential of a packet p in step t, includes the distance from p to its destination, denoted by $dist_p(t)$, and some additional amount of potential.

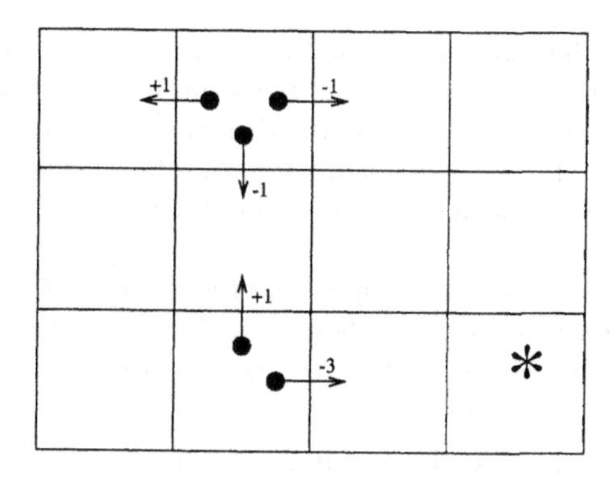

Figure 6 Changes in the potential of packets in one step

Initially, every packet p has $2n$ additional units of potential. As long as p is not of type A, its additional potential remains fixed. When p becomes a type A packet, it "drops" two units of its additional potential in every step.

If a type A packet is deflected by some other packet, then the two packets "switch" their additional potential, and the advancing packet drops 2 units of the potential it has just received from the deflected packet.

Formally, we denote the amount of additional potential of the packet p after step t by $C_p(t)$ (see Figure 6).

1. Initially, for every p, $C_p(0) = 2n$.
2. If after step t the packet p is not restricted, or if it is a restricted packet of type B, then $C_p(t) = 2n$.
3. If after step t the packet p is a restricted packet of type A, then there are two cases:
   (a) p did not deflect any restricted packet of type A in step t. In this case $C_p(t) = C_p(t-1) - 2$.
   (b) p deflected a restricted packet of type A. Denote this packet by q (there is exactly one such packet). In this case $C_p(t+1) = C_q(t) - 2$.
4. If p has reached its destination by step t then $C_p(t) = 0$.

From the properties of algorithms which give priority to restricted packets, it follows that in case 3.b, the packet p was a type B packet at the beginning of step t. Therefore, p had $2n$ units of additional potential. At the end of step t, the packet q had $2n$ units of additional potential (since it was deflected), and the packet p has $C_q(t-1)-2$ units of additional potential. Thus, the sum of additional potential of the packets p and q is $2n+C_q(t-1)-2$. This is exactly the same amount we would have got if q would have deflected p.

Thus, as far as the potential function is concerned, it does not matter whether p deflects q or vice-versa. Therefore, when we analyze the changes in the potential of a node, we can always assume that the type A packets are advancing.

The potential of a packet p at the beginning of step t is $\phi_p(t) \stackrel{\text{def}}{=} dist_p(t)+C_p(t)$. The potential of the mesh at the beginning of step t, is $\Phi(t) \stackrel{\text{def}}{=} \sum_p \phi_p(t)$. We proceed to show that the potential function Φ satisfies Property 3.2.

Lemma 3.11 *Consider an algorithm which gives priority to restricted packets, and let S be a node in the mesh. If S contains $\ell \leq 2$ packets in step t then it looses at least ℓ units of potential in that step, and if S contains $\ell > 2$ packets in step t then it looses at least $4-\ell$ units of potential in that step.*

The proof goes by a case analysis on whether S contains a restricted packet or not. Since the potential function satisfies Property 3.2, and $0 \leq \phi_p(t) \leq 4n$, we can use Theorem 3.9 with $d=2$ and $M=4n$ to obtain

Theorem 3.12 *Every greedy routing algorithm that give priority to restricted packets solves every routing problem with k packets within at most $8\sqrt{2}\cdot n\sqrt{k}$ steps.*

Remark: A hot-potato routing problem in the mesh can be split into two independent problems according to the parity of the packets origin. These problems do not interfere with one another. Therefore, if every node is the origin of a single packet (that is $k=n^2$), we can strengthen the result a little, getting a bound of $8n^2$. Also, when every node is the origin of 4 packets, we get a bound of $16n^2$, which is only 8 times the lower bound.

4 PRIORITY TO INERTIAL PACKETS

The method that is described in this section, sometimes called "chain of deflections", achieve improved performance for general requests on a subset of the algorithms considered in the previous section. Namely, we consider algorithms which give priority to restricted packets, but also prefer packets which already advanced towards their destinations in the previous step using the same direction [8]. The idea in the analysis is to show that each deflection in the network can be accounted for one of the other packets in the system, and in this way to bound the maximal number of deflections that a single packet experiences. Variants of this argument appear in [8, 11, 14]. We start by presenting the algorithm.

Algorithm General_Request. *The algorithm is described in terms of the behavior of packets.*

Priority to restricted packets: *Packets that are restricted (i.e., they have only one good direction) always get priority. (Among several restricted packets and a certain direction, arbitrarily one of them advance.)*

Priority to inertial packets: *Packets that are already moving horizontally in rows towards their target columns (i.e., they made such a move in the previous step), or are already moving vertically towards their rows, get priority over (non-restricted) packets which would like to join this direction.*

Greedy principle: *In all other cases packets are sent according to the greedy principle.*

Note that applying the above rules may involve conflicts which are not resolved by the specified rules. This does not disturb the analysis, i.e., the following claims hold for any algorithm which is consistent with the above rules.

Theorem 4.1 *Algorithm* GENERAL_REQUEST *completes the routing of any routing request of size k in at most $2(k-1)+d_{max}$ steps.*

We remark that the theorem is not sensitive to the initial number of packets in a node in the sense that the same result holds even if up to four packets originate at each processor. In this case, of course, there might be some additional conflicts to resolve in the first step. Again, the theorem holds regardless of the way these conflicts are resolved provided it is consistent with the algorithm (i.e., restricted packets get priority).

Another remark is that there are routing problems which require $k + d_{max}$ steps for any algorithm. For instance, if all k packets are destined to the upper left corner and all of them have the same distance to that corner. For such single-destination routing problems Algorithm GENERAL_REQUEST can be refined in order to utilize the additional structure so that this lower bound is matched, see Section 5. For general routing problems on the mesh Theorem 4.1 is stronger than the related results in [18], [12] and [9], and its proof is simpler.

For the proof of Theorem 4.1 we use a "blaming" mechanism. In other words, we let the packets "blame" each other for being deflected. For each deflection of a packet we create a *token* which will be carried by advancing packets until the carrier packet arrives at its destination with a load of tokens. This packet will be "blamed for" the deflections for which it carries tokens. At this stage the tokens are removed. We make all this formal in the following rules governing the creation, delivery and removal of tokens.

The Token_Rules.

- ▷ *If a packet p is deflected at step t of the algorithm then it produces a token $T(p,t)$.*
  - *If p was advancing at $t-1$ or if p is restricted at the end of that step, then there is a restricted packet q which advances at t from the deflecting node in the direction p was advancing, and $T(p,t)$ is given to q.*
  - *Otherwise, $T(p,t)$ is given to the packet that proceeds in the direction opposite to the direction in which p is deflected. Clearly, there is such a packet that advances in a direction good for p, otherwise p would have not been deflected.*
- ▷ *A packet that carries tokens and is deflected will move these tokens to the advancing packet which gets the newly created token.*
- ▷ *When a packet arrives at its destination, all the tokens that are carried by it are removed.*

Thus, a token is never carried in a certain step by a packet that is deflected at that step. Furthermore, it can be shown that once a token is created, it takes a minimal path to the point where it is removed.

Lemma 4.2 *Two tokens $T1$ and $T2$ that are created by the same packet p never "meet" at the same node at the same time.*

Proof: Let us assume w.l.o.g. that $T1$ was created by p before $T2$ (note that a packet may create at most one token at each step).

Suppose $T1$ is carried from the time of its creation by a restricted packet. Then, at the step following its creation, $T1$ advances one step in some direction and p is deflected in some other direction. Since $T1$ is carried by a restricted packet, it will not change directions until it will eventually disappear, and it will make progress at every step. Hence, no token which is produced by p after the creation of $T1$ may overtake it.

Suppose at the step following its creation $T1$ is carried by a non-restricted packet q. q is advancing in a direction, say X^-, which is also good for p. At the end of that step q is located two steps ahead of p in X^-, and is located at the same point with respect to all other directions. Consider $T2$ which is created at a moment following this situation. If $T1$ and $T2$ ever meet then it means that $T2$ overtook $T1$ in all the directions good for q, including X^-. However, we also know that $T1$ goes along a minimal path to the meeting point, which is a contradiction. ∎

Proof of Theorem 4.1: From Lemma 4.2 we conclude that when a packet arrives at its destination it carries at most one token produced by each other packet. On the other hand, a packet produces a token at each deflection and all tokens are always carried by packets until they are removed.

We thus conclude that the number of deflections which a packet experiences is at most the number of other packets in the system, i.e., $k-1$. Each deflection delays a packet by exactly two steps. Hence the total number of steps that it will take to complete the routing is at most $d_{max} + 2(k-1)$. ∎

In the proof of Theorem 4.1, the most important fact is that out of every node from which a packet is deflected there is another packet that advances and has either the same or a subset of the good directions of p. This fact implies the shortest path observation, which, in turn, implies the theorem. It can be easily verified that for a strongly greedy algorithm which gives priority to restricted packets this fact always holds and so we get the following corollary [9, 14].

Corollary 4.3 *If in Algorithm* GENERAL_REQUEST *the* priority to inertial packets *rule is eliminated, and the* greedy rule *is replaced by a* strongly greedy rule *requirement, then the resulting algorithm completes the routing in* $d_{max} + 2(k-1)$ *steps.*

5 OPTIMAL SINGLE-TARGET ROUTING ON THE 2-DIMENSIONAL MESH

In this section single-target routing on the two-dimensional mesh is considered [8]. The idea in the algorithm is to send packets on the "diagonal" towards the target, thus trying to avoid the packet from becoming restricted. The intuition is that as long that the packet is not restricted, it preserves a choice of two good directions. The analysis shows that the intuition in this case is correct, i.e., the diagonal policy leads to an optimal result, which seems to give a clue for the design of greedy routing algorithms. Nevertheless, we are not aware of algorithms that attempt to handle routing requests that are more general, and which make use of this approach.

Note that since all packets in the system are destined to the same place, each arc of every node is either always good for all the packets or bad for all the packets. A greedy routing algorithm can be seen as giving priority to arcs. When there are less packets than good arcs, the algorithm decides which good arcs to use. When there are more packets than good arcs, it decides which arcs will be used for deflections (since all the packets are going to the same destination all that matters is which arcs are used regardless of the identity of the packets).

For a node z, let $dx(z)$ be the horizontal distance from z to the target, let $dy(z)$ be the vertical distance from z to the target, and let $dist(z)$ be the distance from z to the target (i.e. $dist(z) = dx(z) + dy(z)$). Note that packet is restricted iff $dx(z) = 0$ or $dy(z) = 0$. We say that a packet is *on the diagonal* if $dx(z) = dy(z)$. We can logically divide the system into four *quarters* relative to the target. I.e., the lower-right quarter consists of all nodes which are not higher nor left to the target node. Each node in the system is in at least one of the quarters. A restricted node is in two adjacent quarters. We say that it

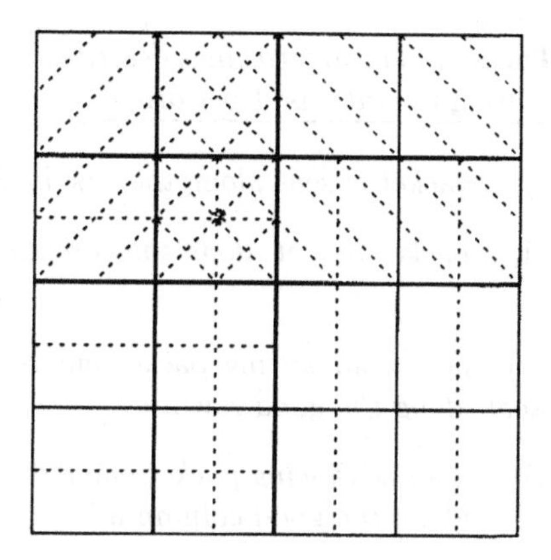

Figure 7 The system can be divided into four quarters with respect to a target node.

is *on the border* between two quarters. The target is in all the quarters (see Figure 7).

Consider the following algorithm.

Algorithm Single_Target. *The algorithm is described from the point of view of a node z and the routing operations it performs.*

- ▷ *If $dx(z) > dy(z)$, a packet is sent along the row in the good direction.*
- ▷ *If $dx(z) < dy(z)$, a packet is sent along the column in the good direction.*
- ▷ *If $dx(z) = dy(z)$ and an advancing packet entered via a row-arc, a packet will be sent along the good row-arc.*
- ▷ *If $dx(z) = dy(z)$ and an advancing packet entered via a column-arc, a packet will be sent along the good column-arc.*
- ▷ *When a restricted packet is deflected, arcs that will make it non-restricted are preferred.*
- ▷ *If $dx(z) = dy(z)$, $dist(z) = 2$ and a deflected packet entered via a row-arc, a packet will be sent along the good column-arc.*
- ▷ *If $dx(z) = dy(z)$, $dist(z) = 2$ and a deflected packet entered via a column-arc, a packet will be sent along the good row-arc.*
- ▷ *The algorithm is greedy – a bad arc will be used only if all the good arcs of the node are used.*

The first observation concerning Algorithm SINGLE_TARGET is that when all packets are destined to a single target its rules do not conflict with those of Algorithm GENERAL_REQUEST from Section 4. Thus, we can augment it so that it terminates fast for multi-destination routing problems. This is done in the following way: In each step, a node that holds packets that are headed to different targets will apply the rules of Algorithm GENERAL_REQUEST, otherwise, if all the packets are destined to the same target, the rules of Algorithm SINGLE_TARGET will be applied. It is obvious that if all the packets in the system are destined to the same target node, then at all times, in all the nodes, only packets that are destined to the same target will be involved. Thus, all the nodes in each step will apply the rules of Algorithm SINGLE_TARGET, and Algorithm SINGLE_TARGET's time bound will hold. If this is not the case (i.e. there are several targets), note that Algorithm GENERAL_REQUEST is given in terms of priority assignment which are not restrictive when the packets are destined to the same location. On the other hand, Algorithm SINGLE_TARGET specifies which direction a packet should take when it has a choice, hence its rules will never conflict with the previously applied priority assignment of Algorithm GENERAL_REQUEST. Although it could be that at a certain step, at a certain node, only packets that are headed to the same target will be involved (and thus, Algorithm SINGLE_TARGET will be applied), still, there won't be any contradiction with the rules of Algorithm GENERAL_REQUEST, therefore it's time bound will hold. We get a "compound algorithm" which, for a single-target routing situation, is guaranteed to terminate in the time bound that is found in the following analysis (see Theorem 5.15). For a multi-target situation the algorithm is guaranteed to terminate in the time bound stated in Theorem 4.1. The crucial point here is that we do not need to know in advance that the routing problem in hand is of the single-target type.

Observation 5.1 *For* any *general routing request, Algorithm* SINGLE_TARGET *can be made to terminate in the same time as Algorithm* GENERAL_REQUEST, *i.e.,* $2(k-1)+d_{max}$, *while preserving its speedup for single-target situations.*

We now return to consider single-target routing. In Algorithm SINGLE_TARGET each packet attempts to reach the diagonal of nodes $\{z : dx(z) = dy(z)\}$, and to pass it. This is reminiscent to the Z^2 policy of Badr and Podar [2] (which is, however, probabilistic and is presented in the context of the store-and-forward routing paradigm). Badr and Podar show that giving higher probability to the path which goes along the diagonal is optimal with respect to the expected routing time of a message. Here we show that trying to stay on the diagonal utilizes the potential of packet pairs to reach the destination at the same time. Both statements give evidence (contrary to the so called "greedy" policies [22]) that every packet should attempt to come as close as possible to its destination without becoming restricted.

Definition 5.2 *The path that a packet p is supposed to take (starting at a certain step at a certain location) is the path that p would take according to the algorithm if it was the only live packet in the system. The path that a packet p is supposed to take with respect to a given set of packets S, is the path that p would take according to the algorithm if the packets in S were the only live packets in the system.*

For example, suppose that two packets p_0 and p_1 are at node $(3,1)$, both headed at node $(1,1)$. Each of them (when disregarding the other packet) is supposed to take the path $(3,1) \longrightarrow (2,1) \longrightarrow (1,1)$. However when both packets are considered, then one is supposed to take the path $(3,1) \longrightarrow (2,1) \longrightarrow (1,1)$ and the other one is supposed to take the path $(3,1) \longrightarrow (3,2) \longrightarrow (2,2) \longrightarrow (1,2) \longrightarrow (1,1)$. See Figure 8.

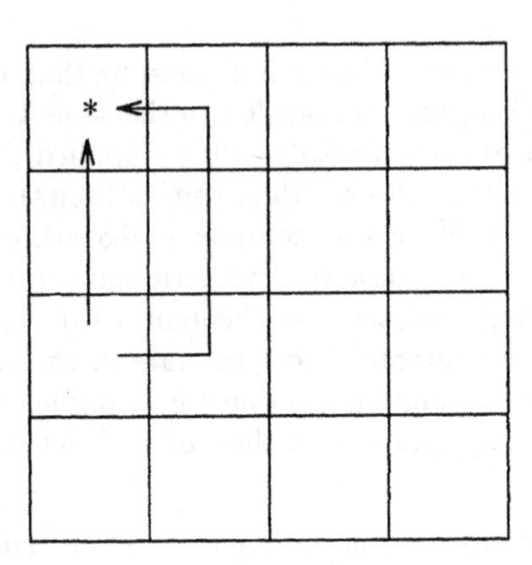

Figure 8 The paths that two packets in node (3,1) that are headed to node (1,1) are supposed to take (when both of them are considered).

Observation 5.3 *In Algorithm* SINGLE_TARGET *for some packet p and an arbitrary set of live packets S, suppose that $e_1, \cdots, e_m$ is the path of arcs that p is supposed to take (with respect to S) starting at time t. Then at time $t+i-1$ for all $1 \leq i \leq m$, a packet will be sent on arc e_i.*

This follows from the fact that the rules of the algorithm are in the form of "if the packet is in a node that has some property, then a packet will use a certain arc". Thus the packet that actually takes the arc is indistinguishable from the packet that was supposed to take the arc.

Observation 5.4 *Suppose that a packet p is the only live packet in the system. Then p does not switch quarters on its way to the destination.*

Note that the path that p will take according to the algorithm if it is the only live packet in the system is equivalent to the path that p is supposed to take (with respect to itself only) (see Definition 5.2). Observation 5.4 follows from the fact that if p is the only live packet in the system, it will never be deflected. In order to pass the border between two quarters p has to be deflected from a restricted node. Therefore, p will never cross the border to another quarter and will stay within the same quarter until it reaches the target. By induction on the distance and case analysis on the packets' locations we can show (See an example in Figure 9):

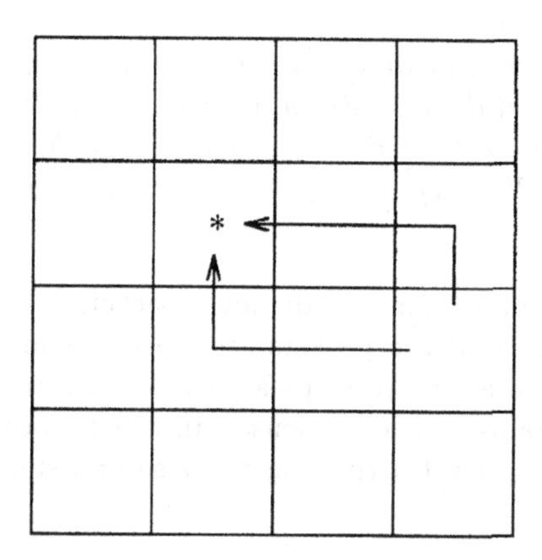

Figure 9 There are two packets in the same node. One of them becomes restricted. The paths that the packets are supposed to take are disjoint and are totally within the same quarter.

Claim 5.5 *If there are two non-restricted packets at distance i from the target and within the same quarter (not necessarily at the same node), then after i steps two packets will reach the target from nodes in that quarter.*

Corollary 5.6 *If at step t there are two non-restricted packets at distance i from the destination and within the same quarter Q, then during step $t+j$ for every $0 \leq j < i$ two packets will advance in Q at distance $i-j$ from the destination.*

Proof: Consider at step t two non-restricted packets at distance i from the destination and within the same quarter Q. By Claim 5.5 they will reach the destination by time $t+i$ from nodes in Q. Consider the paths that these packets are supposed to take (with respect to themselves). By Observation 5.3 there will be two packets moving on these paths with the timing that is claimed. ∎

We distinguish between packets with odd origin-to-destination distance and those for which this distance is even. After odd steps, the odd initial distance packets have even distance to destination and vice versa, and similarly for even initial distance. Since the two dimensional mesh is bipartite, packets from the two classes never meet at the same node. Thus, from now on, unless otherwise stated, we restrict the discussion to one of them, say, to those of even initial distance.

Definition 5.7 *The even distance packets destined to a node z of a network are m-*dense around *z if there exists an integer $j > 0$ such that for all $1 \leq i \leq j$ there are at least m packets at distance $2i$ to z and there are no packets at a distance greater than $2(j+1)$.*

A similar definition applies for odd distance packets. When $m > 0$ is unimportant we simply say that the packets are *dense around z*. Note that if the packets are dense, then after two steps a packet will arrive at the destination, and the network remains at least 1-dense. In other words, when the packets are dense, at least one packet arrives on every second step.

Lemma 5.8 *For any bipartite network with bi-directional links, and* any *greedy algorithm, if all packets are destined to a single node z, then after d_{max} steps the packets will be 1-dense around z.*

Proof: Consider some packet p_0 during the first d_{max} steps. If the packet did not arrive during that time, let i be its distance to the destination. Since the packet did not arrive after d_{max} steps, it must have been deflected at least once during the d_{max} steps. By simple induction it can be proved that from the step after the last deflection till the d_{max}-th step, there is always a packet which is two steps closer to the destination than p_0 (unless p_0 came as close as distance two to the destination). ∎

Theorem 5.9 ([18]) *For* any *bipartite network with bi-directional links, and* any *greedy algorithm, if all k packets are destined to a single node z, then after $2(k-1) + d_{max}$ steps, all packets will reach the destination.*

This theorem was given originally by Hajek [18] for a slightly different type of algorithm. The notion of *greedy routing* makes it easier to prove.

Proof: Consider a packet that is initially the closest to the destination and which defeats all other packets in any conflict (it is easy to see that such a packet exists in a non-empty system). Then this packet arrives by step d_{max}. By Lemma 5.8, after d_{max} steps, all other packets will be 1-dense around the destination and thus, from that time on, at least one packet will arrive at every second step. ∎

We proceed to show a stronger result for the two dimensional mesh. Let $d_{max\ restricted}$ be the maximum origin-to-destination distance of any initially restricted packet. If no packet is initially restricted, then $d_{max\ restricted} = 0$.

Claim 5.10 *Let p_0 be a restricted packet at a distance of at least 2 from the destination after at least $d_{max\ restricted}$ steps of Algorithm* SINGLE_TARGET. *There exists another packet p_1, within one of the two quarters that p_0 is in, and with the same distance to the destination as p_0, such that p_0 and p_1 are supposed to take minimal paths of the same length that are disjoint.*

Proof: A packet that was restricted when the algorithm started and was never deflected, would arrive after $d_{max\ restricted}$ steps. Thus, every packet that is restricted at time $d_{max\ restricted}$ was at some time non-restricted, or was only deflected to restricted nodes. Consider the last deflection of p_0 from node z. If p_0 was deflected to a restricted node, then by the rules of Algorithm SINGLE_TARGET, we know that another packet p_1' was deflected from z to a non-restricted node z' at the same time. Consider the paths that p_0 and p_1' are supposed to take (with respect to themselves). It can be easily seen that the paths which p_0 and p_1' are supposed to take are minimal, disjoint and are totally within the same quarter. See an example on Figure 10. Recall that this is the last deflection of p_0. From that time on p_0 will advance on each step. By Observation 5.3 at each step there will be a packet that will take the arc that p_1' was supposed to take at that step. Thus, at each step there will be a packet at the same distance and in the same quarter as p_0. Otherwise, p_0 was non-restricted and became restricted. p_0 will become restricted at a distance greater than one from the destination only if it meets another packet at the node that sent it to be restricted. This is the case of Claim 5.5. ∎

Putting together Claim 5.5 and Claim 5.10 we get:

Claim 5.11 *After at least $d_{max\ restricted}$ steps of Algorithm* SINGLE_TARGET, *if there are two packets within the same quarter at a distance i ($i \geq 2$) from the target , then after i steps two packets will reach the target.*

Proof: If neither of the packets is in a restricted node, then it is the case of Claim 5.5. Otherwise, at least one of the packets, say p_0, is in a restricted node. By Claim 5.10 there is a non-restricted packet, say p_1, within the same quarter as p_0, such that p_0 and p_1 are supposed to take minimal, disjoint paths of the same length. By Observation 5.3 there will be two packets moving on

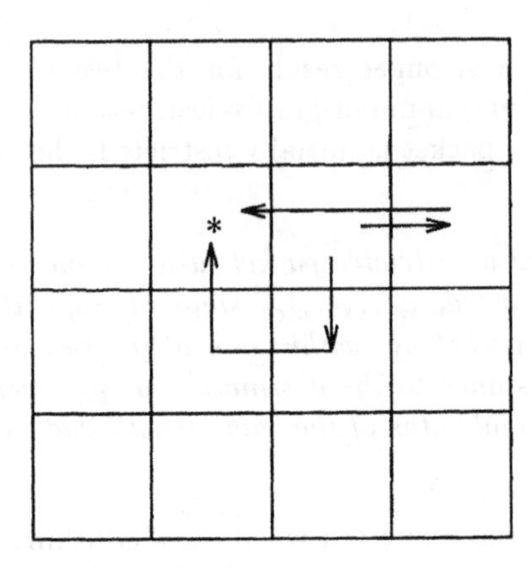

Figure 10 A packet is deflected to a restricted node. The algorithm guarantees that there is another packet that is deflected from the same node to a non-restricted node and that the paths that the packets are supposed to take are minimal, disjoint and are totally within the same quarter.

these paths with the correct timing. Thus, two packets will reach the target after exactly i steps of the algorithm. ∎

Corollary 5.12 *If there are two packets within the same quarter, at a distance of at least two from the target, after at least $d_{max\ restricted}$ steps of Algorithm* SINGLE_TARGET, *then for all $1 \leq j < i$, two packets advance from distance $i - j$ after j steps.*

Proof: The case of two non-restricted packets is the case of Corollary 5.6. Otherwise, at least one of the packets, say p_0, is restricted. By Claim 5.10 there is a non-restricted packet p_1, such that the two packets are supposed to take minimal, disjoint paths of the same length (and that are totally within the same quarter). By Observation 5.3 there will be two packets moving on these paths with the timing stated by the corollary. ∎

Lemma 5.13 *If after $t \geq d_{max}$ steps of Algorithm* SINGLE_TARGET, *there is at least one quarter with at least two packets, then there exists $j > 0$ such that for all $1 \leq i \leq j$, there are at least two packets within the same quarter, that are at distance $2i$ from the target, and there are no packets at distance greater than $2(j + 1)$.*

Note that the lemma claims not only that after step d_{max} (if there is a quarter with two packets) the packets will be 2-dense around the target, but also that if there are packets at distance $2i \leq 2j$, then there are two packets within the same quarter, at distance $2i$.

Proof: Suppose that after $t \geq d_{max}$ steps of Algorithm SINGLE_TARGET, there is at least one quarter with at least two packets and that there is a packet at distance $2j$. We show that for every $0 < 2i < 2j$ there are two packets at distance $2i$ within the same quarter.

Let p_0 be a packet at distance $2j$ after t steps. Consider the last time that p_0 was deflected. If p_0 was deflected from a non-restricted node, then there were two other packets at that node which advanced to nodes in the same quarter. By Corollary 5.12, from that moment and until time t, there are two packets within the same quarter which are two steps closer than p_0 to the destination unless p_0 reached the distance of two or less steps from the destination. Otherwise, if p_0 was deflected from a restricted node, then there was another packet p_1 at that node that advanced at that step. Consider the path that p_1 is supposed to take and let z be the node p_1 is supposed to reach at time t. By Observation 5.3, there will be a packet at that node at time t. Since $t \geq d_{max} \geq d_{max\ restricted}$, then by Claim 5.10, there is another packet within the same quarter as z with the same distance to the destination. ∎

Corollary 5.14 *If after d_{max} steps there is a packet p_0 at distance greater than two from the target, then there are two packets in the same quarter as p_0 or in an adjacent quarter which are two steps closer to the target than p_0.*

The proof uses similar considerations as these made in the proof of Lemma 5.13.

Let t_{first} be the first time, at or after the d_{max}-th step of the algorithm, that all of the quarters have less than two packets. Note that during the first $d_{max} - 1$ steps of the algorithm it could happen that none of the quarters has more than a single packet. In that case, t_{first} is equal to d_{max}. Using the above observations and the previous claims it can be shown that at step t_{first} of the algorithm, if any packets are left in the system, then all of the packets are at distance two from the target, and none of the packets are restricted.

Let *last* be the number of packets in the system at time t_{first}. Note that the value of *last* is at most four. This follows from the fact that all of the packets at time t_{first} are of distance two from the target and none of the packets are

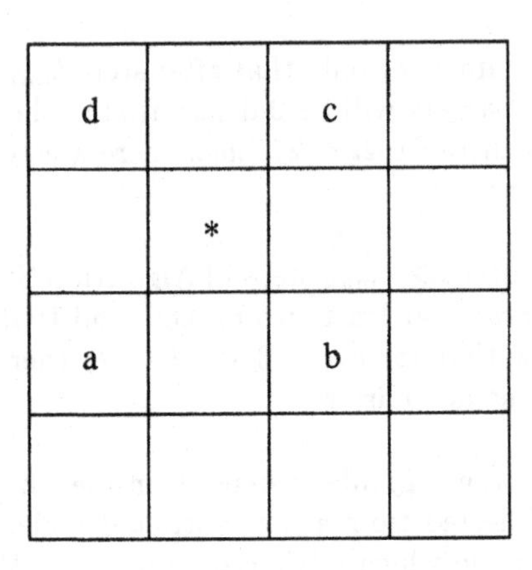

Figure 11 At time t_{first} live packets could reside only in nodes a, b, c or d. There could be no more than one packet at each of these nodes. Thus, there are at most four live packets in the system.

restricted. Thus, there are only packets at distance two on the diagonal, and there is no more than one packet at each such node. See Figure 11.

Using Lemma 5.13 and Claim 5.11 we know that at least two packets will arrive at the destination in every second step, starting from step t_{max} and until t_{first}. Thus, we can show that t_{first} is reached after no more than $d_{max} + k - last$ steps of Algorithm SINGLE_TARGET.

Hence, the worst case scenario is the one in which until step d_{max} only one packet has reached the target, and from that step until t_{first}, only two packets will reach the target, every two steps. We will say that t_{first} *is reached sooner than expected* if before step d_{max} more than one packet has reached the target or if before t_{first}, in one of the (even) steps, more than two packets have reached the target. As we remarked before, t_{first} is even. Thus, if t_{first} is reached sooner than expected, the upper bound on the number of steps until t_{first}, is at least two steps smaller than $d_{max} + k - last$. In fact, if $k - last$ is even, t_{first} will be reached after no more than $d_{max} + k - last - 2$ steps, otherwise $k - last$ is odd and t_{first} will be reached after no more than $d_{max} + k - last - 3$ steps. Note that if until time t_{first} an even number of packets have reached the target, then t_{first} is reached sooner than expected.

Theorem 5.15 *Algorithm* SINGLE_TARGET *will route k packets to a single target in no more than $k + d_{max}$ steps.*

The proof goes by case analysis, considering each of the five possible values of *last* [8].

We remark that if the k packets origins are split to k_{even} of even distance and k_{odd} of odd distance with corresponding maximal distances $d_{max\ even}$ and $d_{max\ odd}$, then the theorem bounds the routing time by $\max\{k_{even} + d_{max\ even}, k_{odd} + d_{max\ odd}\}$.

5.1 Routing to a Small Set of Targets

Theorem 4.1 gives an upper bound of $2(k-1) + d_{max}$ on the termination time for multi-target situations. In this section we use Theorem 5.15 to tighten the bound when k is large relative to the number of targets M. Unfortunately, the resulting algorithm is only weakly greedy.

Let k packets be destined for each one of M targets. We begin by ordering the M destinations in an arbitrary fashion and assume that this ordering is known to each node. Let k_j, $j = 1, 2, \cdots, M$ be the number of packets initially admitted and destined for the jth destination, denoted z_j, so $k = \sum_{j=1}^{M} k_j$. We refer to packets destined to node z_j as *class* j packets. We can logically divide the system into four quarters with respect to one of the targets, say z_j. We refer to each of these quarters as a j*-quarter*. The following greedy algorithm was suggested by Brassil and Cruz [12].

Algorithm Greedy_Multi_Target.

Let each node order arriving packets by class from the lowest to the highest numbered class. Starting from the lowest numbered class, packets are greedily assigned to good exits, if available, in an arbitrary fashion. Then other packets are deflected in an arbitrary fashion.

Theorem 5.16 ([12]) *Suppose a batch of k packets is to be delivered to a set of M targets. If the destinations are ordered so a walk connecting them in order has length P, then Algorithm* GREEDY_MULTI_TARGET *terminates in* $d_{max} + P + 2(k-1)$ *steps on any network.*

Note that for the mesh, we already have a better bound by Theorem 4.1. Using Algorithm SINGLE_TARGET, however, we may get further acceleration by using the following algorithm.

Algorithm Multi_Target.

Let each node of the mesh order arriving packets by class from the lowest to the highest numbered class. Starting from the lowest numbered class, packets are assigned to a preferred exit, if available, according to Algorithm SINGLE_TARGET.

The algorithm may not follow the greedy principle when packets from a prioritized class must be deflected in directions which are good for a lower priority class. Notice, however, that it is weakly greedy. As the following theorem shows, it has a good termination time under certain conditions.

Theorem 5.17 *Suppose a batch of k packets is to be delivered to a set of M targets. If the destinations are ordered so a walk connecting them in order has length P, then Algorithm* MULTI_TARGET *terminates in $d_{max}+P+k+2(M-1)$ steps.*

We remark that this bound is somewhat strange, as it *improves* when the number of destinations M becomes *smaller*. Moreover, it is somewhat unrealistic to assume that the ordering of the destinations minimizes P. In the more realistic case, the destinations are not known in advance. We elaborate on this point below in Section 5.1. Nevertheless, when P and M are small relative to k, this bound becomes almost twice as good as that of Algorithm GENERAL_REQUEST which, in turn, is better than the bound of Theorem 5.16.

The full proof of Theorem 5.17 can be found in [8]. Interestingly, we remark that one may deduce a proof for Theorem 5.16 with respect to any network by repeating the considerations in the proof of Theorem 5.17 with respect to the two dimensional mesh.

Bounding the Length of a Walk Connecting the Destinations

As already mentioned, a natural question to which Theorem 5.17 gives rise is how badly can the value of P influence the bound? When the set of destinations is known in advance, it may be that a short path connecting them, together with a corresponding priority assignment, can be computed. In the more realistic case, the destination set is not known in advance and the priority scheme is a-priori "simple" and fixed.

Fortunately, it is possible to use results by Feige [14] which present an indexing scheme g for the 2-dimensional mesh. If a batch of k packets is delivered to a set of M targets using Algorithm MULTI_TARGET, and the destinations are given priority according to g, then the routing terminate in at most $d_{max} + k + 6n\sqrt{M} + 2(M-1)$ steps [8].

6 RANDOMIZED SINGLE-TARGET ROUTING ON HIGHER DIMENSIONAL MESHES

In this section we present randomized hot-potato routing algorithms on d-dimensional meshes and on the n-dimensional hypercube. The algorithms are designed for routing many packets to a single destination, or a relatively small number of destinations. The important feature which was obtained using this method is that the algorithms utilize the higher in-degree of the nodes and are asymptotically optimal.

Due to some modulo operations and for the sake of notation simplicity we slightly change it for this section only, as follows. The d-dimensional mesh of side-length N, is a graph with N^d nodes. Each node corresponds to a tuple $(x_0, \cdots, x_{d-1})$ where $0 \leq x_i \leq N-1$.

The n-dimensional hypercube is a graph with 2^n nodes and $n2^{n-1}$ edges. Each node corresponds to a binary string of length n. Two nodes are connected by an edge if their corresponding strings differ in exactly one place. Every node in the hypercube has exactly n neighbors. We sometimes refer to edges by their dimension, that is, by the index of the place for which the two end points of the edge differ. The dimensions are indexed by $0, \cdots, n-1$.

As the degree of the n-hypercube with 2^n nodes is n, one might expect that during most steps, as many as n packets will enter the destination. As the hypercube is bipartite, if all packets start with even distance to the target then one may expect packets to arrive during even steps only. Thus, if there are k packets in the system (where k can be as big as $n2^{n-1}$), an immediate lower bound is $2k/n$. However, the bound of [12, 18] is worse by asymptotically a factor of n. Indeed there exists a randomized algorithm that bridge this gap, routing k packets to a single target in $\frac{2k}{n} + o(\frac{k}{n}) + \text{poly}(n)$ steps with high probability. This algorithm is presented here and is generalized in two ways.

First, following [8, 12] it is shown that this can be extended for routing problems with more than one (but rather a small number of) destinations. Then it is shown that the algorithm can be generalized to d-dimensional meshes (with d not necessarily constant). The algorithms are all greedy, (in fact strongly greedy) and have the two-pass property.

The main problem in single target routing is to make equal number of packets reach the destination via each entry. The difficulty with a deterministic approach is the "symmetry" that is inherent in the problem. This difficulty is avoided here by using randomization. However, a simple-minded oblivious random-choice does not suffice for this purpose: The initial distribution of the packets is not known in advance, hence choosing at random during steps in which the packet advances does not guarantee the goal. On the other hand, when the packet is deflected, it gets a chance to add a random "bad" dimension to the set of address bits which need to be switched. This choice is used to ensure that when the deflected packet starts advancing again it will take a path which will end up entering the destination via a random neighbor.

As in the single-target algorithm for the two-dimensional mesh, we distinguish between packets with odd distance to the destination and those for which this distance is even, and restrict our discussion to the packets that their initial distances are even (unless otherwise stated).

Recall also that for greedy algorithms, if the packets are dense, then as long as packets remain dense, at least one packet arrives on every second step. Also, if all packets are destined to a single node z, then after d_{max} steps the packets will be 1-dense around z (Lemma 5.8), and at least one packet will reach the destination at every other step (Theorem 5.9). However, here too we want to achieve considerably more. For instance, for the n-degree hypercube, one expects that as many as n packets will reach the destination at every other step (after possibly a short transition time).

6.1 Cyclic Advancing and its Properties

For a node z denote by $G(z)$ the set of indices in which it differs from the destination node (i.e., the set of good directions). For a node z and a dimension i we define the next dimension in *the cyclic order* to be

$$n(i, z) = (i + \min\{j \geq 1 : (i + j \bmod n) \in G(z)\}) \bmod n$$

For instance, if $z = (0,1,0,1,0)$ and the destination is $(0,0,0,0,0)$, then $n(i,z) = \begin{cases} 1 & \text{if } i = 0,3,4 \\ 3 & \text{if } i = 1,2 \end{cases}$.

Definition 6.1 *An algorithm is called* cyclic advancing *if whenever a packet p enters a node z via a good arc i (i.e., p was not deflected in the previous step), then during the next step a packet will advance from z via the arc $n(i,z)$.*

The *cyclic advancing* condition can be viewed as if the same packet p which enters z via arc i is supposed to be sent out via $n(i,z)$. However, when several packets enter the node it is possible that another packet will be sent via $n(i,z)$, rather than p. If all packets are destined to the same node, the only property which distinguish between them is the arc through which they entered. The packet which advances using $n(i,z)$ "replaces" p in the analysis.

Definition 6.2 *Let z be a vertex and e a good arc entering z. The path that a packet is supposed to take from z with respect to e is the path that a packet would take according to the algorithm assuming that it entered z through e and it is the only packet in the system.*

Observation 6.3 *Let $\mathcal{A}$ be a cyclic advancing single-target algorithm. Consider an instance of the execution of $\mathcal{A}$. Assume that in that instance a packet p advanced to z at time $t-1$ and let $(e_1, \cdots, e_m)$ be the path which p is supposed to take from z (with respect to the arc it used to enter z). Then for any $0 \leq i \leq m-1$ a packet will be sent on e_i at time $t+i$.*

The degree of the destination node is n. This fact is fully utilized if, starting from a certain step, n packets will reach the destination at every even step. When several packets arrive to a neighbor of the destination at a certain step, only one of them will reach the destination at the next step. We thus should try to make n packets reach n different neighbors of the destination node at every even step. We do that by associating a color in the range $\{0, \cdots, n-1\}$ with every packet. This coloring is for analysis purposes only.

Definition 6.4 *The* color of a packet *p is the index $c(p)$ of the last bit which p would flip on its supposed path to the destination.*

Note that for a cyclic advancing algorithm and a single target, this definition sets the color of the packet according to the initial choice of advancing direction. In a multi-packet system this color may change as a result of the packet not following the cyclic advancing rule. However, in such a case another packet, which is identical to the first from the algorithm viewpoint, maintains this color by advancing according to the cyclic advancing rule. We summarize the discussion on the relation between $c(p)$ and the arc which p is supposed to take at the next step in the following observation.

Observation 6.5 *Suppose a packet p at a node z with good arcs along dimensions $d_0 < d_1 < \cdots < d_i$ is colored with $c(p) = d_j$. If the algorithm is cyclic advancing then a packet will exit z during the next step along $d_{(j+1) \bmod (i+1)}$. That packet will have the color $c(p)$ when it leaves.*

Proof: $d_{(j+1) \bmod (i+1)}$ is the arc that p is supposed to take according to the cyclic advancing rule. Now the observation is implied by Observation 6.3 and the definition of $c(p)$. ∎

Consider a set of packets P of equal distance to the destination. The following claim shows that there is a relation between the number of packets in P which arrive to the destination with no deflection, and the number of different colors in P.

Claim 6.6 *Consider an execution of a greedy and cyclic advancing algorithm $\mathcal{A}$. Let P be a set of m packets that are at distance r at time t and suppose that they have distinct colors, that is, the set of colors they have is of size m. Then for each of the following steps $l = 1, ..., r$, a set of m packets will be at distance $r - l$ from the destination. Moreover, these m packets will have the same set of m colors.*

Proof: By induction on r. For $r = 1$, if there are m packets of distance 1 having different colors, then they must be at different nodes. Hence, all packets will enter the destination at the next step.

For $r > 1$ consider a site that contains k of these packets with k different colors. By Observation 6.5, at the next step k packets with the same k colors leave this site and are at distance $r - 1$. As this is true for each of the sites containing the packets with distinct m colors, there will be m packets with the same set of m different colors at distance $r-1$ and by the induction hypothesis we are done. ∎

In particular, in the above setting, at least m packets will arrive to the destination at step $t+r$.

In order to make use of the properties of cyclic advancing algorithms the number of different colors of packets which have the same distance to the destination should be maximized. Since advancing packets keep their colors, the coloring of packets is done in the first step, and during steps in which the packet is deflected or following a deflection step. Random coloring is used to ensure a high variety of colors.

6.2 Routing to a Single Target

We now describe the algorithm for routing to a single target on the hypercube of dimension n. Each node knows the target, hence its distance and the good and bad directions. Each node keeps an out-queue on each arc incident to it. The algorithm is described as the collection of operations performed by a node z of distance r from the destination.

Algorithm Hypercube_Single_Target.

0. *At the first step, packets start advancing arbitrarily according to the greedy rule.*
1. *Add each packet which* advanced *into* z *via arc* i *to the out-queue of arc* $n(i,z)$.
2. *For each packet which* was deflected *into* z *via arc* i,
   (a) *with probability* $\frac{r-1}{n}$ *choose uniformly at random any good arc other than* $n(i,z)$, *and add the packet to its out-queue.*
   (b) *with probability* $\frac{n-r+1}{n}$ *add the packet to the out-queue of arc* $n(i,z)$.
3. *If there are arcs with two or more packets in their out-queue move the extra packets arbitrarily to good arcs with empty out-queues (if there are any).*
4. *As long that there are still arcs with two or more packets in the out-queue, move the extra packets to the out-queues of bad arcs which are chosen at random, i.e., in an arbitrary order, for each such packet, choose uniformly at random a bad dimension with an empty out-queue and move this packet to the queue.*

For the sake of the analysis, the color that is associated to a packet is the color it would have if it would exit according to its choice at Step 2, namely the color it would take if it were the only packet in the node. Note that if a packet is associated with a color, then there is a packet which actually advances from this node and has this color.

It is easy to see that Algorithm HYPERCUBE_SINGLE_TARGET is both greedy and cyclic advancing. There is no restriction on the initial number of packets at each node, as long that this number is no more than the node out-degree, i.e., n.

The number of different colors of packets which are at the same distance to destination, and which were deflected at the previous step, is large. In fact, for a sufficiently large number of such packets, with high probability there will be exactly n different colors. The reason is that when a packet is deflected, the color it will next be associated with (after Step 2 in Algorithm

HYPERCUBE_SINGLE_TARGET) is random with probability $\frac{1}{n}$ for each of the n possible colors.

Suppose that the algorithm could choose independently for every deflected packet a random color out of the set $\{0, \cdots, n-1\}$. The question "how many choices should be made until we get n different colors?" is known as the "coupon collectors problem". It can be analyzed in various ways. It is dealt with here in a slightly more general form, so to be suitable for our needs in Section 6.4 too. Assume we choose out of n colors the color i with probability $p_i \geq p$ and assume that this process is repeated, independently for m choices. Thus the probability that in m choices there will be a color that is not chosen is at most $n(1-p)^m$.

Note that in collecting coupons, the choices are taken randomly and independently. In Algorithm HYPERCUBE_SINGLE_TARGET there is a dependency. The color of each previously deflected packet is determined at Step 2 of the algorithm, and is indeed independent for different packets, and for each packet it is $\frac{1}{n}$ for each of the colors. However, this choice may change at steps 3 or 4 of the algorithm. The color is finally determined by the actual arc through which the packet is sent (if the packet advances), or it is irrelevant if the packet will again be deflected. In particular, two packets at the same node may not choose the same color. However, it is quite clear that the dependency above is in our favor. Formally, assume that the colors were chosen sequentially in an arbitrary order. Let X_i' denote the number of colors after i trials in the random process defined by the algorithm and let X_i be the corresponding number in the coupon collector process. Then for any number t, $\Pr(X_i' < t) \leq \Pr(X_i < t)$. This is because initially each color is being chosen with probability p_i for both processes. In the algorithm this color may later be changed if it was chosen by another packet in the same node. In this case the choice is repeated but is not counted. Thus for any instance of the process defined by the algorithm there is a corresponding instance in the coupon collector process for which for each step i the number of colors collected is at most the same as in the algorithm's.

In our case the bound p is $\frac{1}{n}$, and we get,

Lemma 6.7 *If at the beginning of step t there are at least $m+n$ packets at distance r, then with probability at least $1 - ne^{-\frac{m}{n}}$ during step $t+r+1$ exactly n packets will enter the destination.*

Proof: During step $t+r-1$ at most n packets enter the destination. Thus, at least m packets will be deflected at least once during the time interval $[t, t+r-$

1]. For each such packet, consider the first deflection during that time interval. In each deflection, the deflected packet chooses a new color, and at least m such choices are made by different packets. As explained $\Pr(X'_i < n) \leq \Pr(X_i < n)$, where X_i and X'_i are as before. With $p = \frac{1}{n}$ the latter probability is bounded by $\Pr(X_i < n) \leq n(1 - \frac{1}{n})^m \leq ne^{-\frac{m}{n}}$. Thus with probability $1 - ne^{-\frac{m}{n}}$ there are n colors among these packets. Claim 6.6 may be applied for each of these deflection steps, each with its own set of colors, which sum up to n. Since by Claim 6.6 colors are preserved, the lemma follows. ∎

We now split the packets to *levels* according to their distance to the destination node.

Definition 6.8 *A node z belongs to level i if it is at distance i from the destination node. Level $i+1$ is said to be higher than level i, etc. A level will be called empty if there is no packet in any of its nodes. A level will be called heavy during a certain step if there are at least $3n \ln n + n$ packets in this level at the beginning of that step.*

Recall that only levels containing packets with even initial distance to the destination are considered.

The following observation holds for any network with bi-directional communication links and any greedy single-target routing algorithm.

Observation 6.9

- ▷ *If x packets are deflected from level d to level $d+1$ during step t, then at least x packets will move from level $d+1$ to level d during step $t+1$. The reason is that for every deflected packet which arrives at a node the node has a good arc.*
- ▷ *If at the beginning of step t all levels above the r'th level are empty, then at the end of step $t+1$ all levels above the r'th level are empty.*
- ▷ *If at the beginning of step t there are x packets at levels higher than r then after step $t+1$ there are at most x packets at levels higher than r.*

This observation, combined with Lemma 6.7 and standard Chernoff bounds, turns out sufficient to prove the main theorem. The crux of the proof is to show that as long that there are heavy levels, there are also steps in which a total of n packets reach the destination with high probability. Intuitively, this happen since all the packets in the heavy level advance together towards the destination. On their way, many deflections happen which make the level above contain packets having all the colors with high probability.

Theorem 6.10 *Let $\alpha(n) = o(n)$ any function. Then, k packets are routed with Algorithm ST to a single target on an n-dimensional hypercube in no more than $(1 + \frac{\alpha(n)}{n})\frac{2k}{n} + O(n^2 \log n)$ steps with probability at least $1 - e^{-\frac{\alpha(n)^2 k}{8n^3}}$.*

Note that the complexity of this algorithm is asymptotically optimal, up to an additive quantity that is independent of k. Namely, the ratio between the complexity of the algorithm and the trivial lower bound approaches 1 as k grows. In particular, for $\alpha(n) = \theta(\sqrt{n/\log n})$ Algorithm HYPERCUBE_SINGLE_TARGET becomes nearly n times more efficient than what is given by Hajek's analysis for $n^2 \log n \ll k$ [12, 18].

6.3 Multi-Target Routing on the Hypercube

Suppose a batch of k packets is to be delivered to an ordered set of M targets. A simple version of the idea of Brassil and Cruz [12] is used here to apply a single target algorithm to multi-target problems. The idea is to define priority between packets according to the order of the targets. Packets destined to earlier targets in the order get higher priority. Unfortunately, the resulting algorithm is only weakly greedy.

The algorithm is as follows. Let k packets be destined for each one of M targets. Begin by ordering the M destinations in an arbitrary fashion and assume that this ordering is known to each node. Let k_j, $j = 1, 2, \cdots, M$ be the number of packets initially admitted and destined for the jth destination, denoted z_j, so $k = \sum_{j=1}^{M} k_j$. Refer to packets destined to node z_j as *class* j packets.

Algorithm Hypercube_Multi_Target.

Let each node of the hypercube order arriving packets by class from the lowest to the highest numbered class. Starting from the lowest numbered class, packets are assigned to a preferred exit, if available, according to Algorithm HYPERCUBE_SINGLE_TARGET.

The algorithm may not follow the greedy principle when packets from a higher priority class must be deflected in directions which are good for a lower priority class. Notice, however, that it is weakly greedy. As the following theorem shows, it has a good termination time under certain conditions.

Theorem 6.11 *Suppose a batch of k packets is to be delivered to a set of M targets. Let $\alpha(n) = o(n)$ be any function. Then, the packets are delivered with Algorithm* HYPERCUBE_MULTI_TARGET *in at most* $(1+\frac{\alpha(n)}{n})\frac{2k}{n}+O(Mn^2\log n)$ *steps with probability at least* $1 - Me^{-\frac{\alpha(n)^2 k}{8n^3}}$.

Proof: The proof is simply by iterating the result of Theorem 6.10 M times, each time for the next lower priority class. ∎

Similar to Theorem 6.10, Theorem 6.11 is useful only when there are many more packets than targets. Algorithm HYPERCUBE_MULTI_TARGET becomes n times more efficient than Hajek's result when $Mn^2 \log n \ll k$.

6.4 Routing on Meshes of High Dimensions

The algorithm developed for the hypercube in Section 6.2 can be modified in order to generalize it for d-dimensional meshes. The algorithm comes close to the lower bound of $\lceil 2k/d \rceil$ steps.

The idea here is the same as in the hypercube case. We wish to assign colors to packets in order to get a claim similar to Claim 6.6. Once a large enough variety of colors is guaranteed, it is ensured that a large number of packets will enter the target at each (second) step and thus achieve the accelerated time.

Define first the concept of *cyclic advancing* for a d-dimensional mesh. The main difference in the definition is that for each vertex, there are usually two directions in each dimension (unless the vertex is on the boundary of the mesh). At most one of the directions may be good, the others are bad. Another difference is that the mesh is not a regular graph. The degree of a node may be between $2d$ for an inner node, to d for a corner node. Still as explained there are at most d good directions, and at most one per dimension, thus the good directions may be referred by their dimensions. Let $G(z)$ denote the set of indices of the good dimensions in a node z. As in the cube define the *desired direction* for a packet that advanced in dimension i in the previous step to be

$$n(i, z) = (i + \min\{j \geq 0 : (i + j \bmod d) \in G(z)\}) \bmod d$$

Note that, in contrast to the cube, it might be that $n(i, z) = i$. This will happen if the i'th coordinate of the target is not equal to the i'th coordinate of z.

Definition 6.12 *A hot-potato algorithm for the d-dimensional mesh is called* cyclic advancing *if when a packet p enters a node z via a good dimension i (i.e., if p was not deflected at the previous step), then a packet will advance from z using the dimension $n(i, z)$ during the next step.*

As in the cube, define the *color* of a packet to be the dimension of the last arc on its supposed path. With these definitions one may verify easily that Observations 6.3, 6.5, 6.9 and Claim 6.6 hold for the case of the mesh too.

Let $gd(z)$ be the number of good dimensions of a node z, i.e., the number of dimensions in which the node has a good arc. Let $bd(z)$ be the number of bad dimensions of the node z, i.e., the number of dimensions in which the node has a bad arc. Since a dimension may contain both a good and a bad arc, for every node z we have $d \leq gd(z) + bd(z) \leq 2d$.

We now describe the algorithm for routing to a single target on the d-dimensional mesh. Each node knows the distance and the good and bad directions of its neighbors. Each node keeps an out-queue of packets on each of its out-going arcs. The algorithm is described as the collection of operations performed by a node z.

Algorithm Mesh_Single_Target.

0. *At the first step, packets start advancing arbitrarily according to the greedy rule.*

1. *Add each packet which advanced into the node through an arc in dimension i to the out-queue of the good arc in the dimension $n(i, z)$.*

2. *For each packet which was deflected by a node z_s into the node z through an arc in dimension i, toss a biased coin and do the following:*

   a. *With probability $\frac{d-bd(z_s)}{d}$ choose uniformly at random a dimension j which is not a bad dimension in node z_s (in particular i is not chosen) and add the packet to the out-queue of the (good) arc in dimension j.*

   b. *With probability $\frac{bd(z_s)}{d}$ add the packet to the out-queue of the arc through which it arrived, i.e., try to send it on the good arc in dimension i, going "back" to z_s.*

   Note that in general, not every dimension may be chosen by z. In particular, if $bd(z_s) = d$ then the packet is "returned" to z_s with probability 1.

3. *If there are good arcs with empty out-queues and there are good arcs with more than a single packet in the out-queue, move the extra packets from the latter to the former.*

4. *If there are (good) arcs left with more than a single packet in the out-queue, then deflect the extra packets by assigning them randomly to out-queues of bad arcs. More precisely, this is done as follows: One by one for each packet, choose uniformly at random a dimension which has a bad arc having an empty out-queue, and add the packet to the out-queue of (any) such arc in the chosen dimension (Thus, for the next packet, the random choice may be taken out of a smaller set of dimensions, unless the chosen dimension contains two bad arcs with empty out-queues).*

First note that the algorithm is valid, namely all probabilities are in the interval $[0, 1]$ and can be computed locally by each node. Moreover, the algorithm is greedy and cyclic advancing, and there is no limitation on the initial number of packets in a node, except for the out-degree d. Here too, all colors are chosen by the packets that are deflected with high enough probability.

Claim 6.13 *A packet that is deflected chooses a color at step 2 of the algorithm above with probability at least $\frac{1}{d}$ for every color which is not yet chosen for another packet.*

Proof: When the algorithm deflects a packet from the node z_s to the node z it may be viewed as if it chooses randomly out of the $bd(z_s)$ bad dimensions of z_s (a bad dimension having two bad arcs does not get a higher chance than a one having a single bad arc). The case that some bad directions (or dimensions) may already be taken is to our favor, since it increases the probability to choose the others. Thus, at the deflecting node z_s, each bad dimension is chosen with probability at least $\frac{1}{bd(z_s)}$. Then, when the packet reaches z, the elected dimension becomes the color of the packet with probability $\frac{bd(z_s)}{d}$, hence the a-priori probability to elect it is at least $\frac{1}{d}$.

If there exist dimensions which are not bad in z_s, they may not be chosen by it for deflecting packets, hence they are chosen by the receiving node, z. In this case there are $d - bd(z_s) > 0$ such good dimensions, each chosen with equal probability at the receiving node z. Thus, each dimension which is not bad in z_s is chosen by z with probability $\frac{d-bd(z_s)}{d} \cdot \frac{1}{d-bd(z_s)} = \frac{1}{d}$. ∎

Theorem 6.14 *Algorithm* MESH_SINGLE_TARGET *routes k packets on the d-dimensional mesh with side-length N to a single target in at most $(1+\frac{\alpha(d)}{d})\frac{2k}{d} + O(Nd^2 \log d)$ with probability which is at least $1 - e^{-\frac{\alpha(d)^2 k}{8d^3}}$.*

According to the theorem, Algorithm MESH_SINGLE_TARGET becomes nearly d times faster than the algorithms in [11, 12, 14] when $Nd^2 \log d \ll k$.

The proof follows the same line as that of Theorem 6.10.

7 GREEDY ROUTING OF PERMUTATIONS CAN BE HARD

How simple can a hot-potato routing algorithm be? How greedy? Consider the analysis of partial permutation routing by simple, greedy, algorithms on the 2-dimensional mesh. Intuitively, the balance of the load throughout the network in partial permutation problems should lead to much better upper bounds for permutation problems than for general routing problems. Furthermore, this intuition is strengthened by simulation results, which commonly suggest that simple schemes for greedy routing of permutations should terminate in d_{max} steps, perhaps up to some additive constant factor. Nevertheless, the best such result, and the only one achieving $o(n^2)$ steps, is by Borodin *et al.* [11] which give a greedy algorithm on the mesh which terminate in $O(n^{1.5})$ steps.

Here we present some negative results for fast hot-potato permutation routing by a class of algorithms that are too simple and too greedy [5,6]. More specifically, consider a class of algorithms that do not take into account the full destinations of the packets when making the exit assignment, but merely the directions that send them closer to their destinations. Such algorithms are called *short-sighted*, as they make decisions based only on local information. One motivation for designing such algorithms is the desire for a simple hardware implementation. Indeed, it will be shown that many of the algorithms in the recent literature are short-sighted. In another class of algorithms, called *a-bounded*, a packet is not allowed to deflect outside a thin stripe of a columns (or a rows) around its destination column (destination row, respectively) once it arrives in that stripe. The motivation for a-bounded algorithms comes from an approach that is stingy in a certain sense: a packet will not give up its achievement of reaching a node near its destination column or row.

A combination of the short-sighted and a-bounded classes is a class of algorithms that are short-sighted outside the bounding stripe and that force packets to stay inside their respective stripes once they enter them. Variants of algorithms in the literature are provided that fall into the category of short-sighted a-bounded algorithms. The result obtained is that if the width of the corresponding stripes is a, then a short-sighted a-bounded algorithm will need $\Omega(n^2/q^2(a+1)^3)$ steps to route some (constructive, worst case) permutation on the $n \times n$ mesh, where q is the number of packets that can occupy a node at any one time (e.g., $q = 4$ for hot potato routing). The method is an extension to that of Chinn *et al.* [13], where $a = 0$.

Even if an algorithm is not by definition short-sighted a-bounded — that is, a packet could in general take a path that greatly deviates from any node of any of its minimal paths, even after it has reached within a nodes of its destination row or column — the technique may still be useful to prove lower bounds for such an algorithm. Recall that according to Theorem 3.12 every greedy algorithm that gives priority to restricted packets routes every batch routing problem with $k \leq n^2$ packets in $O(n\sqrt{k})$ steps on the $n \times n$ mesh. The method presented here yields an $\Omega(n^2)$ lower bound for permutation routing by a particular algorithm that is greedy and gives priority to restricted packets.

This implies that Theorem 3.12 is the best one can prove for that entire class of algorithms, even when the involved set of problems consists of permutations only.

Call the method by which a node determines which packets will be accepted the *inqueue policy* of the node. The inqueue policy of a node must ensure that the node does not accept more packets than it is capable of storing. Inqueue policies are not necessary for algorithms that assume unbounded buffer space, since any packet that enters a node will be accepted. However, a real machine will have some limited buffer capacity, q, and so some mechanism is needed to enforce this physical constraint. For hot-potato routing we may think of $q = 4$, however the basic methodology outlined in this section apply also to non hot-potato algorithms, where q may be larger.

A conservative inqueue policy would be as follows: if there are $q - i$ packets in a node at the end of step t, then the node accepts at most i packets in step $t + 1$. This approach is conservative because a node might be able to send one or more of its $q - i$ packets, allowing other packets to be accepted. In the hot-potato domain, the inqueue policy is to accept all incoming packets. Hot-potato algorithms can do this because nodes send all packets they currently store by the next step, allowing room for all incoming packets.

Assume that each node has a central queue of size q. Note that we could instead define a node to have four queues, one for each outlink or one for each inlink. However, a central queue of size $4q$ can simulate these four queues of size q, so we adhere to the more general queue model.

7.1 Short-sighted and a-bounded Algorithms

Each node of the network has a *state*, which may be taken into consideration when making the routing decisions. The algorithm may change the state of the node at the end of every step as a function of the node's current state and the packets that were in the node at the beginning of the step.

The algorithm may also store information in the header of the packets and change this information during algorithm execution. This information is the *state of the packet* and may contain, for example, the number of times the packet was deflected and the number of steps since the beginning of the algorithm. The state of the packet may even contain an infeasible amount of information, such as the collection of all states of the nodes the packet visited (at the time of the visit), and the states of all packets that it met on its way.

Short-sighted Algorithms

An algorithm is called *short-sighted* if the only part of the destination addresses of packets used in its routing decisions is the good directions of the packets. Thus, except for the good directions of packets, a short-sighted algorithm does not consider the destinations of the packets when deciding on the exit assignment. However, the algorithm is allowed to use other information available (e.g., the source addresses of packets in the node).

We refer the reader to [5, 6] for the formal definition of short-sighted algorithms.

Note that in an implementation of a routing algorithm, the direction from which a packet enters a node can be encoded in the packet's state and updated each time it moves. In fact, the origin and destination addresses of packets could also be encoded in its state, but it is convenient for the argument to distinguish these parts of the packet header.

a-bounded Algorithms

The lower bound in this work applies to deterministic routing algorithms on the two-dimensional mesh that force packets to remain near the surroundings of their destination row or destination column when they get close enough. Let a be some positive integral number, $0 \leq a \leq n-1$. An algorithm is called a-bounded if the following condition holds:

▷ Suppose a packet p reaches a node of distance a to its target row (column). The algorithm then guarantees that from that step until it reaches its destination, p will not enter a node of distance greater than a from its destination row (column, respectively).

Alternative definition: An alternative condition, which may be easier to enforce by a routing algorithm, is as follows. The algorithm guarantees that a packet that is deflected from a node of distance a to its destination row (column) into a node of distance $a+1$ to the destination row (column, respectively) returns to the node from which it came in the next step it makes a move.

Short-sighted a-bounded Algorithms

Finally, define algorithms that are a combination of the above definitions. Call a packet *area-bounded* if it already reached within a rows of its destination row or a columns of its destination column. Call all other packets *distant packets*. A *short-sighted a-bounded algorithm* is an a-bounded algorithm that is also short-sighted for distant packets. Note that by the definitions, a short-sighted a-bounded algorithm is not necessarily short-sighted.

We refer the reader to [5, 6] for the formal definition of short-sighted a-bounded algorithms.

For the proof of the main theorem, the unaltered definition of short-sighted a-bounded is used because it is a "clean" definition and the proof is easier to present with the unaltered definition. However, in the sequel, variants of routing algorithms from the literature are presented, that are short-sighted a-bounded under the alternate definition. The alternate definition is used because it is easier to describe the algorithms. After reading the proof of the main theorem, it should be easy for the reader to see that the same asymptotic bounds hold for algorithms that satisfy the alternate definition.

7.2 An Example of a Short-sighted a-bounded Algorithm

Bar-Noy *et al.* [3] gave a simple hot-potato algorithm for routing on the two-dimensional mesh that routes every permutation in $O(n^{3/2})$ steps. The algorithm is as follows.

Each packet moves horizontally (initially to the right) along its origin row, changing direction when it reaches the end of the mesh. Whenever a packet reaches its destination column, it tries to enter it in the up or down directions. A packet fails to enter the column if there are already two packets using the up and down links at that step. (Packets moving straight have priority over turning packets.) If a packet fails to enter its destination column, it proceeds in its current direction in its origin row until it reaches the end of the mesh. Then the packet changes direction, repeating this process until it succeeds in entering its destination column. When the packet succeeds in entering its destination column, it moves until it reaches its destination or until it reaches the end of the mesh, in which case it moves in the opposite direction.

In the Bar-Noy *et al.* algorithm, a packet that moves along a row or a column proceeds in a direction until the number of good directions change (from $\{+X, +Y\}$ to $\{+Y\}$ when it reaches its destination column, or from $\{+X\}$ to $\emptyset$ when it reaches its final destination), or until it arrives at the end of the mesh. Notice that when a packet fails to enter its destination column it may take more than n steps until it has a second opportunity to enter. Even when the packet eventually succeeds in entering its target column, it may take as many as $2n-3$ ($2n-4$ steps in permutation routing) steps until it reaches its final destination. Clearly, the routing decisions taken by each node depend solely on the good directions of the packets in it, the directions from which they came, and whether the node is on the end of the mesh. Thus the algorithm is short-sighted.

Let us attempt to improve the algorithm so that a packet that moves along its row and moves to within a nodes of its destination column will not be moved farther than this distance in all subsequent steps. Similarly, restrict a packet that moves to within a nodes of its destination row. It is easy to verify that this change can be made so that it does not involve unresolvable conflicts in the local routing decisions. For example, the following rules may be applied. Let the *destination stripe* of a packet be its destination column together with a columns to its right and a columns to its left. A packet that is in its destination stripe will be called an *column-bounded* packet.

"Improved" Bar-Noy *et al.* Algorithm.

> ***(1)*** *A packet starts moving to the right in its origin row. If it reaches the end of the mesh, it switches directions.*
>
> ***(2)*** *Whenever a packet moving horizontally reaches its destination column, it tries to enter it in either direction, preferably in the good one.*
>
> ***(3)*** *If a packet reaches its destination stripe, it starts moving from the first column of that stripe to the column at its "other side" and back. When switching directions conflicts with rule **(1)** above (for a different packet), then rule **(3)** takes priority.*
>
> ***(4)*** *Suppose a packet p changes directions as a result of applying rule **(3)** — that is, a packet p' was forced by rule **(3)** to switch directions, causing p to switch. If before the change p was moving towards its destination column, then it keeps trying to return to this direction at every step, giving priority to any other packet that attempts to move in that direction (unless this conflicts with rule **(3)**). On the other hand, if p was moving away from its destination column before the change, then it keeps moving in the new direction.*
>
> ***(5)*** *The vertical movement rules are revised in a similar manner to the horizontal ones.*

The resulting algorithm is short-sighted a-bounded (using the alternate definition of a-bounded). Notice that a packet that reaches its destination stripe may now get two chances to start the vertical movement every $4(a+1)$ steps rather than every $2(n-1)$ steps in the original algorithm. In fact, the analysis of Bar-Noy *et al.* depends strongly on the number of these chances [3]. One might argue that each application of rule **(3)** in which a packet is deflected allows an column-bounded packet another chance, at the worst case expense of the deflected packet missing such a chance. Thus, the worst case performance of the algorithm seems to be no worse, and it may appear that the change can speed up the original algorithm. However, a packet might be frequently deflected by column-bounded packets (by rule **(3)**) and thus rarely have a chance to turn into its destination column. The Bar-Noy *et al.* result depends on the fact that each packet frequently attempts to turn into its destination column (twice every $2(n-1)$ steps). By the lower bound in Section 7.5 (Theorem 7.4), for all $a \in o(n^{1/4})$, the "improvement" actually decreases the performance of the algorithm in the worst case.

7.3 Notation and Definitions

Let c be some constant $0 < c < 1$, so that cn is an integer. Square i is a submesh of size $(cn + i - 1) \times (cn + i - 1)$ at the lower left corner of the mesh.

Column N_i is the $(cn + a + i - 1)$-th column. Row E_i is the $(cn + a + i - 1)$-th row.

The stripe $\widetilde{N_i}$ is the collection of columns $N_{i-a}, N_{i-a+1}, \ldots, N_{i+a}$ (i.e., the column N_i and the a columns to its left and the a columns to its right). Similarly, the stripe $\widetilde{E_i}$ is the collection of rows $E_{i-a}, E_{i-a+1}, \ldots, E_{i+a}$ (i.e., the row E_i and the a rows above it and a rows below it). See Figure 12.

An *N_i-packet* is a packet that originates from square 1, and is destined for a node in column N_i above row E_i. An *E_i-packet* is a packet that originates from square 1, and is destined for a node in row E_i to the right of column N_i. Using the terminology of Section 7.1, note that an N_i-packet or E_i-packet becomes area-bounded when it enters the stripe $\widetilde{N_i}$ or $\widetilde{E_i}$.

In the algorithms to which the bound corresponds, if an N_i-packet enters stripe $\widetilde{N_i}$, it will remain in that stripe until it reaches its destination. If an E_i-packet enters stripe $\widetilde{E_i}$, it will remain in that stripe until it reaches its destination.

A packet is *in square* i if it is in a node that belongs to square i. A packet is *outside square* i if it is not in square i. A packet is said to *exits square* x when it is sent from a node in square x to a node outside that square.

An *exchange of two packets* π *and* π' is an exchange only of their destinations, so that the rest of the information they hold (state, origin, data) remains unchanged.

7.4 The Construction

A permutation is now constructed that will require a short-sighted a-bounded algorithm $\Omega(n^2/q^2(a+1)^3)$ steps to deliver all of the packets, where q is the maximum number of packets that can reside in a node at any given time. The result is proved for the store-and-forward model. To apply the result to hot-potato algorithms, simply observe that the hot-potato model is a special case of the store-and-forward model and that $q = 4$ for hot-potato algorithms.

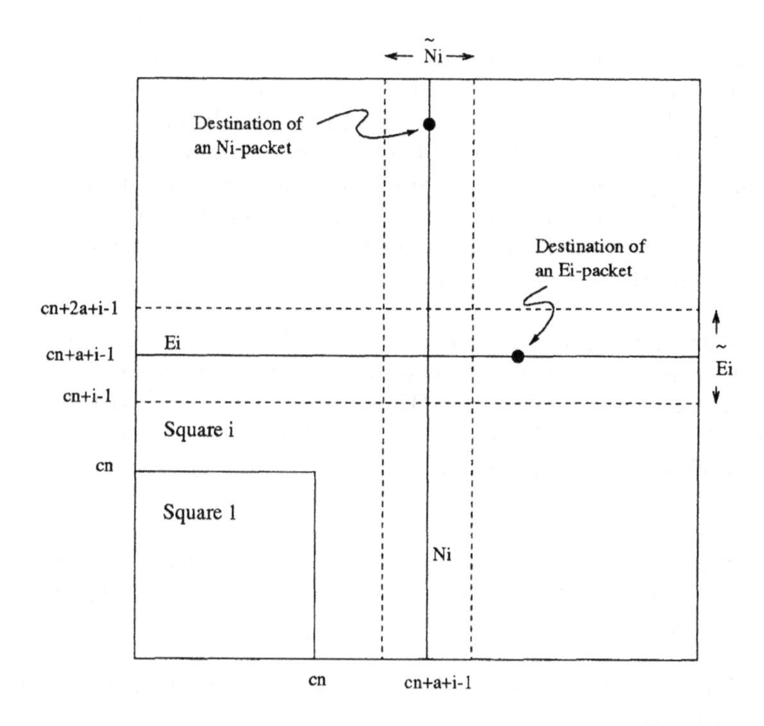

Figure 12 The $n \times n$ mesh

The construction of the bad permutation (or, the algorithm's *bottleneck permutation*) proceeds in phases. Define first an initial partial permutation, then execute the routing algorithm with this permutation as its input, and exchange destinations between packets during execution in a way that will cause the routing algorithm to prevent packets from advancing.

Specifically, the construction begins by placing one packet in each of the nodes of a $cn \times cn$ submesh in the lower left corner of the mesh. Their destinations are nodes above and to the right of this submesh. Throughout the construction, the invariants are maintained that only E_i-packets occupy the north boundary of square i and that only N_i-packets occupy the east boundary of square i. The node at the northeast corner of square i is allowed to contain both E_i- and N_i-packets. Since the routing algorithm is a-bounded, the easternmost $2a+1$ nodes on the north boundary of square $i+2a$ and the northernmost $2a+1$ nodes on the east boundary of square $i+2a$ are the only nodes through which these packets (and hence any packet in square $i+2a$) can exit square $i+2a$.

The invariant can be maintained because no packet in square i has reached a node within a nodes from its destination row or column, and since the routing algorithm is short-sighted for these packets, it cannot distinguish any of them. In the construction, this fact is used to exchange packets that would violate the invariant with packets that do not violate the invariant. For any particular i, the construction can maintain the above invariant for only $\Theta(n)$ steps. When it can no longer maintain the invariant for some value of i, the construction then maintains the invariant for the next higher i.

To summarize, there are $\Omega(n^2/q^2(a+1)^2)$ packets in the $cn \times cn$ submesh destined for nodes outside the submesh. Only $O(a+1)$ packets can exit square $1+2a$ during each step of the first $\Theta(n)$ steps, only $O(a+1)$ packets can exit square $2+2a$ during each step of the next $\Theta(n)$ steps, etc., up to square l, where $l = \Theta(n/q^2(a+1)^2)$. Thus, it takes $\Omega(n^2/q^2(a+1)^3)$ steps to deliver all packets.

Finally, it is shown that the routing algorithm routes the resulting permutation in the same number of steps as the construction did on the original permutation (with exchanges).

We now present the construction. Theorem 7.1 states the desired properties of the construction.

▷ Let $p = \lfloor (2a+1)dn + ((2a+1)q+1)(cn + c^2n + 2a) \rfloor$, where c and d are constants that will be chosen later so that cn and dn are integers. Let $l = c^2n^2/2p$. For every $1 \leq i \leq \lfloor l \rfloor$, put p N_i-packets that have different destinations, and p E_i-packets that have different destinations in square 1 so that

- in every node there is at most one packet,
- in every node of the rightmost column of square 1 (that is, nodes in column N_{1-a} that belong to square 1), there is an N_1 packet, and
- in every node of the top row of square 1 (that is, nodes in row E_{1-a} that belongs to square 1), there is an E_1 packet (except for the node at the upper right corner of square 1).

If $p \leq (1-c)n - l - a$, then such an arrangement is possible, since there will be at least p different destinations for the N_i- and E_i-packets for all $1 \leq i \leq \lfloor l \rfloor$. c and d will be chosen so that the inequality holds. Notice that l is chosen such that $2\lfloor l \rfloor p \leq c^2n^2$, and so there are enough origins for the packets that are placed in square 1.

▷ It is possible to add arbitrarily more packets in order to get a full permutation.

▷ Execute the routing algorithm for $\lfloor l \rfloor dn$ steps. During execution perform the following exchanges, depicted in Figure 13. (Essentially, there are always packets suitable for the exchanges in Theorem 7.1, part 2.)

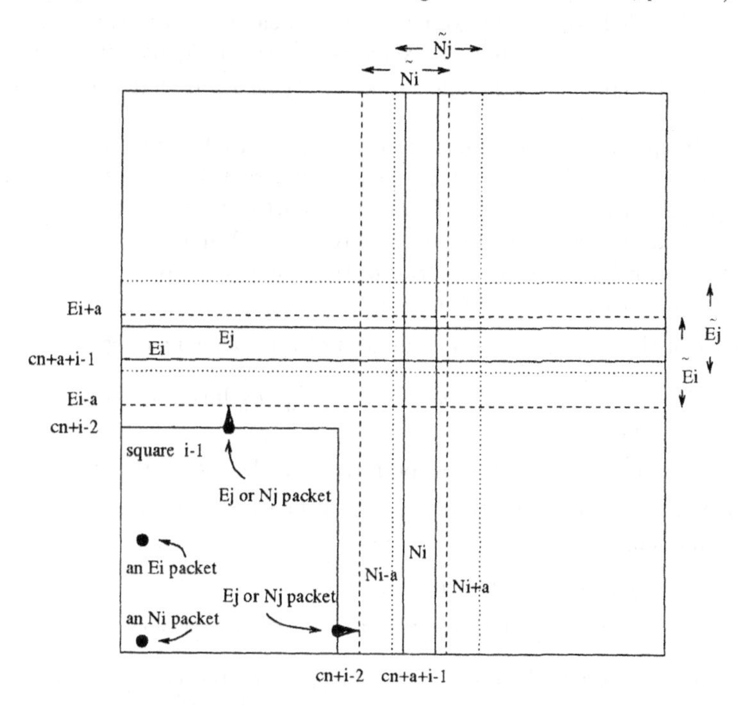

Figure 13 Exchanges during the construction: E_j-packets and N_j-packets (for $j > i$) that exit square $i-1$ during steps 1 to idn are exchanged with E_i-packets and N_i-packets.

- For every $j > i, i \geq 1$, during steps 1 to idn, if an E_j-packet is scheduled to be sent to a node that belongs to the stripe $\widetilde{E_i}$ left of column N_{i-a}, then it is exchanged with an E_i-packet from square $i-1$ that is not scheduled to enter stripe $\widetilde{E_i}$ at that step.
- For every $j > i, i \geq 1$, during steps 1 to idn, if an N_j-packet is scheduled to be sent to a node that belongs to the stripe $\widetilde{N_i}$ and that is below the row E_{i-a}, then it is exchanged with an N_i-packet from square $i-1$ that is not scheduled to enter stripe $\widetilde{N_i}$ at that step.

- For every $j \geq i, i \geq 1$, during steps 1 to idn, if an E_j-packet is scheduled to be sent to a node that belongs to the stripe $\widetilde{N_i}$ below row E_{i-a}, then it is exchanged with an N_i-packet from square $i-1$ that is not scheduled to enter stripe $\widetilde{N_i}$ at that step.
- For every $j \geq i, i \geq 1$, during steps 1 to idn, if an N_j-packet is scheduled to be sent to a node that belongs to the stripe $\widetilde{E_i}$ left of column N_{i-a}, then it is exchanged with an E_i-packet from square $i-1$ that is not scheduled to enter stripe $\widetilde{E_i}$ at that step.

Loosely speaking, these exchange rules ensure that during steps $(i-1)dn$ to idn, only N_j-packets and E_j-packets, where $j \leq i$, are in nodes at the north and east boundaries of square i. Furthermore, E_i-packets are not allowed at the east boundary of square i, and N_i-packets are not allowed at the north boundary of square i during these steps (except for the node at the northeast corner of square i).

Step t of the construction is composed of the following operations:

(a) The algorithm assigns packets to outgoing links, at most one packet per link.

(b) If required, exchanges are performed as described above.

(c) Nodes decide which incoming packets to accept (according to their inqueue policies).

(d) Accepted packets are sent.

(e) The states of the nodes and the states of the packets are updated.

Part **(b)** is executed only for the purposes of the construction and the lower bound argument, The actual routing algorithm ignores part **(b)**. Note that part **(c)** is omitted in hot-potato algorithms, since all incoming packets are accepted.

▷ The collection of pairs determined by the source and destination of the packets after $\lfloor l \rfloor dn$ steps of the construction is called the *bottleneck permutation*. Note that for different short-sighted a-bounded algorithms, the construction produces possibly different bottleneck permutations.

7.5 Relating the Construction and the Bottleneck Permutation

We now show that every short-sighted a-bounded algorithm will need at least $\Omega(n^2/q^2(a+1)^3)$ steps to complete the routing of its bottleneck permutation. Basically, some properties of the construction itself are stated, then it is claimed that these properties also hold when the algorithm is executed with its bottleneck permutation as its input.

The construction is divided into phases, each of dn steps. In the i-th phase (that is, steps $(i-1)dn+1$ through idn), stripes $\widetilde{E_i}$ and $\widetilde{N_i}$ contain E_i-packets and N_i-packets (and possibly N_k- and E_k-packets, where $k < i$), respectively. As a result, these stripes become congested, so that E_j-packets and N_j-packets for $j > i$ remain below stripe $\widetilde{E_i}$ and to the left of stripe $\widetilde{N_i}$. As is shown in Theorem 7.1 below, the only N_i-packets and E_i-packets that exit square $i+2a$ are the ones in the stripe $\widetilde{N_i}$ and $\widetilde{E_i}$, respectively. Since the width of these stripes is $2a+1$, then in every step at most $2a+1$ E_i-packets and at most $2a+1$ N_i-packets exit square $i+2a$.

Formally, by induction on t the following theorem can be proved.

Theorem 7.1 *During the construction, for every $1 \leq i \leq \lfloor l \rfloor$, the following properties hold:*

1. *During step t, for all $(i-1)dn < t \leq idn$, at most $2a+1$ N_i-packets and at most $2a+1$ E_i-packets exit square $i+2a$.*
2. *For any step t, $1 \leq t \leq idn$, whenever the construction determines that an exchange is to occur, there is always a suitable packet for the exchange, provided $l \leq c^2 n$.*
3. *E_j- and N_j-packets, for $j \geq i+1$, do not leave square $i+1$ during step t, for all $1 \leq t \leq idn$.*
4. *Until the end of the i-th phase, i.e., for all t, $1 \leq t \leq idn$, there are no N_i-packets above stripe $\widetilde{E_i}$ or within it, that are also to the left of stripe $\widetilde{N_i}$. Also, there are no E_i-packets to the right of stripe $\widetilde{N_i}$ or within it, that are also below the stripe $\widetilde{E_i}$.*

Corollary 7.2 *After $\lfloor l \rfloor dn$ steps of the construction there is still a packet in the system that has not reached its destination.*

We seek properties similar to those in Theorem 7.1 and, in particular, to those in Corollary 7.2. However, we are actually interested in the execution of the algorithm (no exchanges) with its bottleneck permutation as its input. The idea is to show that if after t steps of the construction packets π and π' are exchanged, then the same situation would be reached (after the exchange) as would have if the exchange would be performed *before* the beginning of the algorithm execution. This property holds because until the exchange, both exchanged packets did not make it to within a rows of the destination row or to within a columns of the destination column. Furthermore, both packets' good directions are up and right, and hence a short-sighted a-bounded algorithm cannot distinguish between them.

It can be shown that exchanging the packets throughout the construction causes the algorithm to behave as though the bottleneck permutation were given as its input (and no exchanges made). An a-bounded algorithm does not allow an area-bounded packet, a packet that reached its target stripe, to "escape" its stripe. In order that this condition be held, the routing algorithm must make decisions that depend on the packets' destinations, and not merely on their good directions. However, the construction performs exchanges only among distant packets. Since the information exchanged in an exchange cannot be used in routing decisions (other than the good directions, which, in this case, are the same), the algorithm cannot distinguish between the exchanged and unexchanged network configuration.

Finally, using the formal version of the above discussion, the main theorem follows.

Theorem 7.3 *Every short-sighted a-bounded algorithm requires at least $\lfloor l \rfloor dn$ steps to complete the routing of its bottleneck permutation.*

The analysis is completed by setting values for c and d which fulfill some constraints. It can be verified that when $n \geq 35 \cdot ((2a+1)q+2)^2$, all the requirements are fulfilled for some c and d where $\frac{0.3}{(2a+1)q+2} \leq c \leq \frac{0.35}{(2a+1)q+2}$ and $\frac{0.3}{2a+1} \leq d \leq \frac{0.35}{2a+1}$. For these values, $p = \Theta(n)$ and $\lfloor l \rfloor dn = \Theta(n^2/q^2(a+1)^3)$. Thus, by Theorem 7.3:

Theorem 7.4 *Every short-sighted a-bounded deterministic algorithm for routing permutations on the two-dimensional mesh with queues of size q requires $\Omega(n^2/q^2(a+1)^3)$ steps in the worst case, for $n \geq 35 \cdot ((2a+1)q+2)^2$.*

7.6 Row-column Algorithms

An algorithm that (**1**) sends a packet on a vertical link only if this link is on the packet's destination column and (**2**) sends a packet only on vertical links once it has crossed a vertical link, is called a *row-column algorithm*. An example is the Bar-Noy *et al.* algorithm from Section 7.2. An algorithm that sends a packet on a vertical link only if this link is in a "stripe" of consecutive columns that include the packet's destination column is called a *row-stripe algorithm*. Clearly, a row-column algorithm is a special case of a row-stripe algorithm with a stripe width of one. Similarly, one may define *column-row algorithms* and *column-stripe algorithms*.

In *dimension order algorithms*, a special case of row-column algorithms, a packet is sent on a vertical link if and only if it reaches its destination column (although the packet can be buffered there until it is sent), and the packet always moves towards its destination whenever it crosses a link. For short-sighted dimension order algorithms, Chinn *et al.* [13] proved that there is a permutation that requires $\Omega(n^2/q)$ steps to route, where q is the maximum queue size. A similar bound is shown in this section for row-column and row-stripe algorithms which are short-sighted and a-bounded.

Define *short-sighted a-bounded column-wise* algorithms in the same way as the more general definition which was given in Section 7.1 for short-sighted a-bounded algorithms, except that here (**1**) consider only row-stripe algorithms, and (**2**) define *column-bounded* packets to be those packets that already reached within a columns of their destination columns.

Observe that the construction of the bottleneck permutation and the resulting bound may be applied for *short-sighted a-bounded column-wise* algorithms. Furthermore in this case the construction and bound may be further improved.

For short-sighted a-bounded column-wise algorithms, it is sufficient to set a bottleneck permutation with N_i-packets, since only column-bounded packets will be sent by it in the vertical direction. In fact, following an observation by Chinn *et al.* [13], a stronger result can be obtained. Let us sketch the main idea. We repeat the construction of the bottleneck permutation, only that this time N_i-packets originate at a $cn \times (1-c)n$ rectangle at the lower left corner of the mesh, destined to a $(1-c)n \times cn$ rectangle at the upper right corner (Figure 14).

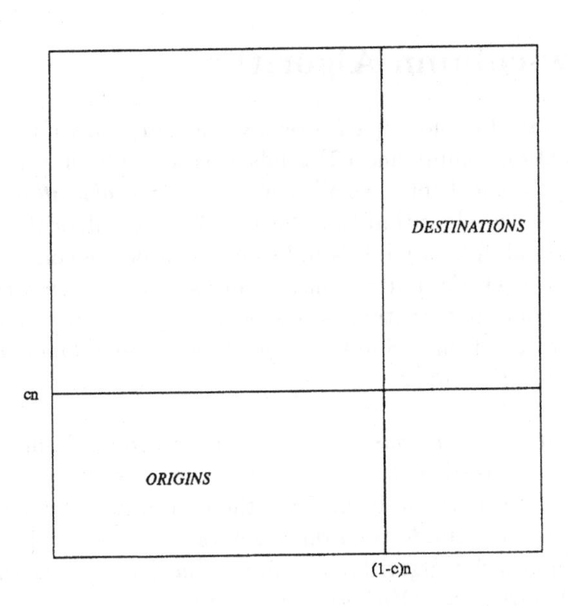

Figure 14 The construction for short-sighted a-bounded column-wise algorithms.

Theorem 7.5 *Every short-sighted a-bounded column-wise deterministic algorithm for routing permutations on the two-dimensional mesh requires*

$$\Omega(n^2/q(a+1)^2)$$

steps in the worst case.

The "improved" Bar-Noy *et al.* algorithm that was presented in Section 7.2 is a row-column algorithm which is short-sighted a-bounded column-wise. Thus we have:

Corollary 7.6 *The "improved" Bar-Noy* et al. *algorithm from Section 7.2 takes $\Omega(n^2/(a+1)^2)$ steps to route its worst case permutation.*

7.7 Algorithms that Perform Well on Average

In this section a row-stripe algorithm is described that routes an overwhelming majority of the batch problems in $2n + O(\log^2 n)$ steps. However, there exists a permutation that will take the algorithm $\Omega(n^2/\log^2 n)$ steps to complete. The algorithm is an adaptation of the Kaklamanis *et al.* hot-potato algorithm [20].

Consider the original Kaklamanis *et al.* algorithm. The mesh is split into vertical stripes, each of $b \log n$ columns. Each stripe is split into blocks of size $b \log n \times b \log n$, where b is some constant.

The Kaklamanis *et al.* Algorithm: (see also [20])

▷ A packet moves right or left along its origin row in the direction of the central column of the stripe to which it is destined.

▷ When a packet reaches its destination stripe, it tries to move upwards at odd columns or downwards at even columns in the direction of the central row of its destination block. A packet fails to turn if sending it conflicts with the transmission of another packet already moving vertically. Out of two packets that attempt to go in the same vertical direction arbitrarily, one is chosen, say, the one coming from the left. A packet that fails to turn proceeds horizontally in the direction it entered and tries to turn at the appropriate columns of its destination stripe. A packet that fails to turn in all of the first $b \log n/4$ appropriate columns, proceeds along the row and changes direction at the end of the mesh.

▷ A packet that succeeds in turning to an appropriate column in its target stripe proceeds until it reaches its target block.

▷ A packet that is already in its target block tries to switch to the opposite direction inside the column. In the block there is a "directed snake" along a Hamiltonian cycle that moves downwards in odd columns and upwards in even columns. If a packet conflicts with another that is already moving along the snake, then it proceeds along the column and again tries to switch directions at the next step. This process goes on as long as the packet is in its destination block. A packet that did not succeed in switching directions inside its destination block will proceed along the column and will switch columns and directions at the end of the mesh.

▷ A packet that succeeds in starting to move along the snake in its target block, proceeds along the snake until it reaches its final destination.

Kaklamanis *et al.* [20] show that for an overwhelming majority of the batch problems, this algorithm will terminate the routing in at most $2n + O(\log^2 n)$

steps. In particular, they show that in most of the permutations, each packet that attempts to turn in its destination stripe will succeed to do so in the first $b \log n/4$ trials. Also, each packet that tries to switch directions in its destination block will succeed to do so before it exits the block.

The "improved" Kaklamanis *et al.* Algorithm:

One attempt to "improve" their algorithm is by making it $b \log n$-bounded column-wise. The required change is a switch in the direction of a packet when it reaches a distance of $b \log n$ steps from the destination column (rather than at the end of the mesh). Notice that an area-bounded packet meets another horizontally-advancing packet $b \log n$ away from its destination column only if the other packet is not area-bounded, or if it is area-bounded but is not at the border of its bounding area.

It is easy to verify that the analysis in [20] holds for the "improved" algorithm as well. However, it is also easy to keep the "improved" algorithm short-sighted. Thus, as with the "improved" Bar-Noy *et al.* algorithm, it is possible to modify the Kaklamanis *et al.* algorithm so that it is a short-sighted $b \log n$-bounded column-wise algorithm. The condition in Theorem 7.5 then holds, and permutation routing takes $\Omega(n^2/\log^2 n)$ steps in the worst case.

Corollary 7.7 *The "improved" Kaklamanis* et al. *algorithm*

- ▷ *routes an overwhelming majority of batch problems in at most* $2n+O(\log^2 n)$ *steps, and*
- ▷ *takes* $\Omega(n^2/\log^2 n)$ *steps to route some worst-case permutation.*

7.8 Algorithms That Are Not a-bounded

One natural question that follows the previous discussion is which features of the algorithm make the lower bound hold? This question is extremely important, as "simple" and "adaptive" behavior of routing algorithms is not yet understood, and as the "nearly minimal" restriction seems somewhat artificial. It turned out that there exists an algorithm which is not a-bounded but for which a similar construction may be applied to get a lower bound of $\Omega(n^2)$ steps.

The specific algorithm that is given tries to keep the routes to destinations minimal in a greedy fashion, tries to keep packets in their destination rows or columns once they reach there, and considers only their good directions in its routing decisions before they reach there. This appears to be a simple, greedy approach that is locally adaptive: a packet takes minimal routes only when it experiences little congestion on its way. Nevertheless, the technique developed in this section shows that the algorithm is slow in routing some bottleneck permutation. Thus, algorithms which follow this "simple" and "greedy" behavior achieve poor performance.

Note that Theorem 3.12 is asymptotically optimal with respect to general routing problems with $k = \Theta(n^2)$ packets, since the case that all k packets have the same destination clearly takes $\Omega(n^2)$ steps. Does the routing time of all algorithms which give priority to restricted packets improve when less general routing requests are involved? The technique presented here gives a negative answer to this question for the set of permutation routing instances, by presenting a specific counterexample.

Consider a greedy algorithm that gives priority to restricted packets and that, among the non-restricted packets, prefers to return a packet that was just deflected from its destination row or its destination column to the node from which it was deflected. Furthermore, except for the above rule and the good and bad directions, the algorithm does not take the destination address into consideration in its routing decisions. Then it can be shown that this short-sighted algorithm behaves as if it were zero-bounded when routing certain bottleneck permutations.

More formally, consider the following greedy algorithm, which is described as the collection of operations taken by a node in the mesh. The rules are given in decreasing order of importance.

Algorithm Greedy_BHS.

(1) *Send the maximal number of restricted packets in good directions. This may be done arbitrarily.*

(2) *Return packets that were just deflected from their destination column or their destination row to the deflecting node, whenever the corresponding links are still free. (If it is not free, such a link may only be taken by a restricted packet.)*

(3) *Send the rest of the packets in any way that is compatible with the greedy rule and the following:*

a. *The exit assignment may not take the full destination addresses into account, only the good and bad directions.*

b. *Deflected packets that are restricted to the right will take the following precedence: If possible, then deflect to the left. If left is taken, then try to deflect downwards. Only when both left and downwards are taken, then deflect upwards.*

c. *Packets being deflected and restricted upwards will take the following precedence: If possible, then deflect downwards. If downwards is taken, then try to deflect to the left. Only when both downwards and left are taken, then deflect to the right.*

It is easy to verify that Algorithm GREEDY_BHS is greedy and gives priority to restricted packets. Thus, according to Theorem 3.12 it will route every routing problem with $O(n^2)$ packets in $O(n^2)$ steps. However, a bound of $\Omega(n^2)$ steps exists for this algorithm for routing a certain worst case permutation.

Theorem 7.8 *There exists a permutation for which Algorithm* GREEDY_BHS *takes $\Omega(n^2)$ steps to route.*

Note that for an algorithm to be greedy it is sufficient to consider the good directions. Thus, a greedy algorithm may be short-sighted. This, however, is not enough for implying Theorem 7.8 by Theorem 7.4, since the algorithm is not necessarily a-bounded: The priority-to-restricted-packets rule (Rule **(1)**) might conflict with this property, and it is of higher importance. In other words, Rule **(1)** is of higher importance than Rule **(2)**, and thus it may happen that a packet will not be able to return to its destination column or row immediately at the step after it is deflected from there. Despite this fact a construction similar to that given in Section 7.4 leads to an $\Omega(n^2)$ bound for this algorithm as well.

In fact it can be shown that when routing certain permutations, Algorithm GREEDY_BHS behaves as if it were zero-bounded. In order to prove this claim, consider in the rest of this section a new construction, where $a = 0$. Thus, an area-bounded packet is either restricted or was restricted in the previous step and will be so in the next step. (We use here the alternative definition of a-bounded in Section 7.1.) The original construction is changed as follows:

▷ The initial routing problem consists of packets of the N_i and E_i types, for all $1 \leq i \leq \lfloor l \rfloor$. No other packets are added (i.e., it is not a full permutation).

▷ Add the following exchange rule to the construction: At the end of the idn-th step, for every E_i- and N_i-packet that is not area-bounded, find a node at the upper-right $cn \times cn$ corner that is not the destination of any other packet, and change the destination of the packet to that node. As far that the exchange rules are concerned throughout the rest of the construction, such a packet will belong to the E_{l+1} type.

Note that c and l were chosen in Section 7.5 such that $2cn + l < n$. Thus there are no packets destined for the upper-right $cn \times cn$ corner at the beginning of the construction, so there are always enough destinations required by this change of destination.

Theorem 7.8 is proved by observing properties of Algorithm GREEDY_BHS. The properties culminate in showing that Algorithm GREEDY_BHS behaves during the construction as if it were zero-bounded. This property turns out to be strong enough for the lower bound analysis from the previous sections to hold.

Property 7.9 *The following hold when Algorithm* GREEDY_BHS *is used in the new construction: Let us call* bad *the event that an E_i-packet or N_i-packet that is area-bounded and attempts to return to its destination row or column is unable to do so because of a prioritized restricted packet. There are no bad events during the construction.*

The proof of this property goes by induction on t. It is somewhat tricky, see [6]. Finally, the proof of Theorem 7.8 is completed by observing that analogues for the types of arguments from Section 7.5 hold in the new construction as well, and by using Property 7.9.

8 CONCLUSIONS

The ultimate routing algorithm should satisfy two basic requirements: it should suit real-world implementations, and it must guarantee a better competitive performance (than that promised by traditional routing theory). These two requirements may in turn imply many others: e.g., that the algorithm must be simple, that it should not be composed of phases, that routing decisions are independent of time, that it is possible to inject new packets into the network throughout the routing, that the algorithm be adaptive to the load distribution, that there is no need for buffers, that there is no need for randomization, that the routing of a permutation terminate in the maximal distance of a packet to its destination plus some small overhead, etc. Although some of these were previously obtained, no algorithm is yet known to exhibit all of them. The research reported in this work was initiated in a desire to pinpoint sets of algorithms and routing policies that will support these requirements.

At an initial point in the research it was observed that algorithms that are implemented in practice are greedy in the sense that they attempt to send packets towards their destinations whenever possible. It was also observed that seemingly-practical algorithms commonly exhibit very good performance in simulations. As a result, the goal was set to find an algorithm which will be suitable for practical purposes, and for which it will be possible to prove a good performance bound. To this end, the target set of algorithms was restricted to include only those having a greedy behavior.

Unfortunately, most of the algorithms that are found in the routing literature are not greedy. One reason is that greedy algorithms adapt to the local loads in the network, and so the movement of the packets become unpredictable, following no a-priori "structure". As a result the paths taken by the packets are very different for different routing requests, which makes the analysis very hard. Thus, as a result of enforcing the greedy requirement on the design of the algorithm, novel analysis techniques need to be devised. The goal of this chapter is to present several techniques that were suggested in this study, and to explain the ideas behind them.

Since the adaptive, greedy behavior is hard to capture, the research focus on simple underlying architectures, and in particular on the two-dimensional mesh. The two-dimensional mesh is a bipartite graph, and so a greedy routing algorithm may result in a livelock situation. The suggested solution, which guided the research, was to give priority to packets that already reached their column or row destinations. There are some positive results to this approach, showing that restricted sub-sets of algorithms are efficient in routing some specific routing problems. In contrast, a negative result shows that the whole set of such algorithms cannot be expected to perform well in the worst case when routing permutations. The main results for the two-dimensional mesh are summarized in Figure 15.

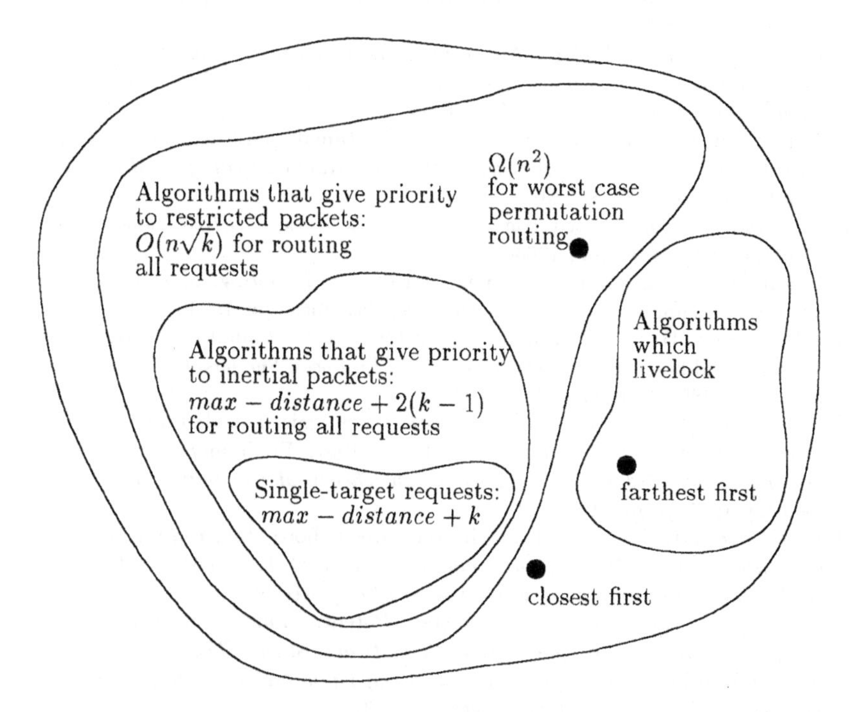

Figure 15 The world of greedy hot-potato routing algorithms on the two-dimensional $n \times n$ mesh. k is the number of packets in the routing request.

Acknowledgements

The author would like to thank his colleagues Ishai Ben-Aroya, Amir Ben-Dor, Donald Chinn, Tamar Eilam, Shai Halevi, and Ilan Newman for taking part in this research, and for their permission to present it here.

REFERENCES

[1] A. S. Acampora and S. I. A. Shah. Multihop lightwave networks: a comparison of store-and-forward and hot-potato routing. In *INFOCOM*, pages 10–19. IEEE, 1991.

[2] H. G. Badr and S. Podar. An optimal shortest-path routing policy for network computers with regular mesh-connected topologies. *IEEE Transactions on Computers*, 38(10):1362–1371, Oct. 1989.

[3] A. Bar-Noy, P. Raghavan, B. Schieber, and H. Tamaki. Fast deflection routing for packets and worms. In *Proceedings 12th Symposium on Principles of Distributed Computing*, pages 75–86. ACM, 1993.

[4] P. Baran. On distributed communications networks. *IEEE Transactions on Communications*, 12:1–9, 1964.

[5] I. Ben-Aroya. Algorithms and bounds for deflection routing. Master's thesis (In Hebrew), Technion, Computer Science Department, Apr. 1994.

[6] I. Ben-Aroya, D. D. Chinn, and A. Schuster. A lower bound for nearly minimal adaptive and hot potato algorithms. In *4th European Symposium on Algorithms*, pages 471–485, Barcelona, Sept. 1996.

[7] I. Ben-Aroya, I. Newman, and A. Schuster. Randomized single target hot potato routing. *Journal of Algorithms*, 23:101–120, 1997. Preliminary version in *Proceedings of the 3rd Israeli Symposium on Theory of Computing and Systems*, January 1995, pages 20–29.

[8] I. Ben-Aroya, E. Tamar, and A. Schuster. Greedy hot-potato routing on the two-dimensional mesh. *Distributed Computing*, 9(1):3–19, 1995. (Also in *Proceedings of the 2nd European Symposium on Algorithms*, Utrecht, 1994).

[9] A. Ben-Dor, S. Halevi, and A. Schuster. Potential function analysis of greedy hot-potato routing. *To appear in Mathematical Systems Theory*, 1997. Preliminary version in *Proceedings of the 13th Symposium on Principles of Distributed Computing*, Los Angeles 1994, pages 225–234.

[10] A. Borodin and J. Hopcroft. Routing, merging, and sorting on parallel models of computation. *Journal of Computer and System Sciences*, 30:130–145, 1985.

[11] A. Borodin, Y. Rabani, and B. Schieber. Deterministic many-to-many hot potato routing. Technical Report RC 20107, IBM Watson Research Report, 1995.

[12] J. T. Brassil and R. L. Cruz. Bounds on maximum delay in networks with deflection routing. In *Proceedings of the 29th Allerton Conference on Communication, Control and Computing*, pages 571–580, 1991.

[13] D. D. Chinn, T. Leighton, and M. Tompa. Minimal adaptive routing on the mesh with bounded queue size. *Journal of Parallel and Distributed Computing*, 34(2):154–170, 1996. Preliminary version in *Proceedings of the 6th Symposium on Parallel Algorithms and Architectures*, June 1994, pages 354–363.

[14] U. Feige. Observations on hot potato routing. In *Proceedings of the 3rd Israeli Symposium on Theory of Computing and Systems*, pages 30–39, Jan. 1995.

[15] U. Feige and P. Raghavan. Exact analysis of hot-potato routing. In *Proceedings of the 33rd Symposium on Foundations of Computer Science*, pages 553–562. IEEE, Nov. 1992.

[16] A. G. Greenberg and J. Goodman. Sharp approximate models of adaptive routing in mesh networks. In O. J. Boxma, J. W. Cohen, and H. C. Tijms, editors, *Teletraffic Analysis and Computer Performance Evaluation*, pages 255–270. Elsevier, Amsterdam, 1986.

[17] A. G. Greenberg and B. Hajek. Deflection routing in hypercube networks. *IEEE Transactions on Communications*, June 1992.

[18] B. Hajek. Bounds on evacuation time for deflection routing. *Distributed Computing*, 5:1–6, 1991.

[19] S. Halevi. On greedy hot potato routing. Master's thesis (In Hebrew), Technion, Computer Science Department, July 1993.

[20] C. Kaklamanis, D. Krizanc, and S. Rao. Hot-potato routing on processor arrays. In *Proceedings of the 5th Symposium on Parallel Algorithms and Architectures*, pages 273–282. ACM, 1993.

[21] M. Kaufmann, H. Lauer, and H. Schröder. Fast deterministic hot-potato routing on meshes. In *Proceedings of the 5th International Symposium on Algorithms and Computation (ISAAC)*, volume 834 of *Lecture Notes in Computer Science*, pages 333–341. Springer-Verlag, 1994.

[22] F. T. Leighton. *Introduction to Parallel Algorithms and Architectures*. Morgan Kaufmann Publishers, 1991.

[23] Y. Mansour and B. Patt-Shamir. Many-to-one packet routing on grids. In *Proceedings of the 27th Symposium on Theory of Computer Science*, pages 258–267, Las-Vegas, Nevada, 1995. ACM.

[24] N. F. Maxemchuk. Comparison of deflection and store and forward techniques in the manhattan street and shuffle exchange networks. In *IEEE INFOCOM*, pages 800–809, 1989.

[25] F. Meyer auf der Heide and C. Scheideler. Routing with bounded buffers and hot-potato routing in vertex-symmetric networks. In *Proceedings of the 3rd European Symposium on Algorithms*, pages 341–354, Corfu, Greece, Sept. 1995.

[26] I. Newman and A. Schuster. Hot-potato algorithms for permutation routing. *IEEE Transactions on Parallel and Distributed Systems*, 6(11):1168–1176, Nov. 1995.

[27] I. Newman and A. Schuster. Hot-potato worm routing via store-and-forward packet routing. *Journal of Parallel and Distributed Computing*, 30:76–84, 1995.

[28] J. Y. Ngai and C. L. Seitz. A framework for adaptive routing in multicomputer networks. In *Proceedings of the 1st Symposium on Parallel Algorithms and Architectures*, pages 1–9. ACM, 1989.

[29] R. Prager. An algorithm for routing in hypercube networks. Master's thesis, University of Toronto, Computer Science Department, 1986.

[30] C. Seitz. The Caltech Mosaic C: an experimental, fine-grain multicomputer. In *Proceedings of the 4th Symposium on Parallel Algorithms and Architectures*, San Diego, June 1992. ACM. Keynote Speech.

[31] B. Smith. Architecture and applications of the HEP multiprocessor computer system. In *Proceedings of (SPIE) Real Time Signal Processing IV*, pages 241–248, 1981.

[32] T. Szymanski. An analysis of hot potato routing in a fiber optic packet switched hypercube. In *IEEE INFOCOM*, pages 918–926, 1990.

[33] Z. Zhang and A. S. Acampora. Performance analysis of multihop lightwave networks with hot potato routing and distance age priorities. In *IEEE INFOCOM*, pages 1012–1021, 1991.

12

MODELS FOR OPTICALLY INTERCONNECTED NETWORKS

Pascal Berthomé and Michel Syska*

LRI, Université Paris-Sud, Orsay, France

**SlooP: joint project I3S-CNRS, INRIA and University of Nice - Sophia Antipolis, France*

ABSTRACT

Switching techniques used in optically interconnected networks differ from those used in classical electronically interconnected networks. This yields new communication models. The aim of this chapter is to survey the results of communication models in three fields: the design of networks, the algorithmics of data communication and the computational models of multiprocessor systems interconnected with optical networks.

1 INTRODUCTION

Massively parallel computers are proposed as the solution for high performance computing. However, parallel computing involves a lot of data communications between the processors that cooperate on the same computation. The amount of time required to perform those communications is prohibitive to the overall performance of the systems considered. As a consequence, dense interconnection network design and fast collective communication protocols are the keys for achieving expected performances. Indeed, multiprocessors systems are made of independent processing units - equipped with a local memory - exchanging data over an interconnection network. Two kinds of topologies are used: point-to-point and multi-stage interconnection networks. For instance, hypercubes and grids are popular point-to-point interconnection networks used in parallel computers. Multi-stage interconnection networks were designed in the case of telecommunication networks but are also relevant for workstation based computing. In this case, a central switch provides a virtual complete

P. Berthomé and A. Ferreira (eds.), Optical Interconnections and Parallel Processing: Trends at the Interface, 355-393.

graph topology. Moreover, as the existence of standard message passing libraries such as PVM and MPI makes it easier today to program such systems, interconnections are getting more and more importance.

The impact of the optical technology on the network modeling is investigated in the following. Three types of models are given: interconnection models (topologies), communication models and computational models.

Usually a N nodes multiprocessor system is represented by a graph $G = (V, E)$, in which V is the set of nodes of the graph representing the processors (and local memory associated to that processor) and E is the set of edges of the graph representing the communication links between processors. The model is accurate for parallel computers where processors are pair-wise connected, each edge representing the connection between two neighbors. The communication of data between two nodes is thus of the one-to-one type. In the case of optical interconnection networks, it is "easy" to implement a one-to-many type of communication, extending the concept of neighbors. This can be represented by hypergraphs and will be developed in Section 3.

When one node has a piece of data to communicate to other nodes, the corresponding message in which the data is encapsulated may have to switch through intermediate nodes, thus introducing delay in the time required to deliver the message to its destination node. Collective communications corresponds to the case when the communication implies more than two nodes.

Two paradigms of elementary collective communications are usually considered: *broadcasting* and *gossiping.* In broadcasting, one node has a piece of data it would like to share with all the other nodes in the network. At the end of the protocol, all the nodes must have that piece of data in their local memory. In gossiping, all the nodes are performing a broadcast simultaneously. At the end of the protocol, all the nodes have pieces of data originated in all the other $N - 1$ nodes of the network considered.

Communication models are required in order to describe the algorithms and the time complexity of communication algorithms. The main results in the case of electronically interconnected networks can be found in [20, 39]. The results in the case of optically interconnected networks are given in Section 4.3.

Finally, two models of computation that take advantage of these communication models are introduced in Section 5. These models can be seen as extensions of the popular PRAM model.

2 SWITCHING TECHNIQUES

The nodes communicate with their neighbors by exchanging messages through *channels.* A channel is a one-way point-to-point connection between two nodes connected by a physical link (arc of the graph). Several channels can share the same physical link in the case of *multiplexing.*

The communication features of the interface between the memory and the communication links should also be characterized for each processor. During a communication, if each node can only send or receive one message on one link at a time, the communication is called 1-*port.* If, on the contrary, each node can simultaneously use all its links, the communications are called Δ-*port*, where Δ refers to the maximum degree of the nodes in the network.

When a message is transmitted between processors that are not directly linked, the message must be routed through intermediate nodes and this routing is done with the help of routers. A router is characterized by its *switching time*, also called *latency.* Switching in a router consists of receiving a destination address, decoding the address in order to determine the appropriate output channel, and sending the message through this channel. Depending on the protocols used, switching can also include physical connection of the input link with the output link determined by the router. The latency can be only a few tens of nanoseconds in case of hardware routing but can go beyond a microsecond in the case of software routing.

2.1 Usual switching techniques

The various usual switching techniques are described by Kermani and Kleinrock in [38]. They are as follows.

Circuit-switching

This is the principle of a telephone: a connection is established first (this means reserving a sequence of channels) and the conversation begins after.

Message-switching

Messages move through the network towards their final destination by passing through intermediate nodes. At each stage, the channel used is immediately freed. This technique is known under the name *store-and-forward* in the context of distributed machines. One flaw of this technique is the necessity of large registers for storing the message on the intermediate processors. In fact, such messages are usually stored in the global memory. However, memory access time, being proportional to the size of the messages, slow down communication dramatically.

Wormhole routing

In the most recent distributed memory machines, the store-and-forward routing mode was displaced by *wormhole* routing.
Contrary to the store-and-forward mode, in which messages (or packets) are entirely stored in the memory of a processor before being transmitted to the next processor, in the wormhole routing mode the messages proceed through the processor network flit by flit (a *flit* — *flow control digit* — is the size of the buffer of a channel), with the first flit containing the destination address. The header, that is, the first flit, progress by a channel each time it is possible. The rest of the message follows, freeing the last channel which contains the end of the message. The last channel then becomes available for another message.

It is very important to distinguish this routing mode from routing by packet-switching. In the latter, each packet contains the destination address in its header and can be routed independently. In wormhole routing mode, only the first flit contains the destination address.

Now the time required to send a message of size L over links having a constant bandwidth $\frac{1}{\tau}$ is considered.

In the case of store-and-forward, the transmission time of the message between two neighboring processors is given by the sum of a *start-up* (or initialization) time, denoted by β, which is the time it takes to initialize the memory registers and the time of receipt procedures, and a propagation time $L\tau$, which is directly proportional to the length L of the message. Thus the cost of sending a message of size L at distance d is $d(\beta + L\tau)$. This cost could be decreased if one allows us to split the message in packets, and pipeline the packets along the path. For an optimal size of packets the cost is decreased to [20]: $\left(\sqrt{L\tau} + \sqrt{(d-1)\beta}\right)^2$.

In the case of wormhole or circuit-switching, one can send a message directly to a node at distance d with time $\alpha + d\delta + L\tau$, where the parameter α is the start-up time of the sending process and the delay δ is the time required to switch the router at each intermediate node.

In order to compare the last two models, observe that a distributed machine with circuit-switching or wormhole routing can also communicate neighbor-to-neighbor. Hence $\beta = \delta + \alpha$.

2.2 Wavelength Division Multiplexing

A new model of commutation is presented here, interested readers should refer to [17, 47] for an excellent introduction to the techniques. Only logical aspect of the communication are considered and we keep the optical implementations details to the minimum.

The large optical spectrum may be divided into numerous different channels, and each is assigned a different wavelength. This approach is known as WDM: Wavelength Division Multiplexing. The limit on the number of wavelengths available depends on the technology of lasers and optical filters. Technical details are out of the scope of this chapter. However, subcarrier multiplexing and electronic Time Division Multiplexing could be used within each wavelength in order to increase the number of different possible channels. This multiplexing will not slow down the communications as the interfaces of the nodes are not able to take full advantage of the bandwidth of optical fibers.

WDM lightwave networks are usually classified into *broadcast-and-select* networks and *wavelength-routing* networks. In both categories, single-hop and multihop networks [45, 46] could be considered. However, the models presented here focus on broadcast-and-select networks.

In broadcast-and-select networks, N stations connected to the same network use N different wavelengths to communicate via a passive network fabric (Star coupler). Each station is equipped with at least one transmitter T and one receiver R. The wavelength of each transmitter is broadcast to all receivers, see Figure 1.

The right side of the figure is an example with 3 stations and each one is built of one transmitter and one receiver. Every station emits its signal on its own wavelength, and receive all the other signals.

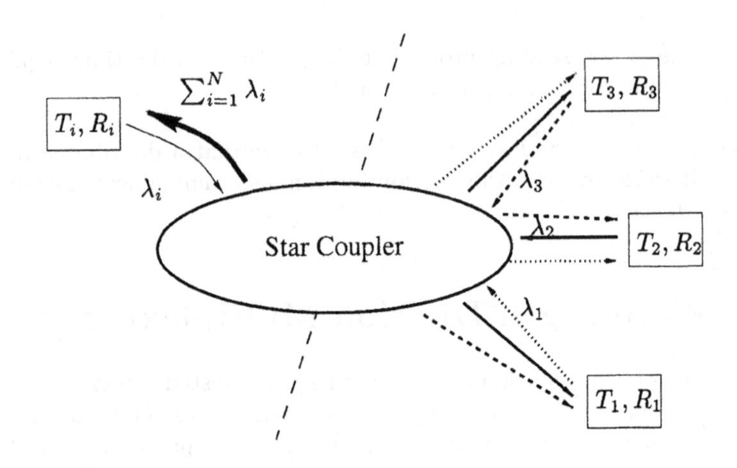

Figure 1 Broadcast-and-select

Two types of transmitters/receivers are available: fixed or tunable wavelength types. When fixed wavelength receivers are used, the transmitters have to be tunable and should be tuned to the appropriate wavelength before each communication. In the case of tunable transmitters and fixed wavelength receivers, the source node selects the wavelength before the communication is established. The last case is when both transmitters and receivers are tunable, the arbitration protocol usually uses a control channel. Fixed wavelength devices are often chosen due to the prohibitive cost of tunable ones.

Most of the results described in the following section concern OPS based networks. Rainbow [36] is an example of a practical implementation of such network.

Usually, the number of wavelengths available to build an interconnection network is limited due to cost reasons, thus all the messages could not be delivered in one hop. Messages transit through switches and networks are said to be multi-hop. In the case of multi-hop networks, conversions from/to electronic or photonic domain are required.

Possible multi-hop network topologies are described in Section 3, and corresponding collective communication issues are given in Section 4.

The case of single-hop networks is presented in Section 4.3. These networks are also known as *all-optical* networks as messages reach their destination in one hop without being converted to electronical representation in between. When the number of wavelengths available is not sufficient to complete the data exchange in one step, the number of additional steps required has to be minimized.

Others limitations could come from the optical power budget, indeed a minimum power is required at each receiver (dividing the signal may introduce loss). The transmitter should have a higher power.

The communication algorithm will have to take into account these different switching techniques.

3 TOPOLOGIES: FROM GRAPH TO HYPERGRAPH MODELS

A graph representation of interconnection networks is considered and new results in the design of topologies motivated by optical devices are given. Indeed, the way nodes are interconnected in a network is driven by technical constraints: complete interconnection is limited to a small number of nodes as each I/O port grows up the complexity of a node. Even optics has limits on the fan-out of switches.

The following definitions will be used in this chapter.

3.1 Definitions

The usual notations are taken from [6].

- ▷ A *directed graph* (or simply a *digraph*) $G = (V, A)$ where V is called the vertex set and A, a multiset whose elements are from $V \times V$, is called the arc set. A *symmetric digraph* is a digraph such that if $(u, v) \in A(G)$ then $(v, u) \in A(G)$.
- ▷ The number of vertices of the graph is called its *order* and is denoted by N.
- ▷ y is said to be a *successor* of x if there is an arc (x, y). The set of successors of a vertex x is denoted by $\Gamma_G^+(x)$ and its cardinality, denoted by $d^+(x)$, is called the *outdegree* of x. The set $\Gamma_G^-(x)$ of *predecessors* of x and the *indegree* $d^-(x)$ are defined similarly.

- ▷ In many cases the distinction between initial and end vertices is irrelevant. Thus the notion of an *undirected graph* is introduced: an arc (x, y) is replaced by the set consisting of the two vertices x and y, called an *edge* of the graph and denoted by $[x, y]$.

- ▷ Two vertices are *adjacent* or *neighbors* if there exists an arc or edge between them.

- ▷ Given a vertex x of a graph G, the number of edges incident with x is called the *degree* of x, denoted by $d_G(x)$ (or by $d(x)$ if confusion is unlikely).

  The maximum over the degrees of all vertices of G is called the *maximum degree* and is denoted by $\Delta(G)$, or simply Δ.

  The minimum over the degrees of all vertices of G is called the *minimum degree* and is denoted by $\delta(G)$ or, simply, δ.

- ▷ A *path* between two vertices x and y (and denoted $P(x, y)$) of a graph G is a sequence $x_1, x_2, \ldots, x_k$ of vertices such that pairs of consecutive vertices are adjacent while $x_1 = x$ and $x_k = y$. A *dipath* from node x to node y is a directed path which consists of a set of consecutive arcs beginning in x and ending in y.

  A path using each vertex at most once is called *elementary*. In the following all the paths considered are elementary and elementary will not be mentioned.

  The *length* of a path (resp. dipath) is the number of edges (resp. arcs) in it.

- ▷ Given two vertices x and y of a graph G, the *distance* between x and y is the length of a shortest path between them and is denoted by $\delta(x, y)$.

- ▷ The *diameter* of a graph G, denoted by $D(G)$ or, simply, D, if the context is clear, is the maximum of the distances $\delta(x, y)$ over all pairs of vertices of G.

- ▷ A *cycle* in a graph G is a path whose initial and end vertices are identified. A cycle is usually meant an *elementary* cycle, that is, one using no vertex more than once.

- ▷ The *Cartesian sum*, often called *Cartesian product* or *box product*, denoted by $G \Box G'$, of two graphs $G = (V, E)$ and $G' = (V', E')$, is the graph whose vertices are the pairs (x, x') where x is a vertex of G and x' is a vertex of G'. Two vertices (x, x') and (y, y') of $G \Box G'$ are adjacent if and only if either $x = y$ and $[x', y']$ is an edge of G', or $x' = y'$ and $[x, y]$ is an edge of G.

3.2 Degree vs Diameter

New optical devices, such as Optical Passive Stars (see Figure 1), bring new interests in network topologies research. Signals are broadcast simultaneously on different wavelengths and these devices could implement what is called a bus network. A bus is a multiple access medium shared among two or more nodes, whether it is based in electronics or optics (see [50] for a description of possible implementations). These networks are modeled by hypergraphs where vertices represent the processors and edges represent the buses. The following construction methods of bus networks that connect a large number of processors with a given maximum processor degree Δ, a maximum bus size r, and a network diameter D are taken from [12]. Hypergraphs are used to represent the underlying topology of the bus interconnection networks.

(Δ, D, r)*-hypergraph problem*

An *(undirected) hypergraph* H is a pair $H = (\mathcal{V}(H), \mathcal{E}(H))$ where $\mathcal{V}(H)$ is a non-empty set of elements, called *vertices*, and $\mathcal{E}(H)$ is a finite set of subsets of $\mathcal{V}(H)$ called *edges*. The number of vertices in the hypergraph is $n(H) = |(\mathcal{V}(H)|$ and the number of edges is $m(H) = |\mathcal{E}(H)|$ where the vertical bars denote the cardinality of the set. The *degree* of a vertex v is the number of edges containing it and is denoted by $\Delta_H(v)$. The *maximum degree* over all of the vertices in H is denoted by $\Delta(H)$. The *size* of an edge $E \in \mathcal{E}(H)$ is its cardinality, and is denoted by $|E|$. The *rank* of H is the size of its largest edge, and is denoted by $r(H)$. A *path* in H from vertex u to vertex v is an alternating sequence of vertices and edges $u = v_0, E_1, v_1, \cdots, E_k, v_k = v$ such that $\{v_{i-1}, v_i\} \subseteq E_i$ for all $1 \leq i \leq k$. The *length* of a path is the number of edges in it. The *distance* between two vertices u and v is the length of a shortest path between them. The *diameter* of H is the maximum of the distances over all pairs of vertices, and is denoted by $D(H)$.

An hypergraph with maximum degree Δ, diameter D, and rank r, is called a (Δ, D, r)*-hypergraph*. An example of a $(2, 2, 5)$*-hypergraph* is given in Figure 2. The problem on bus networks considered in the introduction is known as the (Δ, D, r)*-hypergraph problem* and consists of finding (Δ, D, r)-hypergraphs with the maximum number of vertices or finding large (Δ, D, r)-hypergraphs. The maximum number of vertices in any (Δ, D, r)-hypergraph is denoted by $n(\Delta, D, r)$.

In the case $r = 2$ (graph case), this problem has been extensively studied and is known as the (Δ, D)-graph problem (see for example [10], [11]). The maximum number of vertices in any (Δ, D)-graph is denoted by $n(\Delta, D)$. See the table [1] and construction of such graphs detailed in [10].

Finally, let us mention that the drawing of hypergraphs can be very complex and therefore it is useful to represent an hypergraph H with a bipartite graph,

$$R(H) = (\mathcal{V}_1(R) \cup \mathcal{V}_2(R), \mathcal{E}(R))$$

called the *bipartite representation graph.* Every vertex v_i in $\mathcal{V}(H)$ is represented by a vertex v_i in $\mathcal{V}_1(R)$ and every edge E_j in $\mathcal{E}(H)$ is represented by a vertex e_j in $\mathcal{V}_2(R)$. An edge is drawn between $v_i \in \mathcal{V}_1(R)$ and $e_j \in \mathcal{V}_2(R)$ if and only if $v_i \in E_j$ in H.

Moore bound

A bound on the maximum number of vertices in a (Δ, D, r)-hypergraph (analogous to the the classical Moore bound [34]) can be easily calculated: Each vertex belongs to at most Δ edges and each edge contains at most r vertices. Thus there can be at most $\Delta(r-1)$ vertices at distance one from any vertex. More generally, the maximum number of vertices at distance i from any vertex can be at most $\Delta(\Delta-1)^{i-1}(r-1)^i$. Therefore

$$n(\Delta, D, r) \leq 1 + \Delta(r-1) \sum_{i=0}^{D-1} (\Delta-1)^i (r-1)^i.$$

This bound is known as the *Moore bound for undirected hypergraphs*, and hypergraphs attaining it are known as *Moore geometries*.

For $D > 2$, Moore geometries cannot exist, with the exception of the cycles of length $2D+1$ (the case $\Delta = 2$ and $r = 2$). For a comprehensive survey on these results see [7]. Even for $D = 2$ and $r = 2$ (graph case), only four Moore graphs can exist.

3.3 Directed hypergraphs

Now let us mention the directed vs undirected question. Indeed, a lot of work has been done with undirected hypergraph models while actual problems deal with either directed topologies or symmetric directed ones. The problem of

[1] The table of the largest known (Δ, D)-graphs is maintained by the group of graph researchers in Barcelona, at URL : `http://www_mat.upc.es/grup_de_grafs/table_g.html`

global communications will be studied under the directed assumption and a formal definition of directed hypergraphs will be given in the following. See Section 5 for examples.

In the directed bus networks a bus is divided into two subsets. One subset of nodes can use the bus only to send messages while the nodes of the other subset can only receive messages from the bus.

Definition

A *directed hypergraph* H is a pair $(\mathcal{V}(H), \mathcal{E}(H))$ where $\mathcal{V}(H)$ is a non-empty set of elements (called *vertices*) and $\mathcal{E}(H)$ is a set of ordered pairs of non-empty subsets of $\mathcal{V}(H)$ (called *hyperarcs*). If $E = (E^-, E^+)$ is a hyperarc in $\mathcal{E}(H)$, then the non-empty vertex sets E^- and E^+ are called the *in-set* and the *out-set* of the hyperarc E, respectively. The sets E^- and E^+ need not be disjoint. $|E^-|$ is the *in-size*, and $|E^+|$ is the *out-size* of hyperarc E. The *maximum in-size* and the *maximum out-size* of a directed hypergraph H are, respectively,

$$s^-(H) = \max_{E \in \mathcal{E}(H)} |E^-| \quad \text{and} \quad s^+(H) = \max_{E \in \mathcal{E}(H)} |E^+|.$$

If $s^- = s^+ = 1$, a directed hypergraph is nothing more than a digraph.

Let v be a vertex in $\mathcal{V}(H)$. The *in-degree* of v is the number of hyperarcs that contain v in their out-set, and is denoted by $d_H^-(v)$. Similarly, the *out-degree* of vertex v is the number of hyperarcs that contain v in their in-set, and is denoted by $d_H^+(v)$. The *maximum in-degree* and the *maximum out-degree* of H are, respectively,

$$d^-(H) = \max_{v \in \mathcal{V}(H)} d_H^-(v) \quad \text{and} \quad d^+(H) = \max_{v \in \mathcal{V}(H)} d_H^+(v).$$

A *walk* in H from vertex u to vertex v is an alternating sequence of vertices and hyperarcs $u = v_0, E_1, v_1, E_2, v_2, \cdots, E_k, v_k = v$ such that $v_{i-1} \in E_i^-$ and $v_i \in E_i^+$ for each $1 \leq i \leq k$. The *length* of a walk is equal to the number of hyperarcs on it. The *distance* and the *diameter* are defined analogously to those in the undirected case.

The incidence relations between the vertices and hyperarcs in a directed hypergraph H are represented using a bipartite digraph,

$$R(H) = (\mathcal{V}_1(R) \cup \mathcal{V}_2(R), \mathcal{E}(R))$$

called the *bipartite representation digraph.* Every vertex v_i in $\mathcal{V}(H)$ is represented by a vertex v_i in $\mathcal{V}_1(R)$ and every hyperarc E_j in $\mathcal{E}(H)$ is represented by a vertex e_j in $\mathcal{V}_2(R)$. An arc is drawn from $v_i \in \mathcal{V}_1(R)$ to $e_j \in \mathcal{V}_2(R)$ if and only if $v_i \in E_j^-$ in H, and an arc is drawn from $e_j \in \mathcal{V}_2(R)$ to $v_i \in \mathcal{V}_1(R)$ if and only if $v_i \in E_j^+$ in H.

3.4 Practical topologies

Due to practical reasons, a good topology for an interconnection network is rarely the largest known (Δ, D)-graph or (Δ, D, r)-hypergraph. Indeed, the network has to be implemented with the available technology, and the design should be scalable for economic reasons.

Parallel computers use to have electronic based interconnection networks. Popular topologies are hypercubic networks such as hypercubes, meshes, tori, or more generally speaking k-ary-n-cubes [39]. They offer regularity, symmetry, high connectivity, fault tolerance, simple routing and also reconfigurability. In the case of the hypercube network, the logarithmic diameter is one attractive feature of the topology. However, a major drawback is the scalability of the networks: the node complexity (degree) increases with the total number of nodes in the network.

By the way, low (and/or constant) degree and small diameter networks are better candidates. This is the case for *de Bruijn* and *Kautz* graphs which are among the best known with respect to the Δ and D parameters.

Once the problem is stated for graph models, one have to deal with hypergraph models. What kind of topology is best suited for bus networks? Different networks were recently proposed that take advantage of the optical issues. Bus networks are considered today because of the advantages of optics over electronics: high bandwidth, large fan-out and low signal crosstalk. In addition to the hypergraphs presented here, interested readers should refer to the table in [8].

Hypermeshes and hypercubes

Mesh is probably the most popular topology in many areas of interconnection networks : VLSI routing, Wafer Scale Integration, arrays of processors, parallel computers, metropolitan networks. Main qualities are: scalability, simple routing (including deadlock-free routing), natural embedding of numerical structures (i.e. vector and matrices) on the topology, and also the easy layout of the network with the current planar technology.

Let first define the n-dimensional mesh denoted by $M(p_1, p_2, \ldots, p_n)$, as the cartesian sum (see definition 3.1) of n paths on p_i vertices, with $i = 1, 2, \ldots, n$ and $p_i \geq 2$.

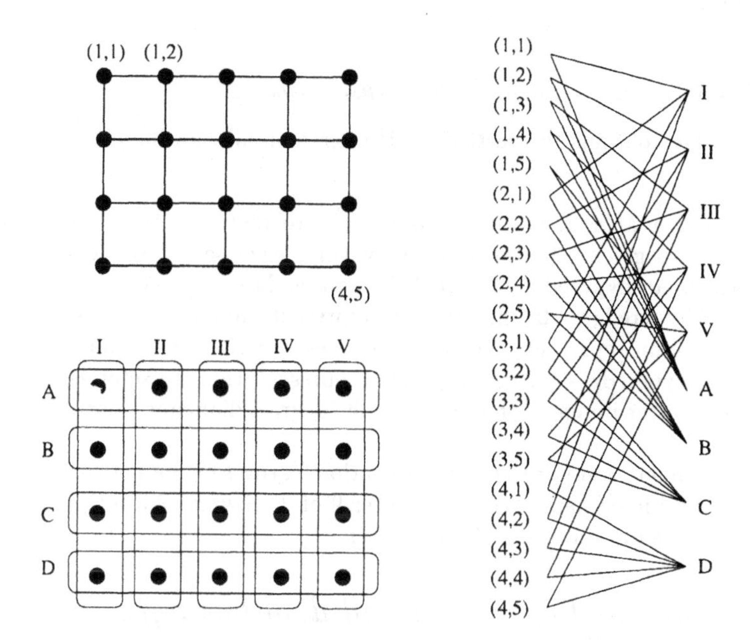

Figure 2 Mesh $M(4,5)$, Hypermesh *HM*$(4,5)$ and its bipartite representation

Each vertex will naturally be denoted by a $n-$*tuple* $(i_1, i_2, \ldots, i_n)$, i.e., $V(M(p_1, p_2, \ldots, p_n)) = \{(i_1, i_2, \ldots, i_n), 1 \leq i_k \leq p_k, k = 1, \ldots, n\}$.

A particular case of this graph is the hypercube, when $p_i = 2$, $i = 1, \ldots, n$. The *hypercube of dimension* n, denoted by $H(n)$, is a graph whose vertices are all words of length n over the two-letter alphabet $\{0, 1\}$ and whose edges connect two words which differ in exactly one coordinate. Vertex $x_1 x_2 \cdots x_i \cdots x_n$ is thus joined to vertices $x_1 x_2 \cdots \overline{x_i} \cdots x_n$ with $i = 1, 2, \ldots, n$.

This graph could be seen as the cartesian sum of n paths of length 2 i.e. a n dimensional mesh $M(2, 2, \cdots, 2)$.

Note that people has also considered the graph based upon the cartesian sum of n cycles which is called an n-dimensional torus.

Hypermeshes

Then, the hypermeshes proposed in [50] are defined with the notation of Section 3.2 by:

▷ $\mathcal{V}(HM(p_1, p_2, \ldots, p_n)) = V(M(p_1, p_2, \ldots, p_n))$

▷ an edge contains the vertices which agree on all coordinates but one

In Figure 2 a $M(4,5)$ is represented, and two pictures of the corresponding $HM(4,5)$ are given. In the bottom picture, each set of vertices (rows and columns) is an hyperedge of $HM(4,5)$, and we label from I to V and A to D the set of hyperedges. In the bipartite representation on the left of the figure, the left column represents the set of vertices and the right column represents the set of hyperedges (crossbars). This representation is useful as it provides us with a better image of the hypergraph (degree, diameter, routing, ...).

The implementation nor the communication features of the network studied in [50] are considered here. See the chapter 9 of T. Szymanski in this book for a detailed presentation of the Hypermeshes.

Spanning bus hypercubes and dual bus hypercubes

Wittíe [53] has defined two hypercube based bus networks, with generalization to W-wide D-dimensional mesh (in the actual generalized hypercube, a W-letter alphabet is used and the graph is defined as the iterated cartesian sum of complete graphs over W vertices).

In a spanning bus hypercube all W nodes aligned in the same dimension are interconnected with a bus. That is every node is connected to a different bus in each dimension. This could be a strong limiting factor for implementation, though P. Dowd [22] has described an efficient multiple access control to implement these topologies. The media access control overcomes the large degree of the graph.

In a dual bus hypercube, some buses are removed from the network in order to have only two (here dual stands for two and not duality) bus connections per node.

Compound techniques

A general technique used to construct large (Δ, D, r)-hypergraphs is to start from good ones for small values of Δ, D and r, then to combine them to build larger ones [12]. A first combination is the cartesian sum, but other graph products can also be used.

Optical Multimesh Hypercube (OMMH)

An OMMH [41] is characterized by a triplet (l, m, n) where l and m represents respectively the row and column dimensions of a torus, and n represents the dimension of a binary hypercube. Recall a torus is similar to a mesh but defined as the cartesian sum of cycles instead of paths ($C_l \square C_m$ in the two dimensional case).

The OMMH is actually the cartesian sum of a n-dimensional hypercube ($H(n)$) and a 2-dimensional torus $TM(l, m)$: $H(n) \square TM(l, m)$. The aim is to find a tradeoff between the hypercube (small $\log_2 N$ diameter but large $\log_2 N$ degree) and the toroidal mesh (large $\sqrt{N}$ diameter but small constant degree).

In Figure 3 we show the example of $H(3) \square TM(2, 3)$. The representation is taken from [41]. Each dashed line represents a cycle in the torus and all the vertices crossing the line belonging to it. A. Louri and H. Sung proposed a 3-dimensional optical implementation of the network which is not a bus network, but a complex interconnection network with an efficient optical implementation. The construction could easily be extended to an hypergraph network by displacing the torus by an hypermesh of same order.

Partitioned Optical Passive Star (POPS)

The interconnection network is built up with several passive star couplers. A (n, d, r) POPS consists in n nodes of degree d linked with a redundancy factor r. The redundancy refers to the number of paths available between any pair of nodes. Each path traverses exactly one coupler and all couplers have equal fan-

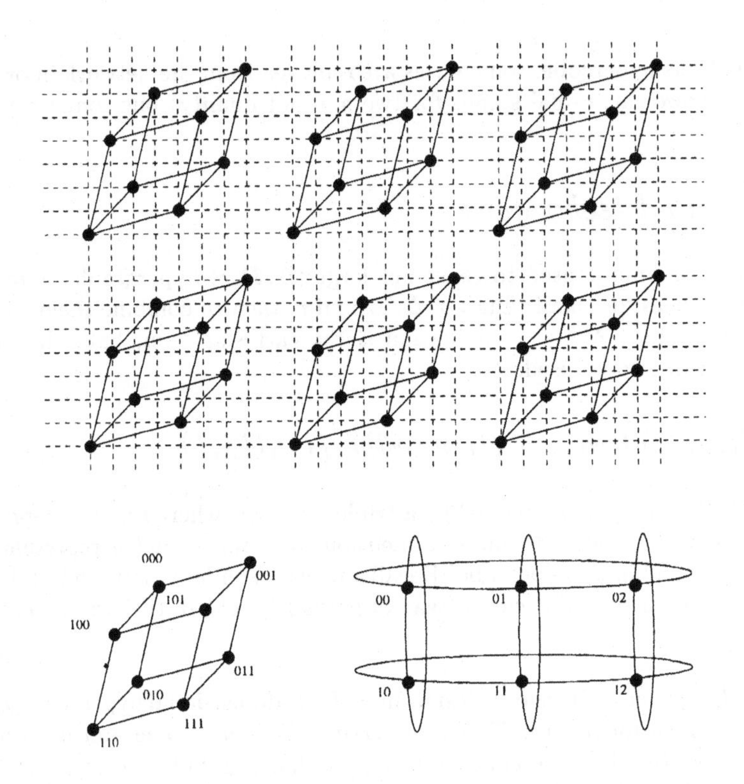

Figure 3 Optical Multimesh Hypercube $(3,2,3)$

in and fan-out d (in-degree and out-degree). The control and the construction of such a network is described in [18].

The n nodes are partitioned in groups. We will present the case of groups of size d with redundancy $r = 1$, and d divides n. Each node is connected to every star coupler in a perfect-shuffle way. In Figure 4, the 9 nodes are partitioned in 3 groups, each group corresponding to 3 star couplers. The graph is represented as a directed bipartite graph, with a repeated set of vertices to the left and to the right of the couplers. One set represents the transmitters while the other set represents the receivers, for every node in the graph.

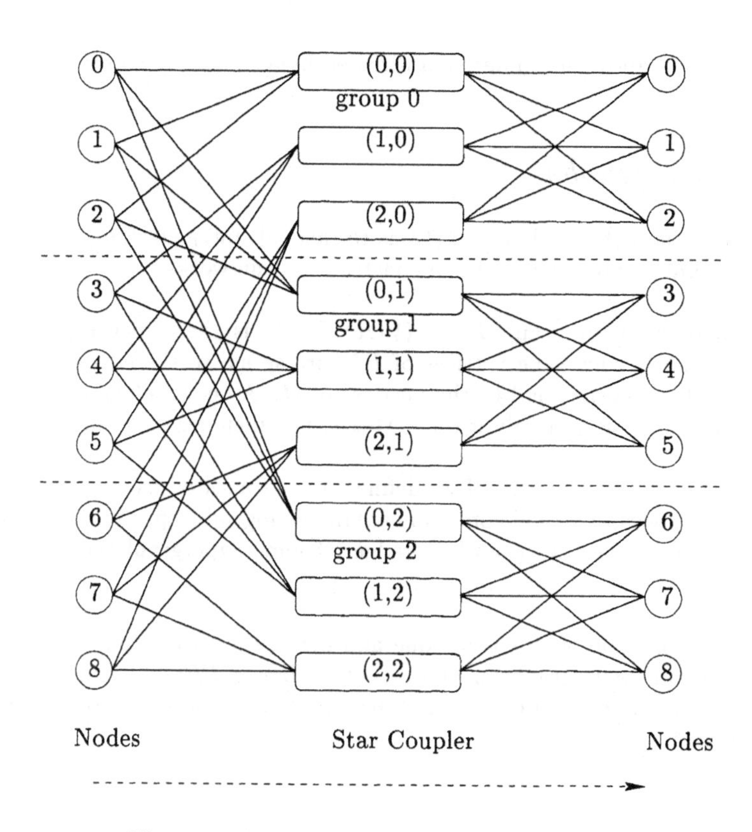

Figure 4 Partitioned Optical Passive Star $(9, 3, 1)$

Stack-Graphs

Stack-graphs [16] are obtained by piling up copies of one original graph, and by replacing each stack of edges by one hyperedge. An hyperedge contains all the extremities of the copies of one original edge.

Let $G = (V, E)$ be the original graph. The corresponding stack-graph $\varsigma(G, m)$ is defined as follows:

- ▷ $\mathcal{V}(\varsigma(G,m)) = \{0, \cdots, m-1\} \times V$, $m \geq 1$
- ▷ $\mathcal{E}(\varsigma(G,m)) = \cup_{(x,y)\in E}\mathcal{E}_{(x,y)}$ where $\mathcal{E}_{(x,y)} = \{0, \cdots, m-1\} \times \{x, y\}$

This definition allows us to derive a directed hypergraph if the original graph is a digraph.

Dual hypergraphs

The basic idea of the duality tool [12] is to take advantage of the properties of the best point-to-point networks to construct bus networks.

The *dual* of an hypergraph $H = (\mathcal{V}(H), \mathcal{E}(H))$ is the hypergraph H^* ($= (\mathcal{V}(H^*), \mathcal{E}(H^*))$) where the vertices of H^* correspond to the edges of H, and the edges of H^* correspond to the vertices of H. A vertex e_j^* is a member of an edge V_i^* in H^* if and only if the vertex v_i is a member of E_j in H.

Consider a graph $G = (V, E)$. If we define a bus as a set of processors in which any pair of processors could communicate in one logical step, then it is natural to think to a node $v \in V$ as an hyperedge -of some hypergraph- containing all the neighbors of v.

We give an example of this technique and detail the two graphs in Figure 5. We take as an input graph an undirected *de Bruijn* $UB(2,3)$ and give as an output its dual hypergraph (see the bipartite representation on the right side of the figure).

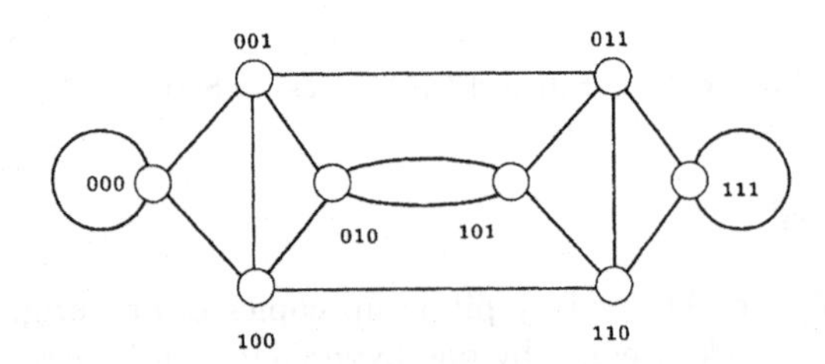

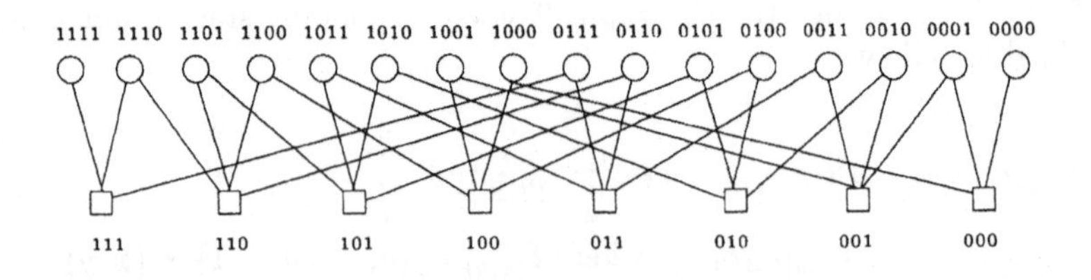

Figure 5 Binary *de Bruijn* network $B(2,3)$ and its dual hypergraph

The *de Bruijn* digraph (resp. graph), denoted by $B(d, D)$ (resp. $UB(d, D)$), has $N = d^D$ vertices with diameter D and in-degree and out-degree d (resp. degree $2d$). The vertices correspond to the words of length D over an alphabet of d symbols. The arcs (or edges) correspond to the shift operations: Given a word $X = x_1 \cdots x_D$ on an alphabet $\mathcal{A}$ of d letters, where $x_i \in \mathcal{A}$, $i = 1, 2, \cdots, D$, and given $\lambda \in \mathcal{A}$, the operation:

- ▷ $x_1 \cdots x_D \longrightarrow x_2 \cdots x_D \lambda$ is called a left shift;
- ▷ $x_1 \cdots x_D \longrightarrow \lambda x_1 \cdots x_{D-1}$ is called a right shift.

In the *de Bruijn* digraph $B(d, D)$, the successors are obtained by left-shift operations, whereas in the *de Bruijn* graph $UB(d, D)$, the neighbors are obtained by either left or right shift operations. An example of a *de Bruijn* digraph is given in Figure 5. The corresponding undirected *de Bruijn* hypergraph is obtained by transforming arcs to edges (i.e., removing the directions of the arcs). Here we do not remove the redundant edges (i.e., those with multiple occurrences in the graph, or those linking the same vertices).

Let us take the $UB(2, 3)$ of Figure 5. If we consider it as an hypergraph of rank 2, then we label each edge of $UB(2, 3)$ with the following construction. For every edge we consider the original arc which is coded on $D = 3$ digits, then we add the suffix 0 or 1 of the corresponding left shift. This set of edges gives the set of 16 vertices of the dual hypergraph $UB^*(2, 3)$. The (hyper)edges of $UB^*(2, 3)$ are made of the 8 vertices of $UB(2, 3)$ and are represented with boxes (□) in the bipartite representation of Figure 5. Note that each vertex belongs to at most 2 (hyper)edges, and that will be the case in any dual construction. The rank of the output hypergraph depends on the degree of the input graph. In the case of the *de Bruijn*, any even r can be chosen.

The feasibility of such network topology with optical technology and the integration of fundamental optical operations to construct the network has been investigated by A. Louri and H. Sung [40].

These hypergraphs (and also *Kautz* hypergraphs) were proposed for virtual topologies of large optical networks by [37].

Bus-Mesh networks

Tong et al [51] have proposed a network architecture which combines Time and Wavelength Division Multiplexing. This network can be represented by a directed hypergraph.

As the number of different wavelengths required for a large system is often larger than what is technically available, and as tunable transmitters are capable of tuning to only a small subset of wavelengths. The idea here is to design a multihop network.

When a node needs to transmit a message to another node which is not directly achievable (not tuned to the transmitter's wavelength), the message is relayed by intermediate nodes like in store-and-forward networks. If the number of wavelengths required by the system is still too high, Time Division Multiplexing can be used combined with each wavelength, by allocation of time slots.

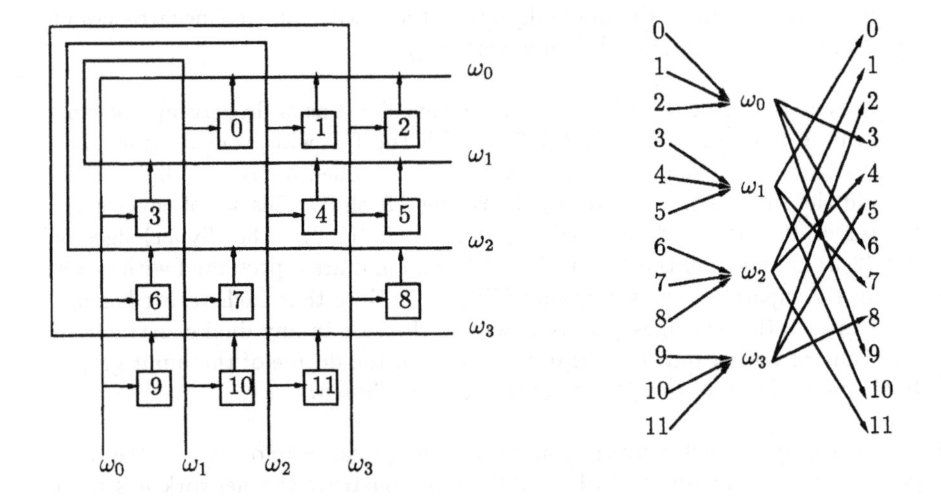

Figure 6 Bus-Mesh network (fixed wavelength transmitter/receiver)

We will illustrate that technique on the Bus-Mesh network. The network is based on a passive star. Each node transmits and receives on two fixed different wavelengths. The time domain is divided into time slots, each slot being large enough to contain a packet of data. We assume all the slots have the same size and are infinitely repeated in a cycle.

Figure 6 shows 12 nodes using 4 wavelengths to communicate. Each node can transmit on one wavelength and receive on a different one. The time domain is divided into 3 slots. The table of transmission cycle indicates for each time slot t_0, t_1, and t_2, which node is authorized to transmit on the corresponding wavelength. The arrow indicates which stations are listening on that wavelength. The bipartite directed hypergraph representation is given on the right side of the figure. For the sake of clarity, we have duplicated the vertices in two sets: transmitters and receivers.

	t_0	t_1	t_2
ω_0	0 $\mapsto$ 3, 6, 9	1 $\mapsto$ 3, 6, 9	2 $\mapsto$ 3, 6, 9
ω_1	3 $\mapsto$ 0, 7, 10	4 $\mapsto$ 0, 7, 10	5 $\mapsto$ 0, 7, 10
ω_2	6 $\mapsto$ 1, 4, 11	7 $\mapsto$ 1, 4, 11	8 $\mapsto$ 1, 4, 11
ω_3	9 $\mapsto$ 2, 5, 8	10 $\mapsto$ 2, 5, 8	11 $\mapsto$ 2, 5, 8

Table 1 Transmission cycle of a 12 nodes, 4 wavelengths Bus-Mesh network

This scheduling of time slots creates a virtual topology. Each packet includes a destination address. If node 4 wants to send a packet to node 6, it has to send the packet at time t_1, and only node 0 will relay the packet at time t_0.

The model of this network is a directed hypergraph. Indeed we can think of 4 hyperedges (associated to the 4 different wavelengths), each edge being made of an input set of vertices, and an output set of vertices.

The generalization of *de Bruijn* and *Kautz* directed hypergraphs is described in [9].

4 COLLECTIVE COMMUNICATIONS

We already stated the problem of collective communications in conventional point-to-point interconnection networks.

Assuming different cost functions of sending a message from an originator node to a destination node at distance d from the originator, the problem is to minimize the total cost of the protocol (broadcasting or gossiping). This problem has been extensively studied in the case of parallel computers and interested readers can refer to [20]. Off-line problems are considered (i.e. the communications patterns are known in advance).

4.1 Communication models for electronic networks

Introduction to broadcasting with store-and-forward switching

We assume that the protocol follows a step by step execution: during one *logical step*, every vertex can send and/or receive one elementary packet of data, according to either model $1 - port$ or $\Delta - port$. The logical step ends when all the transfers are done.

In the following, the hypothesis of store-and-forward switching is used. It is easy to see that the broadcast time of an arbitrary vertex x of a graph G of order N under the one-port full duplex constraint (denoted by F_1) satisfies $b_{F_1}(x) \geq \lceil \log_2 N \rceil$. Indeed, the number of vertices informed at time $t+1$ is at most twice the number of vertices informed at time t.

Proposition 2 *For* $N \geq 2$, $b_{F_1}(K_N) = \lceil \log_2 N \rceil$.

PROOF. The vertices of K_N are numbered from 0 to $N-1$, and we take the vertex 0 as the source of the broadcast. Consider the following broadcast protocol. At time $i \geq 1$, an informed vertex p sends the message to the vertex $2^{i-1} + p$ (if $2^{i-1} + p < N$). It is easy to see, by induction, that at time i all the vertices from 0 to $2^i - 1$ are informed. This guarantees broadcasting in $\lceil \log_2 N \rceil$ time units. □

Broadcasting in specific networks

Unfortunately, there is no general method for computing the minimum broadcast time of a graph in this model. Each new graph is a new special case. Most of the graphs used as models of distributed architecture (meshes, tori, hypercubes, butterfly graphs, *de Bruijn* graphs, cube-connected-cycles graphs, star-graph, and so on) have been studied and their broadcast times are known, sometimes up to a constant. The following proposition and its corollary help in finding an upper bound on broadcast time of any graph.

Proposition 3 *In a p-ary tree of depth h, the broadcast time of the root, under the constraint F_1, is at most $p \times h$.*

Corollary 2 *Let G be a graph and h an integer. If there is, for each vertex x of G, a p-ary tree spanning G, with root x and of depth h, then $b_{F_1}(G) \leq p \times h$.*

In particular, if we can find, for each vertex in a graph of diameter D, a binary spanning tree of depth D, then we obtain for this graph an upper bound of $2D$ on its broadcast time, that is, twice the lower bound.

Δ-*PORT*

Under the Δ-port constraint (often called *shouting* and denoted F_* when links are full-duplex) each processor can communicate with all its neighbors at the same time and so there is no problem in finding a broadcast algorithm for any graph. When a vertex receives a message, it sends it to all its neighbors. Thus, for a graph of diameter D we have

$$b_{F_*}(G) = D.$$

General techniques

A lot of work has been done on the problem of broadcasting and gossiping in point-to-point interconnection networks. However we could point out two general techniques.

In the case of store-and-forward, the general problem consists in finding the maximum number of arc-disjoint spanning trees. Then the message is cut in parts of equal sizes, and each part is "pipelined" on a different spanning tree [20].

In the case of wormhole, one could use coding theory to find optimal covering sets of the graph [21]. These sets represent the vertices that get the information at every logical step of the algorithm. The minimum number of steps is $\lceil \log_{\Delta+1} N \rceil$ when Δ concurrent communications can occur at each step.

We will see in the following that the problem is one more time different in the case of bus networks and WDM routing.

4.2 Collective communications in bus networks

We limit the study of collective communications in bus networks to the communication models used in the literature.

Three types of buses are considered:

- ▷ One-to-one (OTO) bus. This is the electronic model of buses. At any time two nodes belonging to the same bus could exchange data in one logical step, as if they were neighbors in a usual graph topology.

  Earlier designs of networks used this model (for instance the spanning bus hypercube or dual bus hypercube described in 2).

- ▷ One-to-all (OTA) bus, or CREW (Concurrent Read Exclusive Write). This is the motivation for bus networks: one node can broadcast its information to all the other nodes of the bus in one logical step. This has been made possible by the specificity of optical components that naturally broadcast optical signals with a large fan-out.

  Most of the models under study - such as hypermeshes - belong to that model.

- ▷ All-to-all (ATA) bus, or CRCW (Concurrent Read Concurrent Write). In this model, we suppose that all the nodes can exchange data concurrently, thus performing a "gossip" in one step. It is possible when each station connected to the bus has enough receivers (like it is the case in Lambdanet [32]), but this could be costly.

Usually, buses are 1-port: a node can communicate with only one bus at the same time even though the node could be connected to Δ buses.

The broadcasting problem is straightforward in the 1-port (OTA) model, and some work has been done on gossiping in meshes by Sotteau and Hillis [33]. Fujita and Yamashita consider the gossiping problem in mesh-bus computers (similar to squared hypermeshes: n^2 nodes arranged on $n \times n$ array are connected by n column-buses and n row-buses. The algorithm completes the gossiping in $\lfloor n/2 \rfloor + \lceil \log_2 n \rceil + 1$ steps. A lower bound on the number of steps for this problem is shown to be $\lfloor n/2 \rfloor + \lceil \log_2 n \rceil - 1$, thus the protocol takes at most 2 more steps than an optimal algorithm.

4.3 Collective communications in WDM switched networks

We assess the gossiping problem in all-optical networks using wavelength division multiplexing access [17]. The problem considered is to minimize the number of wavelengths required to perform a given communication pattern between the nodes of the network using only one step (One hop problem). Indeed, under the WDM switching assumption several messages can go through a link until they do not use the same wavelength. Using the graph model, we can think of permutations with color-disjoint channels. This could be seen as a generalization of previous models in the way that wormhole in the case of WDM when the number of wavelengths (colors) equals 1, and store-and-forward in the case when the path are restricted to edges.

Routing (one-to-one communication instance of the problem) is developed in chapter 11 and the presentation here is restricted to the off-line problem, in which communications are known before we decide which wavelength has to be allocated.

Note also that we do not consider switches that use wavelength converters. When converters are available, the color of an incoming path on the switch could be changed to a different one at the output port (i.e. paths are multicolored).

In Figure 7 we give the example of five paths which have to be routed in a ring. On the left side of the figure, it is easy to check that we need 3 colors while only 2 colors are required on the right side, if we use two colors for path $(4, 1)$.

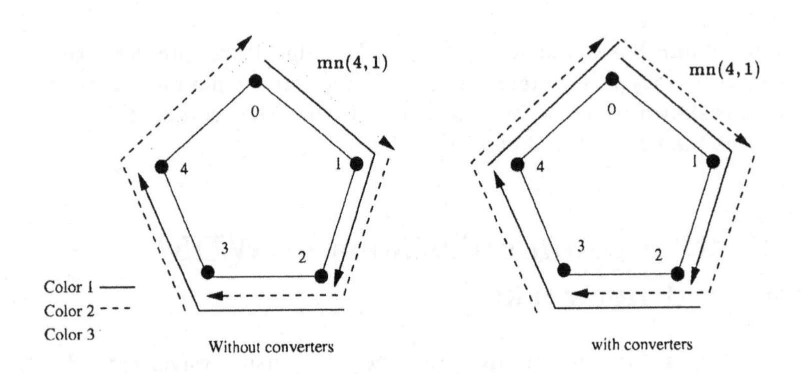

Figure 7 Routing in a ring

Broadcasting and Gossiping in WDM networks

We take the notation of [4] from which the following results are quoted. The network is modeled as a *symmetric digraph.*

Definition of the wavelength-routing problem

- ▷ A *request* is an ordered pair of nodes (x, y) in G (corresponding to a message to be sent from x to y).
- ▷ An *instance* I is a collection of requests. Note that a given request (x, y) can appear more than once in an instance.
- ▷ A *routing* R for an instance I in G is a set of dipaths $R = \{P(x,y) \,|\, (x,y) \in I\}$.
- ▷ The *conflict graph* associated to a routing R is the undirected graph (R, E) with vertex set R and such that two dipaths of R are adjacent if and only if they share an arc of G.

In the example depicted in Figure 7, $G = C_5$ (ring of rank 5), $I = \{(i, i + 2 \mod 5), i = 0..4\}$, and R is the set of paths represented with arrows in the figure.

Let G be a digraph and I be an instance. The *problem* (G, I) asks for a routing R for the instance I and assigning each request $(x, y) \in I$ a wavelength, so that no two dipaths of R sharing an arc have the same wavelength.

If we think of wavelengths as colors, the problem (G, I) is to find a routing R and a vertex coloring of the conflict graph (R, E), such that two adjacent vertices are colored differently.

We denote by $\vec{w}(G, I, R)$ the chromatic number of (R, E), and by $\vec{w}(G, I)$ (or briefly just $\vec{w}$ if there is no ambiguity) the smallest $\vec{w}(G, I, R)$ over all routings R. Thus $\vec{w}(G, I, R)$ is the minimum number of wavelengths for a routing R and $\vec{w}(G, I)$ the minimum number of wavelengths over all routings for (G, I).

Any routing by undirected paths induces a routing by directed paths, and a coloring of the undirected paths is also a coloring of the directed paths, as two edge-disjoint paths will become two arc-disjoint dipaths. Hence $\vec{w}(G, I) \leq w(G, I)$ for any problem (G, I), and every upper bound on w is an upper bound on $\vec{w}$.

Off-line communication problems

As in the case of store-and-forward and wormhole models, the following special instances of the routing problem are considered:

- ▷ The *All–to–All* instance, or *gossiping* instance, denoted I_{ATA}: $I_{\mathrm{ATA}} = \{(x, y) \mid x \in V(G),\ y \in V(G), x \neq y\}$.
- ▷ The *One–to–All* instance, or *broadcasting* instance, denoted I_{OTA}: $I_{\mathrm{OTA}} = \{(x_0, y) \mid y \in V(G), y \neq x_0\}$, where $x_0 \in V(G)$. A *One–to–Many* instance, or *multicasting* instance is a subset of some instance I_{OTA}.
- ▷ A *k-relation* is an instance I_k in which each node is a source and a destination of no more than k requests. A 1-relation is also known as a *permutation* instance. Note also that the instance I_A is an $(N-1)$-relation.

The load parameter of a network

- ▷ Given a network G and a routing R for an instance I, the *load* of an arc $\alpha \in A(G)$ in the routing R, denoted by $\vec{\pi}(G, I, R, \alpha)$, is the number of dipaths of R containing α. The *load* (also called *congestion*) of G in the routing R, denoted by $\vec{\pi}(G, I, R)$, is the maximum load of any arc of G in the routing R, that is, $\vec{\pi}(G, I, R) = \max_{\alpha \in A(G)} \vec{\pi}(G, I, R, \alpha)$.
- ▷ The *load* of G for an instance I, denoted by $\vec{\pi}(G, I)$, or $\vec{\pi}$ if there is no ambiguity, is the minimum load of G in any routing R for I, that is,

$\vec{\pi}(G,I) = \min_R \vec{\pi}(G,I,R)$. For the All–to–All instance I_{ATA}, $\vec{\pi}(G,I_{\text{ATA}})$ (respectively $\pi(G,I_{\text{ATA}})$) is called the *arc forwarding index* (resp. *edge forwarding index*, see [33,49]) of G.

The relevance to the problem of this parameter is shown by the following lemma:

Lemma 1 $\vec{w}(G,I) \geq \vec{\pi}(G,I)$ *for any instance* I *in any network* G.

In other words, to solve a given problem (G,I) one has to use a number of wavelengths at least equal to the maximum number of dipaths having to share an arc.

In general, minimizing the number of wavelengths is not the same problem than realizing a routing that minimizes the number of dipaths sharing an arc (congestion). Indeed, the problem is made much harder due to the further requirement of wavelengths assignment on the dipaths. In order to get equality in Lemma 1, one should find a routing R such that $\vec{\pi}(G,I,R) = \vec{\pi}(G,I)$, for which the associated conflict graph is $\vec{\pi}(G,I)$–vertex colorable.

Question 4 *Does there always exist a routing* R *such that the two conditions hold simultaneously:* $\vec{\pi}(G,I,R) = \vec{\pi}(G,I)$ *and* $\vec{w}(G,I,R) = \vec{w}(G,I)$*?*

Theorem 5 ([5]) *Determining* $\vec{\pi}(G,I)$ *in the general case is NP–complete.*

Sketch of proof. First observe that determining $\vec{\pi}(G,I)$ is equivalent to solving the integral multicommodity directed flow problem with constant capacities. It is shown in [25] that this problem is NP–complete even for two commodities and all capacities equal to one. □

For some special cases, $\vec{\pi}(G,I)$ can be efficiently determined. This is obviously the case for trees, where routing is always unique. This is also the case of the One–to–Many instances where the problem can be reduced to a flow problem (in the graph obtained from G by considering the sender node as the source, giving a capacity $\vec{\pi}$ to each arc of G, and joining all the vertices of G to a sink t with arcs of capacity 1).

The load $\pi(G,I)$ can be defined analogously for an undirected graph and it is proven that $\vec{\pi}(G,I) \leq \pi(G,I) \leq 2\vec{\pi}(G,I)$. For One–to–Many instances I, one can also show that $\vec{\pi}(G,I) = \pi(G,I)$.

Question 6 *Does the equality $\vec{\pi}(G, I_{\mathrm{ATA}}) = \lceil \pi(G, I_{I_{\mathrm{ATA}}})/2 \rceil$ always hold ?*

Arbitrary network topologies

Arbitrary instances

For a general network G and an arbitrary instance I, the problem of determining $\vec{\mathrm{w}}(G, I)$ has been proved to be NP–hard in [23]. In particular, it has been proved that determining $\vec{\mathrm{w}}(G, I)$ is NP–hard for trees and cycles. In [24] these results have been extended to binary trees and meshes. NP–completeness results in the undirected model were known much earlier (actually, well before the advent of the WDM technology). In particular, in [31] it is proved that the problem of determining $\mathrm{w}(G, I)$ is NP–complete for trees. This result has been extended in [23] to cycles, while in [24] it has been proved that the problem is efficiently solvable for bounded degree trees.

In view of this last result and of the NP–hardness of determining $\vec{\mathrm{w}}(G, I)$ for binary trees, it might seem that the problem of computing $\vec{\mathrm{w}}(G, I)$ is harder than that of computing $\mathrm{w}(G, I)$. This is not true in general. For instance, the determination of $\mathrm{w}(G, I)$ remains NP–complete when G is a star network, whereas $\vec{\mathrm{w}}(G, I)$ can be efficiently computed. Indeed, in the undirected model this problem corresponds to an edge-coloring a multigraph [35], each node of which corresponds to a branch in the star network. In the directed case, the same problem becomes equivalent to an edge-coloring of a bipartite multigraph, and the problem is efficiently solvable by König's theorem.

In [1] an upper bound in the undirected model is given, the same holds also in the directed case:

Theorem 7 *(Aggarwal et al. [1]). For any problem (G, I), where G has m arcs, $\vec{\mathrm{w}}(G, I) \leq 2\vec{\pi}(G, I)\sqrt{m}$.*

Let R be a routing for an instance I in a network G. Let L be the maximum length of its dipaths and Δ the maximum degree of its conflict graph. It is clear that $\Delta \leq L\vec{\pi}(G, I, R)$. By a greedy coloring we know that $(\Delta + 1)$ wavelengths are sufficient to solve the problem (G, I). Thus $\vec{\mathrm{w}} = O(L\vec{\pi})$ and similarly $\mathrm{w} = O(L\pi)$. A set of critical undirected problems which reach asymptotically this upper bound (and that of Theorem 7) has been given in mesh-like networks (see [1]). By adapting their examples (orienting alternately the vertical links up and down), the same result is obtained for the general case (not symmetric) digraphs:

Theorem 8 *For every π and L, there exists a directed graph G and an instance I such that $\vec{\pi}(G,I) = \pi$, $L = \max_{(x,y)\in I} \delta(x,y)$ and $\vec{w}(G,I) = \Omega(\pi L)$.*

Question 9 *Does Theorem 8 hold for symmetric digraphs?*

Specific instances

The following theorem gives the exact value of $\vec{w}(G, I_{\text{OTA}})$ for a worst case instance I_{OTA} in various classes of important networks, namely the *maximally arc connected* digraphs, including the wide class of vertex transitive digraphs. A digraph G is maximally arc connected if its minimum degree is equal to its arc connectivity.

Theorem 10 *(Bermond et al. [13]). For a worst case One–to–All instance I_{OTA} in a maximally arc connected digraph G of minimum degree $d(G)$,*

$$\vec{w}(G, I_{\text{OTA}}) = \vec{\pi}(G, I_{\text{OTA}}) = \left\lceil \frac{N-1}{d(G)} \right\rceil .$$

In addition, an efficient network flow based algorithm is given to solve the problem (G, I) with $\vec{w}(G, I)$ wavelengths, for any One–to–Many instance I in any network G.

Theorem 11 *(Beauquier et al. [5]). $\vec{w}(G,I) = \vec{\pi}(G,I)$, for any One–to–Many instance I in any digraph G.*

Many other instances are relevant of interest, however a lot of practical communication problems have to deal with the on-line hypothesis, and efficient algorithms have to be designed. The techniques used are different and are not presented in this chapter.

5 INTRODUCING OPTICAL COMMUNICATIONS IN GENERAL MODELS OF PARALLELISM

In this section, we review the different models of parallelism involved by using optical technologies. These models are mainly derived from the PRAM model. In the following, we recall what is the PRAM model and its optical extensions, namely the OCPC model in Section 5.2 and the OPS model in Section 5.3.

5.1 The Parallel Random Access Machine

The PRAM model (Parallel Random Access Machine) was one of the first model for designing parallel algorithms. It presents the parallel extension of the sequential RAM model [2] in order to design parallel algorithms. The **PRAM** model consists of P processors and M memory modules. In a unit time step, all the processors can access the data in one of the memory modules (Read operation), perform a simple computation on their local registers, and store the result in one of the memory module (Write operation). Problems arise when several processors want to access the same memory location. In 1982, Snir proposed a classification of the PRAM models in terms of the possibility of multiple reads and writes [48], namely, *Exculsive*, when only one processor can access to a Read or Write ressource at the same time, and *Concurrent* in the other case. In case of concurrent multiple write, the result has to be clearly specified [27]. Let note, for example, the COLLISION rule, where a special character is put in a memory cell after a write conflict.

The PRAM model is indeed a theoretical model for a parallel machine. However, it does not take into account the physical network the parallel machine uses to communicate. Several extensions such as the XRAM model solve partially the problem [19].

5.2 The Optically Connected Parallel Computer

In 1988, Anderson and Miller proposed another way of resolving the Write conflicts, based on optical assumptions. The general model is as follows. The **Local Memory PRAM** [3] consists of a collection of processors and a collection of memory modules. Each processor and each memory module has one optical receptor and one optical transmitter. Each transmitter can be focused on a receptor in one unit time. Two light beams do not interfere unless they are focused on the same receptor.

Based on these assumptions, Valiant proposed the **seclusive-PRAM** [52] (S*-PRAM). This model is now called the **completely connected Optical Communication Parallel Computer**. Using this model, it is very easy to implement an acknowledgment process. All the processors that have received a message without error can send back an acknowledgment message to the originator. This latter message will reach its destination without any conflict since processors send messages to a single destination. Thus, in a unit time, it can be assumed that all the processors know whether they have succeeded in accessing a memory module. The acknowledgment can also include the data that the processor was asking for.

The power of this model has been widely discussed [3, 26, 28–30, 52] and we will review in the following some of the principal results of this model.

Relations with the PRAM models

The first problem when defining a new model is to compare it to the others and to know where it stands in the model hierarchy. In this extent, MacKenzie and Ramachandran [43] show that the OCPC is equivalent to the Exclusive Read, Concurrent Write PRAM model, whenever the contention resolutions are solved in the same way. This latter model has not been considered by Snir in his classification [48], since the Read operation in a memory cell is usually less constrained than the Write operations.

Valiant [52] described a simulation of an EREW PRAM on an OCPC. He gave a constant delay simulation of BSP (Bulk Synchronous Parallel) computer on the OCPC, connected to a randomized simulation of a $n \log n$ processor EREW PRAM on a an n processor BSP. A direct simulation was given by Geréb-Grauss and Tsantilas [26] with delay $O(\log n \log\log n)$. Different works have provided such a simulation in expected delay $\Theta(\log\log n)$, but they require $n^{\Omega(1)}$ storage at each processor [42, 44]. Goldberg, Matias and Rao found a randomized simulation of an $n \log\log n$ processor EREW PRAM on an n processor OCPC in $O(\log\log n)$ expected delay [30].

The h-relation problem

In the h-relation problem, each processor has at most h messages to send and h messages to receive. A 1-relation is indeed a (partial) permutation, and can be easily realized in one step on the OCPC model. When the communication pattern is known *a priori*, the h relation can be decomposed into h partial permutation and thus can be performed within h steps.

When studying the *on-line* problem, and probabilistic algorithms have to be used in order to derive good bounds. In [29], Goldberg, Jerrum and MacKenzie show that the expected number of communication steps required to route an arbitrary h relation is $\Omega(h + \sqrt{\log\log n})$ on a n-processor OCPC. In [28], Goldberg et al. solved this problem in time $O(h + \log\log n)$.

5.3 The Optical Passive Star

As shown previously, the OCPC model does not take into account all the capabilities of the optical devices. Especially, this model is always a point-to-point model, i.e., in a single step, a processor can send a message to only one receiver. Using Optical Passive Stars, this constraint can be eliminated by defining the OPS model [14]. Some variations on this model have been given, e.g., see [15].

Each processor in a OPS-based network consists of a receiver, a transmitter, and a processing element with a memory unit. The transmitters and receivers may tune to any one of a range of wavelengths. All the receivers and transmitters are connected to an optical coupler, which is an all-to-all communication device with broadcast capability. Thus broadcast becomes an elementary operation, whereas in the point-to-point models such as the OCPC, this is an elaborated operation. If several messages are transmitted simultaneously on the same wavelength, then detectable "noise" is received. The abstract OPS model comprises a finite number of processors and a finite number of wavelengths.

Readers are referred again to [45] for a comprehensive survey on the implementation of such a model. In the remainder of the section, we briefly present some results on this model. Details can be found in [14].

In the following, we consider the OPS as a variant of the PRAM model with a global memory of linear size. A OPS is said to be balanced if it has as many available wavelengths as processors. Scaling problems towards realistic models is the point of Section 5.3, since we have seen in the previous sections that the number of wavelengths is the critical resource of such a system.

The OPS in the PRAM hierarchy

This new abstraction of a PRAM, dealing with new means of communication can be compared to usual models of CRCW PRAM. More precisely, it can be stated the equivalence with the COLLISION CRCW PRAM. This shows that the OPS is completely different from the OCPC model.

Theorem 12 ([14]) *For integers n and m with $1 \leq m \leq n$, the n processor OPS with m wavelengths and the n processor COLLISION PRAM with m global memory cells are equivalent, i.e., each machine can simulate the other one in a step-by-step fashion with a constant slowdown. In particular, the balanced n processor OPS an the balanced n processor COLLISION PRAM are equivalent.*

Self simulation of the OPS model

The self-simulation of a parallel model is directly related to the efficiency and ease of algorithm design: it is desirable that the algorithm designer may assume that as many processors as required by his algorithm are simultaneously available for his program. Once the algorithm designed for kn processors is executed on a given machine with only n processors, a simulation of the kn processor program should be done by the actual n processors. Moreover, considering the OPS model, the set of wavelengths should also be reduced, since the wavelengths are considered as a scarce resource. Similarly, the communication carried out by the larger virtual machine has to be scaled down to the smaller real machine.

The self-simulation is a step-by-step simulation, so that each (physical) processor simulates the operations of k fixed simulated (virtual) processors. Thus the simulation of the tuning and computation phases of each step can be trivially done in k steps of the simulating machine. The main problem that is addressed in the following is to simulate the communication steps.

The main results in [14] show that the OPS does exhibit scalability properties, i.e., a balanced kn processor OPS can be simulated by a balanced n processor OPS in a step-by-step manner with a slowdown of $O(k + \log^* n)$ with high probability. The algorithm consists essentially of a randomized solution to a distributed load balancing problem, added to a deterministic self-simulation algorithm that takes $O(k^2)$ steps in the worst case. From this randomized solution, we can derive a deterministic off-line algorithm which is able to complete the self-simulation problem in $O(k)$ steps.

Note that these solutions use some redirection of the messages in order to balance the requests. If we consider only direct self-simulation, i.e., a message must reach its final destination with no intermediate stop-overs, then the slowdown is more important, even in the off-line case. In this particular case, $\Omega(\min\{k^2, k + \log n/\log\log n\})$ is a lower bound, even in the off-line case, or when the number of wavelengths in unbounded. A simple algorithm with slowdown $O(k^3)$ can be established by scheduling all the possible combinations independently.

6 CONCLUSION

Since the speed of parallel algorithms is mainly constrained by interprocessor communications, models for communications must be established first in order to provide abstract models of computation for parallel architectures. These models are dictated by the underlying network topology and a great deal of work has been undertaken in this area, motivated by the existence of parallel computers.

Three aspects of interconnection networks models have been introduced in this chapter: topologies, communications and computations. Our goal was to state what changes in optical communication systems with respect to electronical ones, and we were able to provide some answers.

Concerning topologies, a hypergraph model was introduced, that gives a better representation of the network than the usual graph model. The communication models presented rely on the use of the wavelength division multiplexing switching technique. The important parameter in all the communication problems under study is the number of different wavelengths required to perform the communication. It is also the case for the computational models derived from these communication models.

Acknowledgements

We wish to thank Bruno Beauquier and Jean-Claude Bermond for helpful discussions and contributions. We acknowledge also the French groups ROI and RUMEUR.

REFERENCES

[1] A. Aggarwal, A. Bar-Noy, D. Coppersmith, R. Ramaswami, B. Schieber, and M. Sudan. Efficient routing in optical networks. *Journal of the ACM*, 46(6):973–1001, November 1996.

[2] A. Aho, J. Hopcroft, and J. Ullman. *The Design and Analysis of Computer Algorithms*. Addison-Wesley Publishing Co., 1974.

[3] R. Anderson and G. Miller. Optical communication for pointer based algorithms. Technical Report CRI 88-14, Computer Science Department, University of Southern California, Los Angeles, CA 90089-0782 USA, 1988.

[4] B. Beauquier, J.-C. Bermond, L. Gargano, P. Hell, S. Pérennes, and U. Vaccaro. Graph problems arising from wavelength–routing in all–optical networks. In *Proc. Conference WOCS97, Geneva*, April 1997.

[5] B. Beauquier, P. Hell, and S. Pérennes. Optimal wavelength-routed multicasting. Discrete Applied Mathematics, to appear.

[6] C. Berge. *Graphs and Hypergraphs*. North-Holland, 1973.

[7] J.-C. Bermond, J. Bond, M. Paoli, and C. Peyrat. Graphs and interconnection networks: diameter and vulnerability. In E. Lloyd, editor, *Surveys in Combinatorics, Invited Papers for the Ninth British Combinatorial Conference*, volume 82 of *London Math. Society Lecture Note Series*, pages 1–30. Cambridge University Press, 1983.

[8] J.-C. Bermond, J. Bond, and C. Peyrat. Interconnection network with each node on two buses. In *Proc. of the Internat. Workshop on Parallel Algorithms & Architectures, Luminy France.*, pages 155–167. North Holland, April 1986.

[9] J.-C. Bermond, R. Dawes, and F. Ergincan. de Bruijn and Kautz bus networks. Technical Report 94-32, I3S (to appear in Networks), 1994.

[10] J.-C. Bermond, C. Delorme, and J.-J. Quisquater. Strategies for interconnection networks: Some methods from graph theory. *Journal of Parallel and Distributed Computing*, 3:433–449, 1986.

[11] J.-C. Bermond, C. Delorme, and J.-J. Quisquater. Table of large (Δ, d)-graphs. *Discrete Applied Mathematics*, 37/38:575–577, 1992.

[12] J.-C. Bermond and F. Ergincan. Bus interconnection networks. *Discrete Applied Mathematics*, (68):1–15, 1996.

[13] J.-C. Bermond, L. Gargano, S. Perennes, A. A. Rescigno, and U. Vaccaro. Efficient collective communication in optical networks. *Lecture Notes in Computer Science*, 1099:574–585, 1996.

[14] P. Berthomé, T. Duboux, T. Hagerup, I. Newman, and A. Schuster. Self-simulation for the passive optical star model. In P. Spirakis, editor, *European Symposium on Algorithms*, number 979 in Lecture Notes in Computer Science. Springer-Verlag, 1995.

[15] P. Berthomé and A. Ferreira. Communication issues in parallel systems with optical interconnections. *International Journal of Foundations of Computer Science*, 1997. To appear in Special Issue on Interconnection Networks.

[16] H. Bourdin, A. Ferreira, and K. Marcus. A comparative study of one-to-many WDM lightwave interconnection network for multiprocessors. In *Second International Workshop on Massively Parallel Processing using Optical Interconnections*, pages 257–264, San Antonio (USA), October 1995. IEEE Press.

[17] C. A. Brackett. Foreword. is there an emerging concensus on WDM networking. *Journal of Lightwave Technology*, 14(6):936–941, June 1996.

[18] D. M. Chiarulli, S. P. Levitan, R. P. Melhem, J. P. Teza, and G. Gravenstreter. Partitioned Optical Passive Stars (POPS) multiprocessor interconnection networks with distributed control. *Journal of Lightwave Technology*, 14(7):1601–1612, July 1996.

[19] M. Cosnard and A. Ferreira. Designing parallel non numerical algorithms. In G. J. D.J. Evans and H. Liddell, editors, *Parallel Computing'91*, pages 3–18. Elsevier Science Publishers B.V., 1992.

[20] J. de RUMEUR. *Communications dans les réseaux de processeurs*. Masson, Paris, 1994.

[21] O. Delmas and S. Perennes. Circuit-Switched Gossiping in 3-Dimensional Torus Networks. In L. Bougé, P. Fraigniaud, A. Mignotte, and Y. Robert, editors, *Proceedings of the Euro-Par'96 Parallel Processing / Second International EURO-PAR Conference*, volume 1123 of *Lecture Notes in Computer Science*, pages 370–373, Lyon, France, Aug. 1996. Springer Verlag.

[22] P. W. Dowd. Wavelength division multiple access channel hypercube processor interconnection. *IEEE Transactions on Computers*, 41(10):1223–1241, October 1992.

[23] T. Erlebach and K. Jansen. Scheduling of virtual connections in fast networks. In *Proc. of Parallel Systems and Algorithms (PASA)*, pages 13–32, 1996.

[24] T. Erlebach and K. Jansen. Call scheduling in trees, rings and meshes. In *Proc. of HICSS*, 1997.

[25] S. Even, A. Itai, and A. Shamir. On the complexity of timetable and multicommodity flow problems. *SIAM J. of Computing*, 5(4):691–703, Dec. 1976.

[26] M. Geréb-Grauss and T. Tsantilas. Efficient optical communication in parallel computers. In *ACM Symposium on Parallel Algorithms and Architectures*, pages 41–48, June 1992.

[27] J. Gil and Y. Matias. Fast and efficient simulations among CRCW PRAMs. *Journal of Parallel and Distributed Computing*, 23(2):135–148, Nov. 1994.

[28] L. Goldberg, M. Jerrum, T. Leighton, and S. Rao. A doubly logarithmic communication algorithm for the completely connected optical communication parallel computer. In *ACM Symposium on Parallel Algorithms and Architectures*, pages 300–309, June 1993.

[29] L. Goldberg, M. Jerrum, and P. MacKenzie. An $\Omega(\sqrt{\log \log n})$ lower bound for routing in optical networks. In *ACM Symposium on Parallel Algorithms and Architectures*, June 1994.

[30] L. Goldberg, Y. Matias, and S. Rao. An optical simulation of shared memory. In *ACM Symposium on Parallel Algorithms and Architectures*, June 1994.

[31] M. C. Golumbic and R. E. Jamison. The edge intersection graphs of paths in a tree. *J. of Combinatorial Theory, Series B*, 38:8–22, 1985.

[32] M. S. Goodman, H. Kobrinski, M. P. Vecchi, R. M. Bulley, and J. L. Gimlett. The lambdanet multiwavelength network: Architecture, applications, and demonstrations. *IEEE Journal on Selected Areas in Communications*, 8(6):995–1004, August 1990.

[33] A. Hily and D. Sotteau. Gossiping in d-dimensional mesh-bus networks. *Parallel Processing Letter*, 6(1):101–113, March 1996.

[34] A. Hoffman and R. Singleton. On Moore graphs with diameters 2 and 3. *IBM J. Research and Development*, 4:497–504, 1960.

[35] I. Holyer. The NP–completeness of edge coloring. *SIAM J. of Computing*, 10(4):718–720, 1981.

[36] F. J. Janniello, R. Ramaswami, and D. G. Steinberg. A prototype circuit-switched multi-wavelength optical metropolitan-area network. *Journal of Lightwave Technology*, May/June 1993.

[37] S. Jiang, T. E. Stern, and E. Bouillet. Design of multicast multilayered lightwave networks. IEEE GLOBECOM'93 Houston Texas, pages 452-457, 1993.

[38] P. Kermani and L. Kleinrock. Virtual cut-through: a new computer communication switching technique. *Computers Networks*, 3:267–286, 1979.

[39] F. T. Leighton. *Introduction to parallel algorithms and architectures.* Morgan Kaufmann, 1992.

[40] A. Louri and H. Sung. Optical binary de bruijn networks for massively parallel computing: Design methodology and feasibility study. Technical Report AZ85721, University of Arizona, 1994.

[41] A. Louri and H. Sung. Scalable optical hypercube-based interconnection network for massively parallel computing. *Applied Optics*, 33(32):7588–7598, November 1994.

[42] P. MacKenzie, C. Plaxton, and R. Rajamaran. On contention resolution protocols and associated probabilistic phenomena. In *ACM Symposium On Theory of Computing*, 1994.

[43] P. D. MacKenzie and V. Ramachandran. ERCW PRAMs and optical communications. In *EUROPAR: Parallel Processing, 2nd International EURO-PAR Conference.* LNCS, 1996.

[44] F. Meyer auf der Heide and C. S. Scheiderler. Fast simple dictionaries and shared memory simulation on distributed memory machines; upper and lower bounds. Pre-print, 1994.

[45] B. Mukherjee. WDM-based local lightwave networks, Part I: Single-hop systems. *IEEE Network Magazine*, 6(3):12–27, May 1992.

[46] B. Mukherjee. WDM-based local lightwave networks, Part II: Multi-hop systems. *IEEE Network Magazine*, 6(4):20–32, July 1992.

[47] R. Ramaswami. Multiwavelength lightwave networks for computer communication. *IEEE Communications Magazine*, pages 78–88, February 1993.

[48] M. Snir. On parallel searching. *SIAM Journal of Computing*, 14(4):688–708, Aug. 1985. Also appeared in ACM Symposium on Principles of Distributed Computing,1982.

[49] P. Solé. Expanding and forwarding. *Discrete Applied Mathematics*, 58:67–78, 1995.

[50] T. Szymanski. "Hypermeshes": Optical interconnection networks for parallel computing. *Journal of Parallel and Distributed Computing*, 26(1):1–23, April 1995.

[51] S.-R. Tong, D. H. C. Du, and R. J. Vetter. Design principles for multi-hop wavelength and time division multiplexed optical passive star networks. *IEEE Journal on Selected Areas in Communications*, 4(2), 1995.

[52] L. Valiant. General purpose parallel architectures. In J. van Leeuwen, editor, *Handbook of Theoretical Computer Science, Volume A: Algorithms and Complexity*, pages 943–971. Elsevier/MIT Press, 1990.

[53] L. D. Wittie. Communication structures for large networks of microcomputers. *IEEE Transactions on Computers*, C-30(4):264–273, April 1981.